HIGH T$_c$ SUPERCONDUCTORS
Electronic Structure

Pergamon Titles of Related Interest

BEVER
Encyclopedia of Materials Science & Engineering,
8-volume set

CAHN
Encyclopedia of Materials Science & Engineering,
Supplementary Volume 1

DHEERE
Universal Computer Interfaces

EMBURY & PURDY
Advances in Phase Transitions

HUTCHINS
Solar Optical Materials

MURPHY & TURNBULL
Power Electronic Control of AC Motors

ROSE-INNES & RHODERICK
Introduction to Superconductivity, 2nd edition

Pergamon Related Journals *(free specimen copy gladly sent on request)*
European Polymer Journal

International Journal of Solids and Structures

Journal of Physics and Chemistry of Solids

Materials and Society

Materials Research Bulletin

Progress in Materials Science

Progress in Solid State Chemistry

Progress in Surface Science

HIGH T$_c$ SUPERCONDUCTORS
Electronic Structure

Proceedings of the International Symposium on the Electronic Structure
of High T$_c$ Superconductors, Rome, 5–7 October 1988

Edited by

A. BIANCONI
University of L'Aquila
L'Aquila, Italy

and

A. MARCELLI
National Institute of Nuclear Physics
Frascati, Italy

PERGAMON PRESS
OXFORD · NEW YORK · BEIJING · FRANKFURT
SÃO PAULO · SYDNEY · TOKYO · TORONTO

U.K.	Pergamon Press plc, Headington Hill Hall, Oxford OX3 0BW, England
U.S.A.	Pergamon Press, Inc., Maxwell House, Fairview Park, Elmsford, New York 10523, U.S.A.
PEOPLE'S REPUBLIC OF CHINA	Pergamon Press, Room 4037, Qianmen Hotel, Beijing, People's Republic of China
FEDERAL REPUBLIC OF GERMANY	Pergamon Press GmbH, Hammerweg 6, D-6242 Kronberg, Federal Republic of Germany
BRAZIL	Pergamon Editora Ltda, Rua Eça de Queiros, 346, CEP 04011, Paraiso, São Paulo, Brazil
AUSTRALIA	Pergamon Press Australia Pty Ltd., P.O. Box 544, Potts Point, N.S.W. 2011, Australia
JAPAN	Pergamon Press, 5th Floor, Matsuoka Central Building, 1-7-1 Nishishinjuku, Shinjuku-ku, Tokyo 160, Japan
CANADA	Pergamon Press Canada Ltd., Suite No. 271, 253 College Street, Toronto, Ontario, Canada M5T 1R5

Copyright © 1989 Pergamon Press plc

All Rights Reserved. No part of this publication may be reproduced, stored in a retrieval system or transmitted in any form or by any means: electronic, electrostatic, magnetic tape, mechanical, photocopying, recording or otherwise, without permission in writing from the publisher.

First edition 1989

Library of Congress Cataloging in Publication Data
International Symposium on the Electronic Structure of High T_c Superconductors (1988 : Rome, Italy)
High T_c superconductors electronic structure: proceedings of the International Symposium on the Electronic Structure of High T_c Superconductors, Rome, 5–7 October 1988/edited by A. Bianconi and A. Marcelli.
p. cm.
1. High temperature superconductors—Congresses.
2. Electronic structure—Congresses.
3. Photoemission—Congresses.
4. X-ray spectroscopy—Congresses.
I. Bianconi, A. (Antonio), 1944– .
II. Marcelli, A.   III. Title.
QC611.98.H54I58   1988 537.6'23—dc20   89-16180

British Library Cataloguing in Publication Data
International Symposium on the Electronic Structure of High T_c Superconductors: 1988 : Rome
High T_c Superconductors.
1. Superconductivity
I. Title II. Bianconi, A. III. Marcelli, A.
537.6'23

ISBN 0-08-037542-1

In order to make this volume available as economically and as rapidly as possible the authors' typescripts have been reproduced in their original forms. This method unfortunately has its typographical limitations but it is hoped that they in no way distract the reader.

Printed in Great Britain by BPCC Wheatons Ltd., Exeter

Preface

This volume arose from an international symposium on the "Electronic Structure of High T_c Superconductors" sponsored by the Associazione Amici dell' Accademia dei Lincei, Europa Metalli-LMI, Commission of European Communities, Consiglio Nazinale delle Ricerche and ENEA. It was held October 5 to 7, 1988, at "La Farnesina" in Rome. The venue was the renaissance villa decorated by Raphael in 1517, and seige of the National Academy of Lincei. The Accademia dei Lincei was founded by Federico Cesi in 1603 and G. Galilei joined the company of Lynceans in 1611.

The most active groups working on the electronic structure of the new cuprate perovskites that exhibit high T_c superconductivity have presented the results of a wide range of spectroscopies such as photoemission, infrared reflectivity and absorption, photoinduced absorption, Raman spectroscopy, magnetic scattering, nuclear magnetic resonance, µSR, electron energy loss, x-ray emission, extended x-ray absorption fine structure and x-ray absorption near edge structure.

The experimental results are compared with the theoretical results of the most distinguished theorists of the electronic structure of these materials. The experimental data are needed today to verify the large number of theories of high T_c superconductivity that have been proposed.

A. Bianconi and A. Marcelli

Prof. Antonio Bianconi
Chair of Physics
Collemaggio
University of L'Aquila,
67100 L'Aquila
Italy

dr. Augusto Marcelli
Laboratori Nazionali di Frascati
Istituto Nazionale di Fisica Nucleare
00044 Frascati,
Italy

Contents

THEORY

A. Fujimori
*Character of doped holes and low-energy excitations
in high T_c superconductors: Roles of the apex oxygen atoms*   3

W.M. Temmerman, G.Y. Guo, Z. Szotek and P.J. Durham
On the validity of the band model for high T_c superconductors   17

H. Krakauer, R.E. Cohen and W.E. Pickett
*Local density approximation (LDA) study of high T_c superconductors:
first principles calculations of lattice statics, lattice dynamics,
and normal state transport properties of La_2CuO_4*   31

J. Ashkenazi and C.G. Kuper
*Correlated electrons, a narrow conduction band,
and interlayer pairing in YBCO*   43

J. Ranninger, R. Micnas and S. Robaszkiewicz
*Thermodynamic and electrodynamic properties in systems
with local electron pairing*   51

OPTICAL PROPERTIES

G.A. Thomas, M. Capizzi, J. Orensten, D.H. Rapkine, A.J. Millis,
P. Gammel, L.F. Schneemeyer and J.V. Waszczak
*The energy gap and two-component absorption in
a high T_c superconductor*   67

M. Cardona and C. Thomsen
Vibrational modes in the CuO_2 planes of $YBa_2Cu_3O_{7-\delta}$ 79

A.J. Epstein, J.M. Ginder, J.M. Leng, R.P. McCall, M.G. Roe, H.J. Ye
W.E. Farneth, E.M. McCarron III and S.I. Shah
*Absorption and photoinduced absorption studies of La_2CuO_4
and $YBa_2Cu_3O_{6+x}$: role of defect states*   87

C. Taliani, R. Zamboni, G. Ruani, A.J. Pal, F.C. Matacotta,
Z. Vardeny and X. Wea
*Photoinduced optical excitations in the $YBa_2Cu_3O_{7-y}$
high T_c superconducting system*   95

D. Mihailovic and J. Solmajer
*Anomalous heating of oxygen vibrations
in Raman spectra of $YBa_2Cu_3O_{7-\delta}$*   103

MAGNETIC INTERACTIONS

Y. Kitaoka, K. Ishida, K. Asayama, H. Katayama-Yoshida, Y. Okabe
and T. Takahashi
Nuclear magnetic resonance in high temperature superconductor   111

M.A. Kastner, A. Aharony, R.J. Birgeneau, Y. Endoh, K. Fukuda,
D.R. Gabbe, Y. Hidaka, H.P. Jenssen, T. Murakami, M. Oda,
P.J. Picone, M. Sato, S. Shamoto, G. Shirane, M. Suzuki,
T. R. Thurston and K. Yamada
Neutron scattering studies of magnetic excitations in $La_{2-x}Sr_xCuO_4$ 123

R. De Renzi, G. Guidi, P. Carretta, G. Calestani and S.J.F. Cox
*$Bi_2Sr_2Y_{1-x}Ca_xCu_2O_{8+\delta}$: The magnetic partner of $Bi_2Sr_2CaCu_2O_8$
A NMR and μSR study.*   141

F. Bordi, S. Onori, A. Rosati and E. Tabet
Low field microwave absorption in ceramic $YBa_2Cu_3O_{7-\delta}$ 149

MATERIALS

F. Celani, L. Liberatori, R. Messi, S. Pace, A. Saggese and N. Sparvieri
*Ozone annealing of YBCO superconductors: Toward the
maximum of diamagnetic T_c and minimum of ΔT_c*   157

PHOTOEMISSION

P. Steiner, S. Hüfner, A. Jungmann, V. Kinsinger and I. Sander
Photoemission on high T_c superconductors   169

F.U. Hillebrecht, J. Fraxedas, L. Ley, J. Trodahl and J. Zaanen
Photoemission spectroscopy of $Bi_2CaSr_2Cu_2O_{8+\delta}$ in the normal and superconducting state   181

A. Balzarotti, M. De Crescenzi, N. Motta, F. Patella, J. Perriere, F. Rochet and Murali Sastry
Electronic structure of $YBa_2Cu_3O_{7-\delta}$ and $Bi_2Sr_2CaCu_2O_{8+\delta}$ by x-ray photoemission and Auger spectroscopies   185

C. Laubschat, E. Weschke, M. Domke, O. Strebel, J. E. Ortega and G. Kaindl
Interface formation between metals and high-T_c superconductors   201

D.D. Sarma, P. Sen, C. Carbone, R. Cimino, B. Dauth and W. Gudat
Resonant photoemission near O 1s threshold n $YBa_2Cu_{2.7}Fe_{0.3}O_{6.9}$ 215

J.-S. Kang, J.W. Allen, B.-W. Lee, M.B. Lee, M.B. Maple, Z.-X. Shen, J.J. Yeh, W.P. Ellis, W.E. Spicer and I. Lindau
Electron spectroscopy studies of high temperature superconductors: $Y_{1-x}Pr_xBa_2Cu_3O_{7-\delta}$ 225

Zhen Xiang Liu, Dian Hong Shen, Kan Xie, Shang Xue Qi, Hong Wei Xu and Nai Juan Wu
Electronic structure studies of Bi-Sr-Ca-Cu-O single crystal superconductor   237

X-RAY SPECTROSCOPY

H. Tolentino, E. Dartyge, A. Fontaine, G. Tourillon, T. Gourieux, G. Krill, M. Maurer and M.-F. Ravet
Oxygen-dependent electronic structure in powder and single crystal of $YBa_2Cu_3O_{7-\delta}$ observed in-situ by x-ray absorption spectroscopy   245

S. Della Longa, M. De Simone, C. Li, M. Pompa and A. Bianconi
    On the variation of the Cu K-edge XANES of $YBa_2Cu_3O_{7-x}$
    with oxygen concentration   259

J. Röhler and U. Murek
    On the temperature dependence of the Cu-O structure in $EuBa_2Cu_3O_{7-\delta}$ 271

A. Bianconi, P. Castrucci, A. Fabrizi, A.M.Flank, P.Lagarde, S. Della Longa,
A. Marcelli, Y. Endoh, H. Katayama-Yoshida and Z.X. Zhao
    Correlation between Tc of cuprate superconductors
    and the energy splitting between in plane and out-of-plane
    polarized Cu 2p->3d transition   281

J. Fink, N. Nücker, H. Romberg and S. Nakai
    Electronic structure studies of high -T_c superconductors
    by valence and core electron excitations.   293

M. De Santis, A. Bianconi, A. Clozza, P. Castrucci, A. Di Cicco,
M. De Simone, A.M. Flank, P. Lagarde, J. Budnick, P. Delogu,
A. Gargano, R. Giorgi and T.D. Makris
    Joint analysis of Cu L_3 XAS and XPS spectra of high T_c
    superconductors for determination of the states induced by doping   313

C.F. Hague, V. Barnole, J.-M. Mariot, C. Michel and B. Raveau
    Soft X-ray emission spectra of high T_c superconductors   325

THEORY

CHARACTER OF DOPED HOLES AND LOW-ENERGY EXCITATIONS IN HIGH-T_C SUPERCONDUCTORS: ROLES OF THE APEX OXYGEN ATOMS

A. Fujimori

Department of Physics, University of Tokyo, Bunkyo-ku, Tokyo 113 Japan

ABSTRACT

Using the results of photoemission spectroscopy and cluster-model calculations, we have studied the location and symmetry of doped holes and the nature of low-energy (0.1-0.5 eV) excitations in the Cu-oxide superconductors. Even though the doped holes may be located predominantly in the CuO_2 planes, appreciable real and virtual transfer of the holes to the apex oxygens is show to occur, which may be important in the hole doping process. The strong antiferromagnetic coupling between the holes and Cu spins, which has been suggested to favor attractive interaction between the holes, is accompanied by this charge transfer and may be coupled to other charge fluctuations such as plasmons and optical phonons.

KEYWORDS

High-T_c superconductivity; photoemission; cluster model; doped hole; phonon softening.

INTRODUCTION

In order to elucidate the mechanism of high-T_c superconductivity in the Cu-oxide systems, understanding of the electronic structure is an essential first step. By means of photoemission (e.g., Fujimori *et al.*, 1987a,b; 1989) and other high-energy spectroscopies (Bianconi *et al.*, 1987; Tranquada *et al.*, 1987), the following basic picture has been established: (a) Electron correlation is strong (U ~ 5-7 eV) for the Cu 3d electron; (d) The Cu 3d and O 2p levels are close to each other and are strongly hybridized; (c) Doped holes are O 2p-like. Although the superconductivity mechanism is still far from understood, at least the ordinary BCS theory based on one-electron band theory is irrelevant to this class of compounds. This situation necessiates the use of correlated two-band models such as extended Hubbard and periodic

Anderson models which include both Cu 3d and O 2p states explicitly. These models have lead to many proposals of spin (Emery, 1987; Imada, 1988) and charge fluctuation mediated (Varma et al., 1987; Hirsch et al., 1988. Tachiki and Takahashi, 1988) superconductivity mechanis. In most theoretical studies based on the two-band models, the properties of the CuO_2 planes have been investigated. The importance of the BaO planes adjacent to the CuO_2 planes was first pointed out by Takayama-Muromachi et al. (1988b) in the study of the $La_{1+x}Ba_{2-x}Cu_3O_{7\pm\delta}$ system, in which a small amount of oxygen vacancies in the BaO planes are shown to suppress T_c dramatically. Bianconi et al. (1988) studied polarized Cu L_3-edge x-ray absorption in $YBa_2Cu_3O_{7-\delta}$ and pointed out the possibility that the doped holes enter the BaO planes. It has also been controversial within the CuO_2 planes whether the holes are in the $p\sigma_{x,y}$ (Emery, 1987; Imada, 1987), $p\pi_{x,y}$ (Aharony et al., 1988), or $p\pi_z$ (Johnson et al., 1988) orbitals (Fig.1). The magnetic coupling between the $p\sigma_{x,y}$ holes and the Cu d^9 spins is antiferromagnetic and extremely strong, whereas the $p\pi$ orbitals as well as the $p\sigma_z$ orbitals in the BaO planes are ferromagnetically coupled to the Cu d^9 spins relatively weakly. The symmetry of the holes is therefore of particular importance in the studies of spin fluctuation mediated mechanisms (Imada et al., 1988; Shiba and Ogata, 1988).

In order to get insight into the character of the doped holes and their coupling to other spin or charge degrees of freedom, we have performed configuration-interaction calculations on CuO_n clusters combined with photoemission spectroscopy. Our results suggest that, even if the holes are predominantly distributed within the CuO_2 planes, there exist significant virtual and real charge transfer between the in-plane and apex oxygens, implying a coupling of the holes to some charge fluctuation such as plasmons and optical phonons.

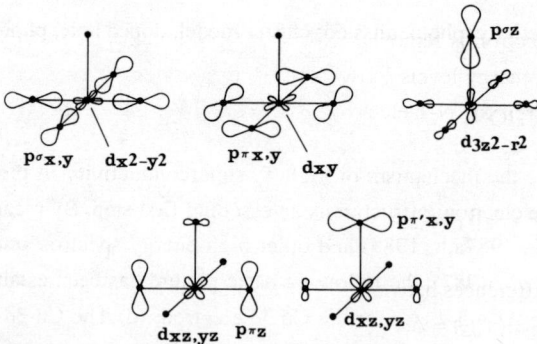

Fig. 1 Various types of oxygen p orbitals for the CuO_5 cluster. Also shown are Cu 3d orbitals which are hybridized with these p orbital

CLUSTER MODEL

We consider a CuO_5 or an elongated CuO_6 cluster depending on the local geometry around the Cu atom. The ground state of the undoped cluster (N-electron system) is represented as

$$\psi_g = \alpha \mid d^9> + \beta \mid d^{10}\underline{L}>, \quad (1)$$

where $\underline{L}$ denotes a ligand hole. The first term in Eq. (1) represents the purely ionic (Cu^{2+}, O^{2-}) configuration, and the second term a p-to-d charge-transfer state. Energy levels of the N-electron system are characterized by the charge-transfer energy, $\Delta \equiv E(d^{10}\underline{L}) - E(d^9)$, and the p-d hybridization, $T \equiv <d^{10}\underline{L}|H|d^9>$. Figure 2 shows schematically these energy levels. The p-d hybridization in the ground state is restricted to the CuO_2 plane and does not involve apex oxygen orbitals because of the x^2-y^2 symmetry of the unoccupied orbital.

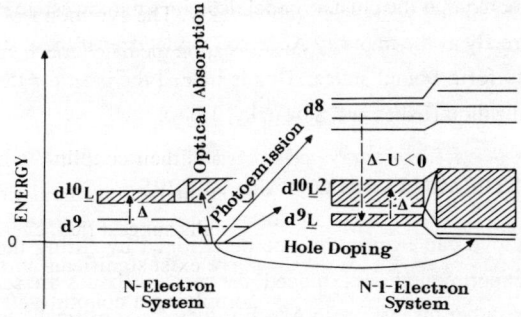

Fig. 2 Schematic energy-level diagram of the CuO_n cluster.

When a valence electron is removed by doping with an extra hole or by photoemission, we explicitly consider energy levels of the N-1-electron system instead of the N-electron system (Fig.2). Eigenstates of the N-1-electron system are given in the form (Fujimori and Minami, 1984)

$$\psi_f = \alpha'_f \mid d^8> + \beta'_f \mid d^9\underline{L}> + \gamma'_f \mid d^{10}\underline{L}^2> \quad (2)$$

Here, the energy differences between the configurations are parameterized as $E(d^9\underline{L}) - E(d^8) = \Delta-U$ and $E(d^{10}\underline{L}^2) - E(d^9\underline{L}) = \Delta$.

As the extra hole may not be necessarily in the $p\sigma_{x,y}$ orbitals, we include p orbitals on the apex oxygen. We introduce a new parameter $\Delta\varepsilon_p$, the energy difference between p orbitals in the

CuO$_2$ plane and on the apex oxygen, i.e., $\Delta\varepsilon_p \equiv \varepsilon(p\sigma_z) - \varepsilon(p\sigma_{x,y})$. In order to evaluate the off-plane p-d and p-p hybridizations, they are assumed to scale with R$^{-3.5}$ and R^{-2}, respectively, where R is the interatomic distance (Harrison, 1980). Thus the relative interatomic distance for the out-of-plane Cu-O bond to the in-plane one, r, has been taken as another independent parameter, which is determined from crystallographic data.

Due to the strong intra-atomic correlation (U) as compared to the interatomic interaction (band effects) for the Cu 3d states, single impurity models such as the cluster and the impurity Anderson models provide a good starting point. As for the band-like O 2p states, their finite bandwidths are explicitly taken into account in the impurty Anderson model, whereas in the cluster model the O 2p band is treated as a set of molecular orbitals derived from the oxygen ligand orbitals. Nevertheless, these two models are virtually equivalent concering the ground states of insulators and also for their spectroscopic properties as far as gross spectral features are concerened. Here, it shoul be noted that the cluster model describes the lowest energy states of the N-1-electron system as correctly as the impurity Anderson model does if these states are split off from the $d^9\underline{L}$ continuum to form bound states. This is indeed the case for the Cu oxides because of the large p-d hybridization (Eskes and Sawatzky, 1988).

SPECTROSCOPIC DETERMINATION OF PARAMETERS

The parameters introduced above can in principle be determined by fitting the calculated photoemission spectra to the experimental ones. Indeed, the spectral shapes are sensitive to T, U, and Δ, but unfortunately are rather insensitive to $\Delta\varepsilon_p$. It is therefore practically impossible to determine $\Delta\varepsilon_p$ by this procedure as can be seen from Fig. 3. Although $\Delta\varepsilon_p$ does not affect the spectral shape appreciably, it does determine the symmetry of the lowest binding energy feature, namely, the symmetry of the extra holes. In Fig. 4, calculated spectra are compared with experiment for La$_{2-x}$Sr$_x$CuO$_4$ and Bi$_2$(Sr,Ca)$_3$Cu$_2$O$_{8+\delta}$ (Fujimori et al., 1989). Best fits have been obtained with U = 6.5 eV, Δ = 1-2 eV, and T = 2.2-2.4 eV.

We have also utilized the Cu 2p core-level photoemission spectra to estimate the parameters. The Cu core-hole state is given by

$$\psi_f = \alpha'_f |\underline{c}d^9\rangle + \beta'_f |\underline{c}d^{10}\underline{L}\rangle \qquad (3)$$

where $\underline{c}$ denotes a core hole. The energy difference between the two configurations is $E(\underline{c}d^{10}\underline{L}) - E(\underline{c}d^9) = \Delta - Q$, where $-Q$ is the Coulomb interaction between the core hole and a d electron.

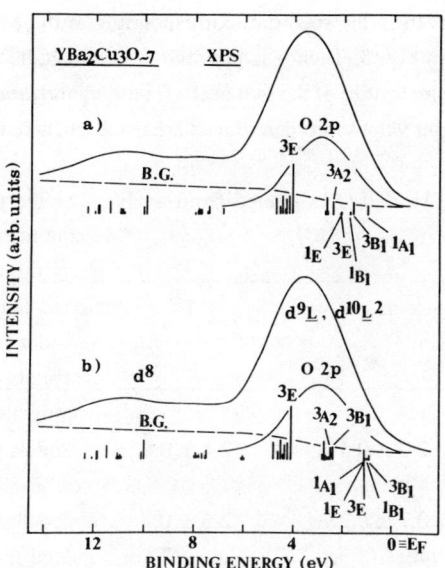

Fig. 3   Calculated valence-band x-ray photoemission spectra of $YBa_2Cu_3O_{\sim 7}$. $U=7.0$ eV, $\Delta=1.2$ eV, and $T=2.3$ eV. (a): $\Delta\varepsilon_p=0.7$ eV; (b): $\Delta\varepsilon_p=1.7$ eV (Fujimori, 1988).

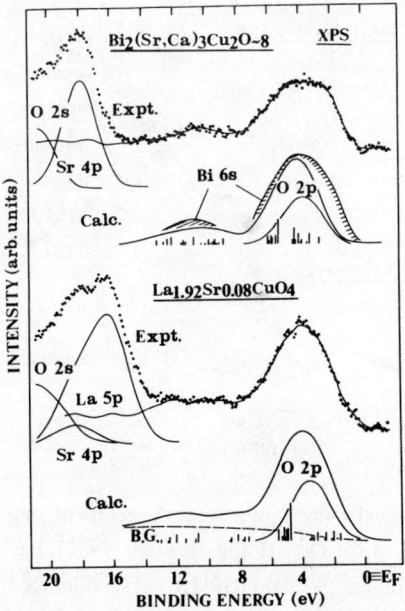

Fig. 4   Valence-band x-ray ($h\nu = 1253.6$ eV) photoemission spectra of single-crystal $La_{2-x}Sr_xCuO_4$ and $Bi_2(Sr,Ca)_3Cu_2O_{8+\delta}$ compared with those calculated using the cluster model (Fujimori et al., 1989).

Thus two peaks are observed in the spectrum corresponding to the $\underline{c}d^9$ and $\underline{c}d^{10}\underline{L}$ final-state configurations as shown in Fig. 5. Then it is possible to obtain Δ, T, and Q from the energy separation and the relative intensities of the two peaks following the procedure given by van der Laan et al. (1981). Parameter values thus determined are listed in Table I (Fujimori et al., 1989).

Table I Parameters Δ, T, U, and Q evaluated from analyses of the photoemission spectra. Energies are in eV.

	Δ	T	Q	U
$Bi_2(Sr,Ca)_3Cu_2O_{8+\delta}$	2.0 ± 0.5	2.4 ± 0.2	8.1± 0.2	7.5± 0.3
$YBa_2Cu_3O_{7-\delta}$	0.7 ± 0.2	2.6 ± 0.2	7.3± 0.1	7.0± 0.3
$La_{1.92}Sr_{0.08}CuO_4$	0.4 ± 0.3	2.7 ± 0.2	6.5± 0.1	6.5± 0.3

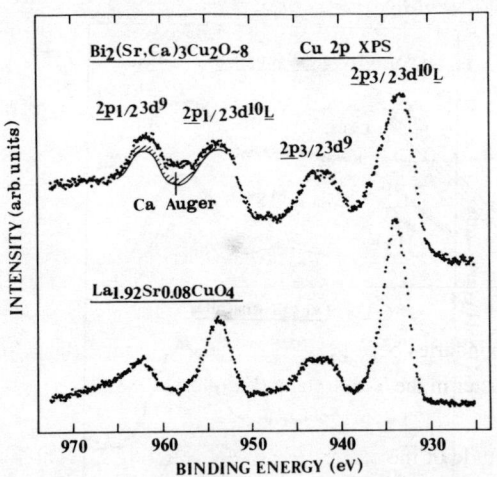

Fig. 5   Cu 2p core-level x-ray photoemission spectra of single-crystal $La_{2-x}Sr_xCuO_4$ and $Bi_2(Sr,Ca)_3Cu_2O_{8+\delta}$ (Fujimori et al., 1989). Each spin-orbit component is further split into the main ($\underline{c}d^{10}\underline{L}$) and satellite ($\underline{c}d^9$) peaks.

We find that Δ is indeed smaller than U, which classifies the Cu-oxide systems in the charge-transfer regime (Zaanen *et al.*, 1985). Also, Δ is small as compared to T, leading to highly covalent character in the Cu-O bond.

CHARACTER OF DOPED HOLES AND LOW-ENERGY EXCITATIONS

The symmetry of the ground state of the hole-doped CuO_5 cluster has been explored, and it is found that only $^1A_1$- and $^3B_1$-symmetry states are possible for a reasonable renge of the parameters as shown in Fig. 6 (Fujimori, 1988). In the $^1A_1$ state, the doped hole is in an x^2-y^2-symmetry molecular orbital consisting of in-plane oxygen $p\sigma_{x,y}$ orbitals and forms a singlet with the Cu spin; In the $^3B_1$ state, the hole is in a $3z^2-r^2$-symmetry molecular orbital consisting of the apex oxygen $p\sigma_z$ as well as of the in-plane $p\sigma_{x,y}$ orbitals (Fig. 1) and is coupled to the Cu spin ferronagnetically.

Low-lying excited states of the doped cluster for realistic T, U, and Δ values are plotted as functions of $\Delta\varepsilon_p$ in Fig. 7. In the case of the $^1A_1$ ground state (small $\Delta\varepsilon_p$), the $^3B_1$-$^1A_1$ splitting

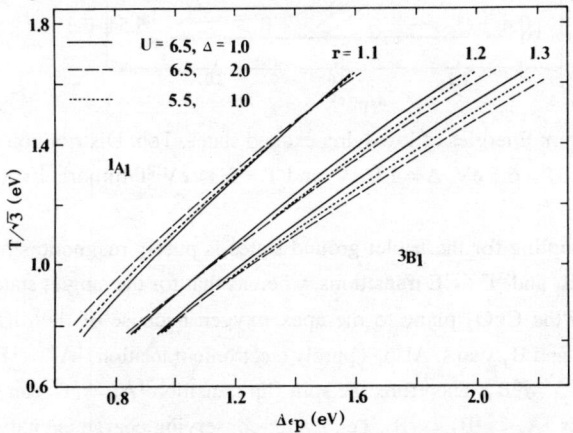

Fig. 6   Boundaries between the $^1A_1$ and $^3B_1$ ground states for the hole-doped CuO_5 cluster in the T-$\Delta\varepsilon_p$ plane (Fujimori, 1988).

represents the magnitude of the antiferromagnetic coupling J_{pd}. This splitting depends on $\Delta\varepsilon_p$ (and on r) and can become huge ($J_{pd} \sim -(0.2-0.5)$ eV) for small $\Delta\varepsilon_p$ (or large r). In the case of the $^3B_1$ ground state (large $\Delta\varepsilon_p$), the $^3B_1$-$^1B_1$ splitting yields the ferromagnetic J_{pd}, which only weakly depends on $\Delta\varepsilon_p$ and r and is relatively small ($J_{pd} \sim 0.07$ eV). It is also noticed form Fig. 7 that a ^3E state is located only ~0.2-0.4 eV above the $^3B_1$ state. As the ^3E state is mainly derived from $p\pi_{x,y}$ orbitals of the apex oxygen ($p\pi'_x$ in Fig. 1), this state can be the ground state of the doped cluster if the crystal field at the apex oxygen site raises the $p\pi'_{x,y}$ orbital energy by 0.2-0.4 eV relative to the $p\sigma_z$ orbital.

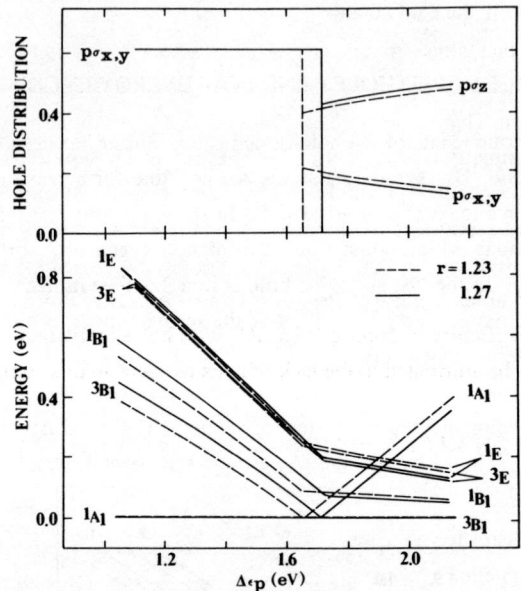

Fig. 7. Bottom: Energies of low-lying excited states. Top: Distribution of the doped hole. $U = 6.5$ eV, $\Delta = 1.0$ eV, and $T = 2.34$ eV (Fujimori, 1988).

The p hole-Cu spin coupling for the triplet ground states is purely magnetic since only spin changes in the $^3B_1 \to {}^1B_1$ and $^3E \to {}^1E$ transitions, whereas that for the singlet state involves a transfer of holes from the CuO_2 plane to the apex oxygen because of the differen p-hole distributions for the A_1 and B_1 states. Also, a purely electronic transition, $^1A_1 \to {}^1B_1$, is located only ~0.1 eV above $^1A_1 \to {}^3B_1$. Therefore, the spin-flip transition $^1A_1 \to {}^3B_1$ can be viewed as a second-order preocess $^1A_1 \to {}^1B_1 \to {}^3B_1$, i.e., a spin-conserving charge excitation followed by a spin flip, and may be dominated by the first-order spin-conserving process $^1A_1 \to {}^1B_1$. These points have not been considered in previous theoretical models but may be important in the studies of superconductivity mechanisms.

The unkmown parameter $\Delta\varepsilon_p$ may be estimated from the mean energies of the partial density of states (DOS) of various oxygens as given by band-structure calculations. Thus $\Delta\varepsilon_p$ is found to be roughly 1-2 eV, making the system in the region where the singlet and triplet states are close in energy. At present, it is controversial whether the doped holes are located in the CuO_2 planes or in the BaO planes, although a recent oxygen core-level electron-energy-loss spectroscopy by

Nücker et al. (1988) has shown that at least for $Bi_2(Sr,Ca)_3Cu_2O_{8+\delta}$ the p-hole orbitals are paralell to the planes, i.e., they are $p_{x,y}$-like.

Here I would like to note that, although the ground state of the N-1-electron system is either A_1 or B_1 (or E) within the cluster model, these states are mixed with each other and are not pure states in the periodic lattice. That is, the pure A_1 states (with a b_{1g}-symmetry hole) and the pure B_1 states (with an a_{1g}-symmetry hole) occur at the M point of the two-dimensional Brillouin zone, but actual hole states occupy a finite k-space volume around this high-symmetry point. Thus, there is a significant mixture of B_1 character in the predominantly $^1A_1$-like ground state even if the cluster calculation indicates a $^1A_1$ ground state for the N-1-electron system. In this sense, part of the doped holes are transferred to the apex oxygen. Since mixing serves to stabilize the system, the difficulty in doping Nd_2CuO_4 with holes by maintaining the structure (Tokura et al., 1988) may be attributed to the lack of apex oxygens in this compound.

IMPLICATIONS ON MECHANISMS OF SUPERCONDUCTIVITY

Magnetic Mechansisms

Magnetic pairing mechanisms have recently been studied systematically on a two-band model consisting of antiferromagnetic Heisenberg Cu spins and itinerant oxygen holes with exchange coupling J_{pd} between them (Imada et al., 1988; Shiba and Ogata, 1988). The results have shown that antiferromagnetic J_{pd} which is strong as compared to the Cu-Cu superexchagne J_{dd} leads to a binding of two holes. This situation corresponds to our $^1A_1$ ground-state case of the doped cluster as far as the spin excitation is concerned. Ferromagnetic J_{pd} does not lead to pairing except for a large J_{pd} (~ 0.8 eV) with very small p-hole transfer, which is not compatible with the small J_{pd} for the $^3B_1$ and ^3E states. Therefore, Cooper pairing seems difficult within the magnetic mechanism in the case of the ferromagnetic coupling.

Charge Fluctuation Mechanisms

The results of our cluster calculations have suggested that the magnetic mechanism is possible only for the antiferromagnetic hole-spin coupling. However, it must be noted that the antiferromagnetic coupling is not purely magnetic but necessarily involves charge transfer between the CuO_2 planes and the BaO planes. This charge transfer is made possible by intercluster coupling or when it is coupled to other charge fluctuations such as excitons, plasmons, and optical phonons. Also, the $^1A_1 \to {}^1B_1$ excitation at slightly higher energy may have stronger coupling to the charge fluctuations because of its purely electronic nature.

Charge fluctuations that are most likely to be coupled to the $^1A_1 \to {}^1B_1$ charge-transfer excitations are optical phonons having B_{1g} symmetry. An out-of-phase vibration of oxygens in the CuO_2 planes along the z direction has this symmetry. The isotope effects which are much smaller than those predicted by the BCS theory are compatible with electron-phonon mechanisms in the extremely strong coupling region (Johnson et al., 1988; Alexandrov, 1988).

The interlayer charge transfer in the pyramidal oxygen coordination induces dipole electric fields also, which may be coupled to plasmon excitations. This effect, however, will be much less significant for the octahedral coordination because of the inversion symmetry at the Cu site, and therefore cannot be a general cause of the high-T_c superconductivity. Nevertheless, this difference between the octahedral and pyramidal systems might explain the different T_c values.

Experimental Evidence for the Roles of the Apex Oxygens

A common structural feature of the Cu-oxide superconductors is the CuO_2 planes and the adjacent BaO (LaO or SrO) planes. Doped Nd_2CuO_4 which has CuO_2 planes but no such "BaO" planes is not superconducting with its original structure, but becomes so when the Cu atoms are coordinated by additional oxygens to form pyramids (Tokura et al., 1989; Takayama-Muromachi et al., 1988a). Also, some metallic Cu oxides having three-dimensional Cu-O networks such as $La_4BaCu_5O_{13}$ and $LaCuO_3$ are not superconducting (Torrance et al., 1988). It therefore seems that the presence of the apex oxygens is a necessary condition for the occurence of superconductivity.

There have been some structural studies reporting anomalous contraction of the distance between the planer Cu and the apex oxygen below T_c in $YBa_2Cu_3O_{7-\delta}$ and related compounds (Ishigaki et al., 1987). This contraction lowers the $A_1 \rightarrow B_1$ excitation energies (see Fig. 7), faciliating the coupling between the $A_1 \rightarrow B_1$ excitations and the B_{1g} optical phonon. Indeed, a softnening of this phonon mode below T_c has been observed by Raman spectroscopy (Cardona et al., 1988; Thomsen et al., 1988), which we may attribuite to the lowering of the $^1A_1 \rightarrow {^1B_1}$ excitation energies and suggests significant electron-phonon coupling for the B_{1g} mode.

Low Carrier Density Aspects of the Oxide Superconductors

We would like to point out that the photoemission spectra of the Bi-oxide (Sakamoto et al., 1987) and Cu-oxide superconductors (Fujimori et al., 1987a; b; Fujimori et al., 1988) have a common characteristic feature that the DOS at the Fermi level (E_F) is unusually low (Fig. 8). The DOS at E_F is low as compared not only to band theory (Park et al., 1988) but also to impurity Anderson-model calculations (Eskes and Sawatzky, 1988), so that this cannot be explained by the strong one-site Coulomb interaction. We therefore suspect that the low DOS at E_F arises from interatomic Coulomb interaction which would be fairly long-ranged due to the low carrier densities in these oxides. Such Coulomb interaction is known to produce a "soft" gap in disordered localized systems and a real gap in Wigner crystals; In general, the long-range Coulomb interaction will reduce the DOS at E_F (Fujimori et al., 1988a). In modeling the electronic structure of the metallic oxides, it would be necessary to somehow take into account this interaction (Takada, 1988), which has been neglected so far in most of theoretical studies. In particular, the long-range Coulomb interaction would be important in the studies of charge fluctuation mediated superconductivity mechanisms.

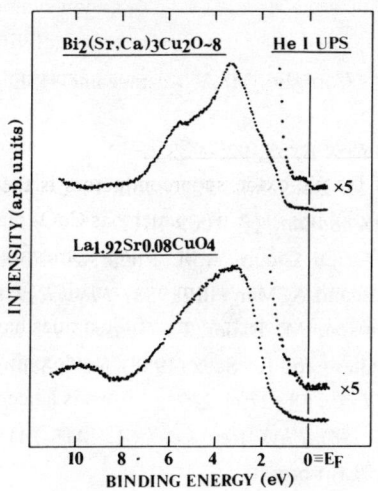

Fig. 8   Ultra-violet (hv = 21.2 eV) photoemission spectra of single-crystal $La_{2-x}Sr_xCuO_4$ and $Bi_2(Sr,Ca)_3Cu_2O_{8+\delta}$ (Fujimori et al., 1989).

CONCLUSION

We have presented results of our photoemission studies and cluster-model calculations on Cu-oxide superconductors, and based on them discussed their electronic structure. In particular, we have studied the location and symmetry of doped holes and the nature of low-energy (0.1-0.5 eV) excitations, which are crucial in the studies of superconductivity mechanisms. Even though the doped holes appear to be located predominantly in the CuO_2 planes, appreciable transfer of the holes to the apex oxygens is shown to occur. Real charge transfer resulting from intercluster interactions is suggested to be important in the hole doping process. Virtual charge transfer is coupled to optical phonons which exhibit a softneing below T_c and might be related to the mechanism of superconductivity. Strong antiferromagnetic coupling between the holes and the Cu spins is shown to be accompanied by simultaneous charge transfer of the same type, and is therefore should be coupled to charge fluctuations.

ACKNOWLEDGEMENT

I would like to acknowledge collaboration and discussions with many people, particularly E. Takayama-Muromachi, S. Takekawa, T. Takahashi, H. Katayama-Yoshida, Y. Tokura, S, Uchida, and M. Imada.

REFERENCES

Aharony, A., R. J. Birgeneau, A. Coniglio, M. A. Kastner and H. E. Stanley (1988). *Phys. Rev. Lett.*, 60, 1330-1333.
Alexandrov, A. S. (1988).*Phys. Rev. B*, 38, 925-927.
Bianconi, A., A.C. Castellano, M. De Santis, P. Rudolf, P. Lagarde, A.M. Flank and A. Marcelli (1987). *Solid. State Commun.*, 63, 1009-1012.
Bianconi, A., M. De Santis, A. Di Cicco, A.M. Flank, A. Fontaine, P. Lagarde, H. Katayama-Yoshida, A. Kotani and A. Marcelli (1988). *Phys. Rev. B*, 38, 7196-7199.
Cardona, M., R. Liu, C. Thomsen, M. Bauer, L. Genzel, W. Konig, U. Amador, M. Barahona, F. Fernandez, C. Otero and R. Saez (1988). *Solid State Commun.*, 65, 71-75.
Emery, V.J. (1987). *Phys. Rev. Lett.*, 58, 2794-2797.
Eskes, H. and G.A. Sawatzky (1988). *Phys. Rev. Lett.*, 61, 1415-1418.
Fujimori, A. (1988). *Phys. Rev. B* (in press).
Fujimori, A. and F. Minami (1984). *Phys. Rev. B*, 30, 957-971.
Fujimori, A., E. Takayama-Muromachi, Y. Uchida and B. Okai (1987a).*Phys. Rev. B*, 35, 8814-8817.
Fujimori, A., E. Takayama-Muromachi and Y. Uchida (1987b). *Solid State Commun.*, 63 857-860.
Fujimori, A., K. Kawakami and N. Tsuda (1988a). Phys. Rev. B, 38, 7889-7892
Fujimori, A., S. Takekawa, E. Takayama-Muromachi, Y. Uchida, A. Ono, T. Takahashi, Y. Okabe and H. Katayama-Yoshida (1989). *Phys. Rev. B* (in press).
Harrison, W.A. (1980). *Electronic Structure and the Properties of Solids*. Freeman, San Fransisco.
Hirsch, J.E., S. Tang, E. Loh, Jr. and D.J. Scalapino (1988). *Phys. Rev. Lett.*, 60, 1618-1621.
Imada, M. (1988). *J. Phys. Soc. Jpn.*, 57, 3128-3140.
Imada, M., Y. Hatsugai and N. Nagaosa (1988). *J. Phys. Soc. Jpn.*, 57, 2901-2904.
Ishigaki, T., H. Asano, K.Takita, H. Kato, H. Akinaga, F. Izumi and N. Watanabe (1987).

Jpn. J. Appl. Phys., 26, L1681-1683.

Johnson, K.H., M.E. McHenry, C. Counterman, A. Collins, M.M. Donovan, R.C. O'Handley and G. Kolonji (1988). *Physica C*, 153-155, 1165-1166.

Nücker, N., H. Romberg, X.X. Xi, J. Fink, B. Gegenheimer and Z.X. Zhao (1989). *Phys. Rev. B* (submitted).

Park, K.-T., K. Terakura, T. Oguchi, A. Yanase and M. Ikeda (1988). *J. Phys. Soc. Jpn.*, 57, 3445-3456.

Sakamoto, H., H. Namatame, T. Mori, K. Kitazawa, S. Tanaka and S. Suga (1987). *J. Phys. Soc. Jpn.*, 56, 365-369.

Shiba, H. and M. Ogata (1988). In: *Proc. 6th Int. Conf. on Crystal-Field Effects and Heavy Fermion Physics*, Fraknfurt (in press).

Tachiki, M., and S. Takahashi (1988). *Phys. Rev. B*, 38, 218-224

Takada, Y. (1988). *Phys. Rev. B*, 37, 155-177.

Takayama-Muromachi, E., Y. Matsui, Y. Uchida, F. Izumi, M. Onoda, and K. Kato (1988a). *Jpn. J. Appl. Phys.* (in press).

Takayama-Muromachi, E., Y. Uchida, A. Fujimori and K. Kato (1988b). *Jpn. J. Appl. Phys.*, 27, L223-226.

Thomsen, C., R. Liu, M. Cardona, U. Amador and E. Moran (1988). *Solid State Commun.*, 67, 271-274.

Tokura, Y. *et al.* (1989) (to be published)

Torrence, J.B., Y. Tokura, A.I. Nazzal, A. Bezinge, T.C. Huang and S.S.P. Parkin (1987). *Phys. Rev. Lett..*, 61, 1127-1130

Tranquada, J.M., S.M. Heald and A.R. Moodenbaugh (1987). *Phys. Rev. B*, 36, 5263-5274.

van der Laan, G., C. Westra, C. Haas and G.A. Sawatzky (1981). *Phys. Rev. B*, 23, 4369-4380.

Varma, C.M., S. Schmitt-Rink and E. Abrahams (1987). *Solid State Commun.*, 62 681-685.

Zaanen, J., G.A. Sawatzky and J.W. Allen (1985). *Phys. Rev. Lett.*, 55, 418-421.

ON THE VALIDITY OF THE BAND MODEL FOR HIGH T_c SUPERCONDUCTORS

W.M. Temmerman, G.Y. Guo, Z. Szotek, P.J. Durham
SERC, Daresbury Laboratory, Warrington WA4 4AD, UK

G.M. Stocks
Oak Ridge National Laboratory, PO Box X, Oak Ridge, TN 37830, USA

INTRODUCTION

The discovery of high T_c superconductivity in cupric oxides has led, amongst other things, to numerous electronic bandstructure studies of these compounds. The foundation of these calculations is Density Functional Theory (DFT) and they are implemented with the Local Density Approximation (LDA) or Local Spin Density (LDS) approximation. In this article we will review what we have learned from these calculations regarding the electronic structure of the family of high T_c materials. To answer this we have to find out how relevant the LSD/LDA is for the electronic properties of cupric oxides. It is well known that the application of LDA DFT for transition metal oxides is fraught with difficulty and controversy (Anderson, 1988). Although the LDA describes the variation of the equilibrium volume through the 3d transition metal oxides, including the volume expansion associated with Mott insulators (Andersen *et al.*, 1979), the theory fails to describe FeO and CoO as antiferromagnetic insulators, and in the case of NiO and MnO where the LDA does yield an insulating ground state (Terakura *et al.*, 1984a,b) the band gap is grossly underestimated. Evidently the high-T_c materials are systems that are on the verge of undergoing a Mott transition; consequently, it is necessary to be circumspect regarding the applicability of LDA DFT both in the metallic (superconducting) and magnetic insulating states.

ELECTRONIC STRUCTURE AND BONDING

The electronic groundstate, and hence the bonding, is described in our self-consistent-field

(SCF) calculations by the Local Density Approximation to the Density Functional Theory. This has been an extremely successful theory of the bonding properties of solids (Andersen et al., 1985). In Figure 1 we show the variation of the total electronic energy as a function of lattice constant. The calculated equilibrium volumes, in the tetragonal structure, are 188.9 Å^3 for La_2CuO_4 and 181.9 Å^3 for $La_3SrCu_2O_8$ (Stocks et al., 1988). The arrows labelled A and B (see Figure 1) mark the corresponding experimentally determined equilibrium volumes (190.Å^3 and 187.7 Å^3 respectively). The differences between theory and experiment are 0.7 and 3.1% respectively. The change in volume upon addition of Sr is overestimated by a factor of 2.6. However, it should be stressed that the experimental volume change is small (1.4%).

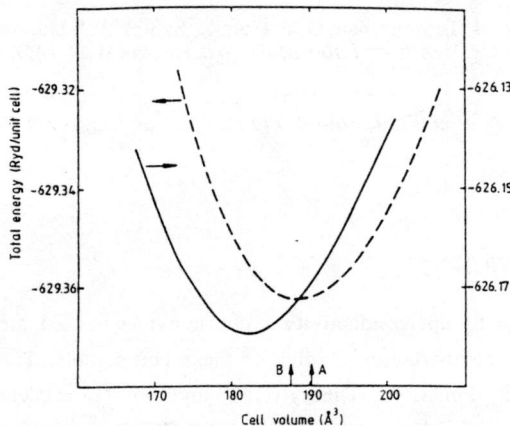

Fig. 1. The calculated volume dependence of the total energy of La_2CuO_4 (broken curve: left-hand scale) and $La_3SrCu_2O_8$ (full curve: right-hand scale). The arrows A and B give the experimentally measured equilibrium volumes of La_2CuO_4 and $La_{1.5}Sr_{0.5}CuO_4$ respectively.

In Figure 2 we plot for La_2CuO_4 the partial pressures associated with the different atomic species as a function of lattice spacing. Since the evaluation of the pressure is not as precise as that of the total energy, the equilibrium volume based on the point at which the total pressure vanishes is at a slightly different point than that obtained from the total energy curves. This notwithstanding the plot presents an interesting picture of the stability of the crystal, which is very similar to that found in the early transition metal oxides (Andersen et al., 1979). The partial pressures resulting from oxygen and from copper are positive and the equilibrium is achieved by balancing them against the negative pressure associated with the lanthanum atoms. The oxygen and copper partial pressures are dominated by the "p" and "d" contributions respectively, while the lanthanum partial pressure comprises a large, essentially volume-independent "df" contribution and a positive "sp" contribution (for the sphere sizes chosen the Madelung contribution is small). In $La_3SrCu_2O_8$ the "df" pressure is less negative than for lanthanum.

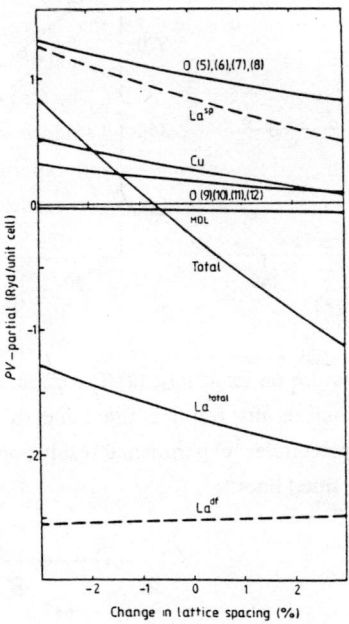

Fig. 2. Total and species decomposed partial pressures in La_2CuO_4. The ordinate is in terms of the percentage deviation in the lattice spacing from a=3.7817 Å, which corresponds to a volume of 189.47 Å^3. The broken curves show the results of further decomposing the partial pressure associated with the La sites into "sp" and "df" angular momentum components. The curve marked MDL gives the Madelung contribution to the pressure (Andersen et al., 1979).

However, this is more than compensated for by a reduction in the lanthanum "sp" pressure and in the "p" pressure of the oxygen at the octahedron apex closest to the plane of Sr atoms. These compensating effects produce the overall slight reduction in volume over that of La_2CuO_4 referred to previously. That the bonding is highly 3 dimensional can be inferred from Figure 3a where the experimental change in c/a as a function of pressure is plotted (Akhtar et al., 1988). In Figure 3b we compare our calculated P-V curve with the measurements. The agreement is excellent. The calculated variation in volume follows experiment almost exactly, and with only a small (about 0.5%) difference between the absolute values of calculated and measured volumes. These results strongly support the quantitative reliability of LDA calculations for the bonding properties of high T_c materials.

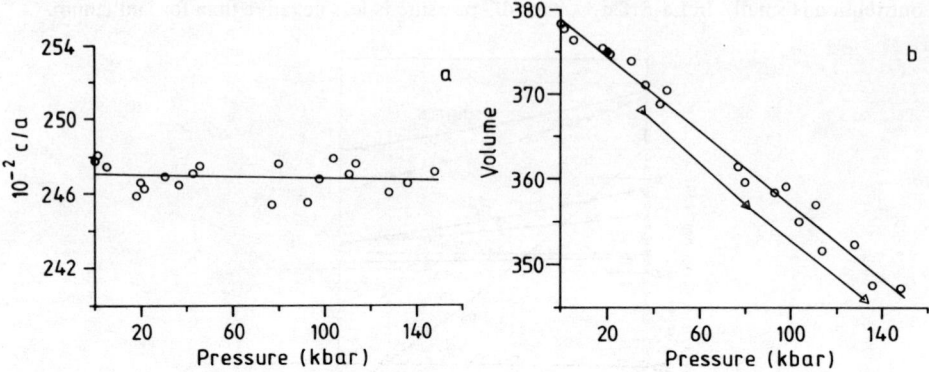

Fig. 3. The effect of pressure on La_2CuO_4. (a) The effect of pressure on c/a. Open circles: experimental results; full line: fitted line. (b) The variation of volume with pressure. Open circles: experimental results; opern triangles: theoretical results; full lines: fitted lines.

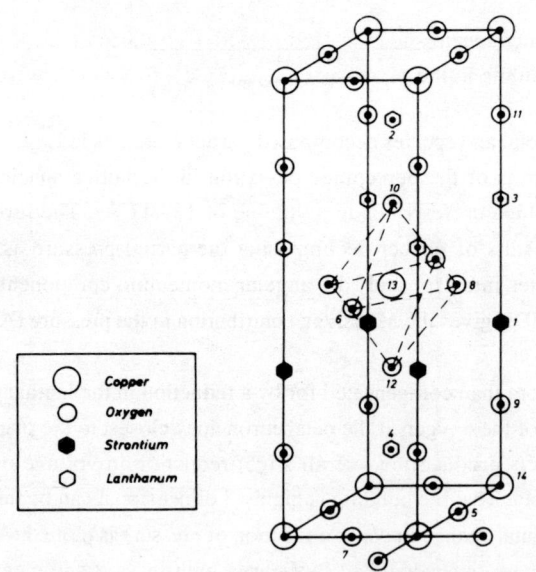

Fig. 4. A schematic representation of the simple tetragonal unit cell used in the $La_3SrCu_2O_8$ calculation. Replacement of the Sr atom (labelled 1) by La gives the unit cell used in the La_2CuO_4 calculation.

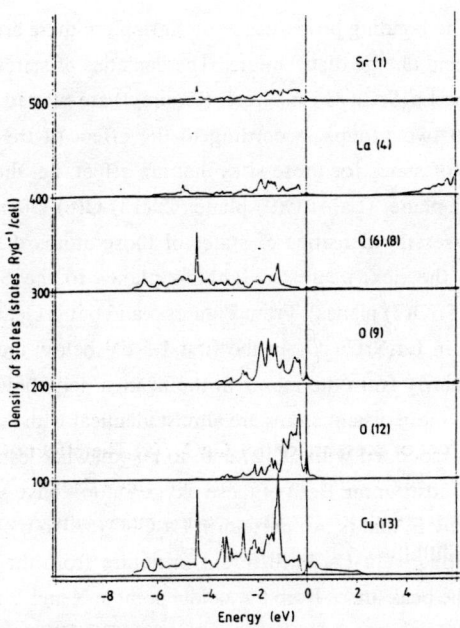

Fig. 5. Atomic-species-decomposed densities of states of $La_3SrCu_2O_8$ for Sr and for atoms in the planes neighbouring the Sr layer. The species labelling is according to Fig. 4.

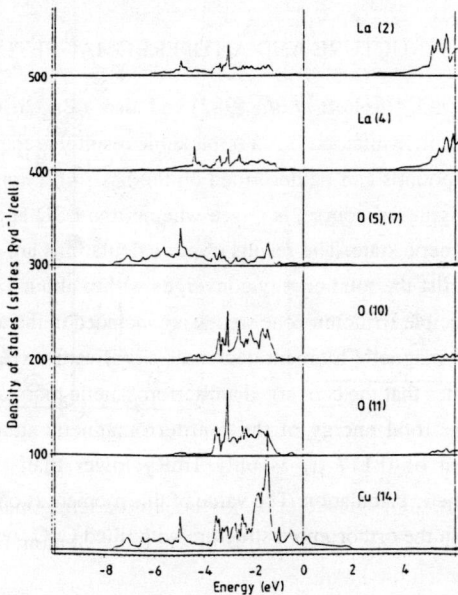

Fig. 6. Atomic-species-decomposed densities of states of $La_3SrCu_2O_8$ for atoms in the next-nearest-neighbour planes to the Sr layer. The species labelling is according to Fig. 4.

Besides the changes in the bonding properties upon Sr doping, there are also major changes to the electronic structure and charge distributions. The densities of states for each of the twelve inequivalent atoms in the $La_3SrCu_2O_8$ supercell (Figure 4) are plotted in Figures 5 and 6. We have divided these into two groups, according to the effect of the Sr addition. Figure 5 comprises the densities of states for those sites that are affect, i.e. those atoms immediately above and below the Sr plane: (La(4)-O(9) plane, Cu(13)-O(6)-O(8) plane, Sr-O(12) plane itself). Figure 6 comprises the densities of states of those atoms that are not substantially affected, i.e. atoms in the next-nearest-neighbour planes to the Sr plane (La(2)-O(11), La(3)-O(10), Cu(14)-O(5)-O(7) planes). From Figures 5 and 6 it is clear that the extra structure in the density of states in $La_3SrCu_2O_8$ in the first 1.5 eV below the Fermi energy E_f is Sr induced and arises primarily from the atoms in the nearest neighbour planes to the Sr. The densities of states for the more distant atoms are almost identical with those of pure La_2CuO_4. It is the density of states associated with O(12) and, to a somewhat lesser extent, O(9) that are most affected by the Sr substitution. Both of these oxygen atoms have structure in the densities of states at and just below E_f that is not present in La_2CuO_4. Also it should be noted that the Fermi energy peak in $La_3SrCu_2O_8$ is different in nature from the peak just below E_f in La_2CuO_4. In La_2CuO_4 the peak arises from the saddle point at X and Y and is dominated by the Cu component; in $La_3SrCu_2O_8$ octahedral apex oxygen atoms (O(9) and O(12)) adjacent to the Sr atoms dominate. Indeed a small saddle-point peak in the densities of states of $La_3SrCu_2O_8$ that results from the unperturbed Cu-O2 plane can still be seen in Figure 6 just above E_f.

ELECTRONIC STRUCTURE AND ANTIFERROMAGNETISM

The cupric oxides La_2CuO_4 (Freltoft *et al.*, 1987) and also $YBa_2Cu_3O_6$ (Tranquanda *et al.*, 1988) are antiferromagnetic insulators. It is a remarkable result that the bonding and the doping properties of these compounds can be described on the basis of a non-magnetic and metallic groundstate. The other issue, of course, is to see whether the LSD approximation to DFT can obtain this antiferromagnetic state. The results of our calculations are summarized in Figure 7 (Guo *et al.*, 1988a). Whilst the total energy converges within about 1mRyd when 80 or more k-points inside the irreducible Brillouin zone wedge are included in the tetrahedral Brillouin zone integration, the antiferromagnetic moment decreases significantly with increasing number of k-points. Figure 7 indicates that the converged antiferromagnetic moment would be non-existent or negligibly small. The total energy of the "antiferromagnetic state" corresponding to 80 k-points and a moment of 0.117 μ_b is only 1mRy lower than the total energy of the corresponding non-magnetic calculation. The value of this moment is only slightly dependent on the structure : 0.117 μ_b in the orthorombic structure with tilted CuO_6 octahedra and 0.109 μ_b in the tetragonal structure.

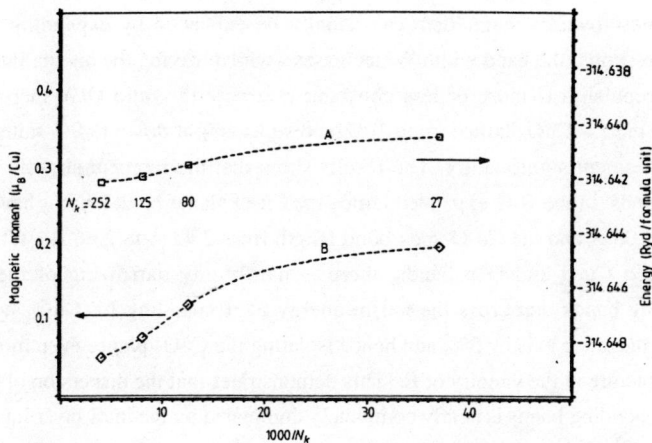

Fig. 7. The self-consistent antiferromagnetic moment on each Cu site (B), and the total energy (A), versus the number of k-points N_k inside the irreducible (1/8) Brillouin zone for the orthorombic structure. The curves are merely a guide to the eye.

Moreover the spin band splitting is always too small to produce a semiconducting gap at the Fermi level. The energy bands of the orthorombic and "anti-ferromagnetic" (80 k-points) compound are shown in Figure 8 (Temmerman *et al.*, 1988). The band splitting across the Brillouin zone varies from D1=0.08 eV to D2=0.15 eV. A periodic lattice distortion (PLD) of a 3.4% modulation of the planar Cu-O bond lengths would give rise to similar band splittings: D1=0.0 eV and D2=0.23 eV.

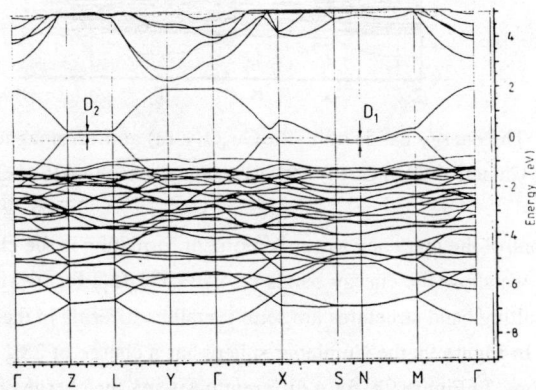

Fig. 8. The energy bands of orthorombic and "antiferromagnetic" $La_4Cu_2O_8$.

The tendency towards magnetism can usually be enhanced by expanding the lattice. This narrows the bands (the band width W decreases), whilst leaving the on-site inter-electronic d-d Coulomb repulsion U more or less constant; therefore the ratio U/W increases. Uniformly expanding the La_2CuO_4 lattice up to 10% and squeezing it down to 9% failed to increase the magnetic moment significantly. The results show that the moment slightly decreases as the lattice expands. In the 10% expanded lattice the Cu-O planar bond lengths have increased from 1.89 Å to 2.08 Å and the Cu-O apex bond length from 2.42 Å to 2.66 Å. Whilst this narrows the occupied Cu d and O p bands, there is hardly any narrowing of the Cu d and O p anti-bonding bands that cross the Fermi energy E_f. Expanding La_2CuO_4 non-uniformly by increasing just the c axis by 5%, and hence isolating the CuO_2 planes even more, does not alter the bandstructure in the vicinity of E_f. This demonstrates that the dispersion of the Cu d_{x2-y2} and O $p_{x,y}$ anti-bonding bands is nearly completely dominated by the intra-layer interactions.

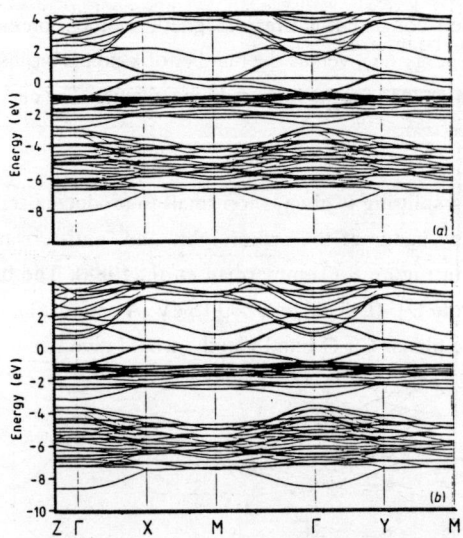

Fig. 9. The energy bands of $Y_2Ba_4Cu_6O_{12}$; (a) antiferromagnetic with up to ℓ=3 on Cu and ℓ=2 on O; (b) antiferromagnetic with up to ℓ=3 on Cu and ℓ=1 on O.

What seems to control the tendency towards moment formation is the charge on the O and Cu sites. In Figure 9 we show the energy bands of $Y_2Ba_4Cu_6O_{12}$ for two different basis sets. In both cases the resulting band structures are semi-metallic, differing in their separation of the Cu d and O p bands. In Figure 9a the Cu planar sphere has a charge of 28.12 electrons, the planar O has 7.72 electrons. In Figure 9b, for a different basis set, the charge on the O site is reduced to 7.67 electrons whilst the charge on the Cu site has remained unchanged. The effect is to separate the O p band from the Cu d bands, increasing the d character at E_f and slightly increasing the antiferromagnetic tendency.

If the antiferromagnetic groundstate is not a LSD solution to DFT, the question arises how far away is the LSD approximation from obtaining an anti-ferromagnetic groundstate? The study of the q and ω RPA-LSD spin susceptibility (Stenzel and Winter, 1985) should provide us with that information. Preliminary results suggest that there is a peak in the spin susceptibility χ (q,0) around the (1,1,0) point (Lueng et al., 1988 and Chui et al., 1988), but reliable estimates of the magnitude of the peak of the interacting susceptibility based on the RPA-LSD formalism are not available yet. Moreover the q and ω behaviour of the spin susceptibility would provide the coherence length of the spin fluctuations. Unfortunately, those calculations are not available yet. However, from LSD banstructure calculations of systems structurally and chemically close to La_2CuO_4, such as K_2NiF_4, K_2CuF_4 and La_2NiO_4, some conclusions regarding the validity of LSD in cupric oxides can be inferred.

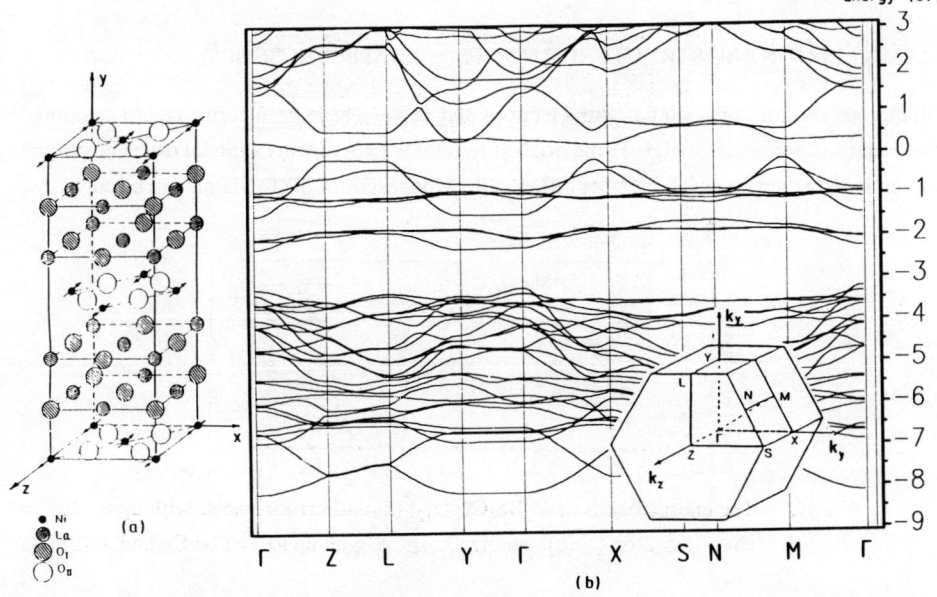

Fig. 10. (a) The antiferromagnetic spin structure of La_2NiO_4 used in the present calculation and (b) the band structure of the antiferromagnetic La_2NiO_4 at the experimental lattice constant, with the Brillouin zone shown in the inset.

Kubler et al., (1988), find K_2NiF_4 to be an anti-ferromagnetic insulator with a moment of 1.5 μ_b and 0.5 eV gap, whilst K_2CuF_4 is ferromagnetic with a moment of 0.4 μ_b. In these fluorides the metal d bands are more separated from the fluoride p bands, and therefore the states at the Fermi energy would be more of metallic d character than in La_2CuO_4. We studied La_2NiO_4 (Guo et al., 1988b and Guo et al., 1988c) which we find to be an anti-ferromagnetic insulator with a moment of $1.0\mu_b$ and a gap of 0.2 eV. In Figure 10 we show this banstructure; first we note that the O p bands are well separated from the Ni d bands. Secondly the metal d bands are split by approximately 0.5 eV: the Ni t_{2g} and both of the Ni e_g bands, i.e. the Ni $3d_{z^2}$-$O2p_z$ and the Ni $3d_{x^2-y^2}$-$O2p_{x,y}$. The majority and minority t_{2g} bands are occupied, whilst only the majority of the e_g bands are occupied, and the energy gap separates the majority from the minority e_g bands. For a gap to open up it is crucial that the e_g orbitals pointing along the z-axis occur at the same energy as the x2-y2 e_g orbitals. We find for example that increasing the ratio of the Ni-O apex to Ni-O planar bond length from 1.04 to 1.08 would take us from an insulator to a metal. This is a consequence of a charge transfer from the Ni to the La which pulls the minority Ni $3d_{z^2}$ - O $2p_z$ below E_f. Also, the tilting of the $Ni-O_6$ octahedra gives rise to charge transfers which in turn affects the magnetic properties, increasing the moment slightly.

ELECTRONIC STRUCTURE AND XPS/EELS SPECTROSCOPY

Numerous spectroscopic studies with electrons and photons have been performed to determine the electronic structure of high T_c materials. From these experiments a model of the electronic structure has emerged which is rather different than the results of the bandstructure calculations.

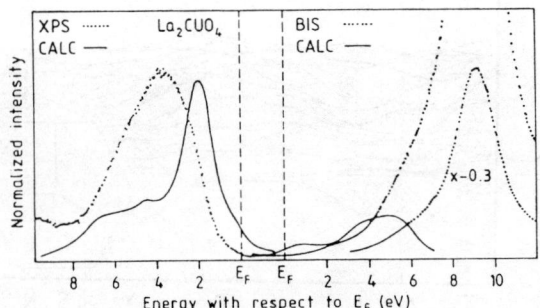

Fig. 11. Experimental and calculated XPS and BIS spectra at 1486.7 eV for La_2CuO_4.

Whilst in the band picture the Cu and O bands are strongly hybridized, from spectroscopic studies it is deduced that the Cu bands are localized and are split in a lower and upper Hubbard Cu d band with the O p band sandwiched in between (Fuggle et al., 1988). In Figure 11 we compare the experimental XPS and BIS with our one electron calculations. The calculated XPS peak is much too narrow and occurs at a binding energy of 2 eV instead of the observed binding energy of 4 eV. The XPS spectra are dominated by the Cu 3d states and therefore the principal reason for the discrepancy between experiment and theory is the strong electron-electron interaction in the Cu 3d states, as is also found in transition metal compounds.

X-ray absorption/emission and EELS experiments provide local probes of the electronic structure. X-ray emission spectroscopy confirmed the conclusion drawn from the XPS experiments regarding the Cu 3d binding energy of 4eV (Barnole *et al.*, 1988). The O p bands on the other hand has been seen in the spectroscopy which probes the unoccupied bands such as the EELS (Nücker *et al.*, 1988) and the X-ray absorption (Kuiper *et al.*, 1988). These experiments demonstrate the presence of O holes on doping the insulating state. One would expect the band picture to give a better description of the band-like O p states. In Figure 12 we show the calculated EELS spectra. At threshold the calculations predict wrongly for undoped La_2CuO_4 a finite intensity due to holes in the two-dimensional Cu $3d_{x2-y2}$ - O $2p_{x,y}$ band. Upon doping with Sr an increase in the intensity is predicted by the calculation for x=0.5 due to states on both the planar and apex O.

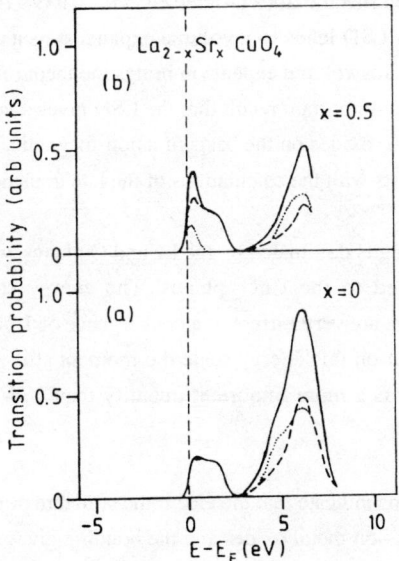

Fig. 12. Calculated oxygen 1s absorption edges for $La_{2-x}Sr_xCuO_4$: (a) x=0 (b) x=0.5. Total: solid curve, dashed curve: planar O, dotted curve: apex O.

This increase in intensity near threshold is caused by large increases in the Fermi energy densities of states on the apex O sites, i.e. O(12) and O(9) closest to the "Sr" doping. Calculations for other concentrations such as x=0.25 and 0.125 together with the symmetry analysis of the O p holes would be extremely useful for further analysis of the EELS and X-ray absorption experiments.

CONCLUSION AND OUTLOOK

As has been mentioned before, the LDA gives a very good description of the bonding properties of the high T_c materials. It describes not only the lattice constant and the P-V curve very well, but it also determines highly accurately the position of the La and apex O atoms in the unit cell (Cohen *et al.*, 1988 and Krakauer *et al.*,1988). Since these results are obtained from a metallic and non-magnetic groundstate which is in marked contrast to the experimental insulating and antiferromagnetic groundstate, one may expect that those two states are energetically not so far away from each other.

It is well known that electron-electron interactions are important in transition metal oxides. The band picture determines the d states to be too band like and this leads to too small a lattice constant in the late transition metal oxides (Andersen *et al.*, 1979). Allowing for these systems to become magnetic within LSD leads to a volume expansion as a consequence of narrowing and localizing the d states. This volume expansion brings the theory in close agreement with the experiment. It is therefore an important result that the LSD gives a very good description of the bonding in the high T_c cupric oxides on the basis of a non-magnetic, strongly Cu-O hybridized, groundstate, and this contrasts with the calculations of the late transition metal oxides.

We also find that the bonding is dominated by the La and O planes, whilst the antiferromagnetic interactions are constrained to the CuO_2 planes. The energy scales of the bonding and antiferromagnetic interaction are very different: a convergence of 1mRyd is more than sufficient to describe the bonding, but on this energy scale the moment still varies by 0.12 μ_b. We also obtain that charge transfer is a more important quantity than U/W to describe the magnetic interactions.

In short, the LSD calculations indicate that the electronic structure of the high T_c cupric oxides is different from the late transition metal oxides and the bonding gives no strong indications, as is the case in transition metal oxides, that the calculated Cu 3d states are too band like. Also, the magnetic properties seem to be different from the late transition metal oxides. In this context, q,ω spin susceptibility studies would be of great interest for high T_c cupric oxides.

ACKNOWLEDGMENTS

The work of GMS was sponsored by the Division of Materials Sciences, US Department of Energy, under contract DE-AC05-84OR21400 with Martin Marietta Energy Systems, Inc.

REFERENCES

Akhtar, M.J., C.R.A. Catlow, S.M. Clark and W.M.Temmerman (1988). *J. Phys. C: Solid State Phys.*, 21, L917

Andersen, O.K., O. Jepsen and D. Glotzel (1985). *Highlights of Condensed Matter Theory* edited by F. Bassani, F. Fumi and M.P. Tosi (North-Holland).

Andersen, O.K., H.L. Skriver, H. Nohl and B. Johansson (1979). *Pure & Appl.Chem.*, 52, 93.

Anderson, P.W., (1988) Proc. Enrico Fermi Int. School, *Frontiers and Borderlines in Many-Particle Physics*, in press.

Barnole V., C.F. Hague, J.M. Mariot, C. Michel and B. Raveau (1989). *Proc. International Symposium on the Electronic Structure of High T Superconductors* (Rome, Italy, 1988).

Chui, S.T., R.V. Kasowski and W.Y. Hsu (1988) *Phys. Rev. Lett.*, 61, 207.

Cohen, R.E., W.E. Pickett and H. Krakauer (1988) submitted to *Phys. Rev. Lett.*

Freltoft, T., J.E. Fischer, G. Shirane, D.E. Moncton, S.K. Sinha, D. Vaknin, J.P. Remeika, A.S. Cooper and D. Harshman (1987.) *Phys. Rev.B*, 36, 826.

Fuggle, J.C., P.J.W. Weijs, R. Schoorl, G.A. Sawatzky, J. Fink, N. Nücker, P.J. Durham and W.M.Temmerman (1988). *Phys. Rev. B*, 37,123.

Guo, G.Y., W.M. Temmerman and G.M. Stocks (1988a). *J. Phys. C: Solid State Phys.*, 21, L103.

Guo, G.Y. and W.M. Temmerman (1988b). *J. Phys.C: Solid State Phys.*, 21, L803.

Guo, G.Y. and W.M. Temmerman (1988c). to be submitted

Krakauer, H., R.E. Cohen and W.E. Pickett (1989). *Proc. International Symposium on the Electronic Structure of High T_c Superconductors* (Rome, Italy, 1988).

Kubler, J., V. Eyert and J. Sticht (1988). *Physica C*, 153-155, 1237

Kuiper, P., G. Kruizinga, J. Ghijsen, M.Grioni, P.J.W. Weijs, F.M.F. de Groot, G.A. Sawatzky, H. Verweij, L.F. Feiner and H.Petersen (1988. *Phys. Rev. B*, 38, 6483.

Lueng, T.C., X.W. Wang and B.N. Harmon (1988). *Phys.Rev. B*, 37, 384.

Nücker, N., J. Fink, J.C. Fuggle, P.J. Durham and W.M. Temmerman (1988). *Phys. Rev. B*, 37, 5158.

Stenzel, E., and H. Winter (1985). *J. Phys. F: Metal Phys.*, 15, 1571.

Stocks, G.M., W.M. Temmerman, Z. Szotek and P.A. Sterne (1988). *Supercond. Sci.Technol.*, 1, 57.

Terakura, K., T. Oguchi, A.R. Williams and J. Kubler (1984a). *Phys. Rev. B*, 30, 4734.
Terakura, K., A.R. Williams, T. Oguchi and J. Kubler (1984b). *Phys. Rev. Lett.*, 52, 1830.
Temmerman, W.M., Z. Szotek and G.Y. Guo (1988).*J. Phys.C: Solid State Phys.*, 21, L867.
Tranquada, J.M., D.E. Cox, W. Kunnmann, H. Moudden, G. Shirane, M. Suenaga, P. Zolliker, D. Vaknin, S.K. Sinha, M.S. Alvarez, A.J. Jacobson and D.C. Johnston (1988). *Phys. Rev. Lett.*, 58, 2802.

LOCAL DENSITY APPROXIMATION (LDA) STUDY OF HIGH T_c SUPERCONDUCTORS: FIRST PRINCIPLES CALCULATIONS OF LATTICE STATICS, LATTICE DYNAMICS, AND NORMAL STATE TRANSPORT PROPERTIES OF La_2CuO_4

H. Krakauer[a], R. I. Cohen[b] and W. E. Pickett[b]

[a] *Dept. of Physics, College of William and Mary,*
Williamsburg, VA 23185
[b] *Complex Systems Theory Branch, Naval Research Laboratory,*
Washington, D.C. 20375-5000

ABSTRACT

The general potential linearized augmented plane wave (LAPW) method is used to determine the structural and vibrational properties of La_2CuO_4 within the framework of the LDA. The theoretically determined volume and c/a ratio are in excellent agreement with experiment. We find that the oxygen in plane breathing mode is stable, in agreement with experiment and contrary to previously published results. The observed tilt mode is predicted to be unstable, with a minimum at about the experimental distortion. Thus the tilt is not due to many-body interactions not present in LDA, as had been suggested earlier. Several X-point displacements have been studied, which allows us to identify the frequencies and eigenvectors of all modes coupled to the pure breathing mode. Several zone center modes are also calculated and found to be in very good agreement with experiment. The calculated normal state transport properties are found to be extremely anisotropic and generally in good agreement with single crystal data. This supports the view that normal Fermi-liquid behavior describes the ground state above T_c, a picture that has been discounted by some. These results demonstrate that the LDA is an excellent starting point for describing the ground state of the high temperature superconductors.

KEYWORDS

High-T_c; lanthanum copper oxides; phonon calculations; LDA

Band theory based on the LDA has proved to be one of the most useful theoretical tools for obtaining accurate quantitative information about many diverse systems. In metals LDA based band structures yield excellent Fermi surfaces and often provide very good approximations to the low-lying quasiparticle excitations. The LDA has also been employed with great success to obtain structural determinations of molecules, surfaces and solids, as well as solid-solid pressure induced phase transformations. In conventional superconductors, which are driven by the electron-phonon interaction, the LDA provides the ingredients for a realistic (virtually) parameter free theory of T_c (Klein *et al.*, 1979, Allen and Mikovic, 1982). The applicability of the LDA to the high T_c superconductors has been strongly questioned, however, because of the presumed dominance of strong correlation effects principally on the Cu ions because of their large Hubbard U. In support of this view several LDA results have been cited as being contrary to experiment. LDA band structures for La_2CuO_4 (see Fig.1) predict it to be a metal, whereas experimentally it is an insulator. Another LDA result (possibly related) is the failure to yield an antiferromagnetic solution (Leung *et al.*, 1988, Temmerman *et al.*, 1988). These are discussed further below. Finally, a parametrized tight-binding calculation of the phonon spectrum of La_2CuO_4 (Weber, 1987) incorrectly predicted an unstable in-plane oxygen breathing mode, and this result has been widely interpreted as indicating the failure of LDA calculations in studying lattice dynamics in the high T_c copper oxides. Our results flatly contradict this, finding instead that this is the highest frequency mode of its symmetry type.

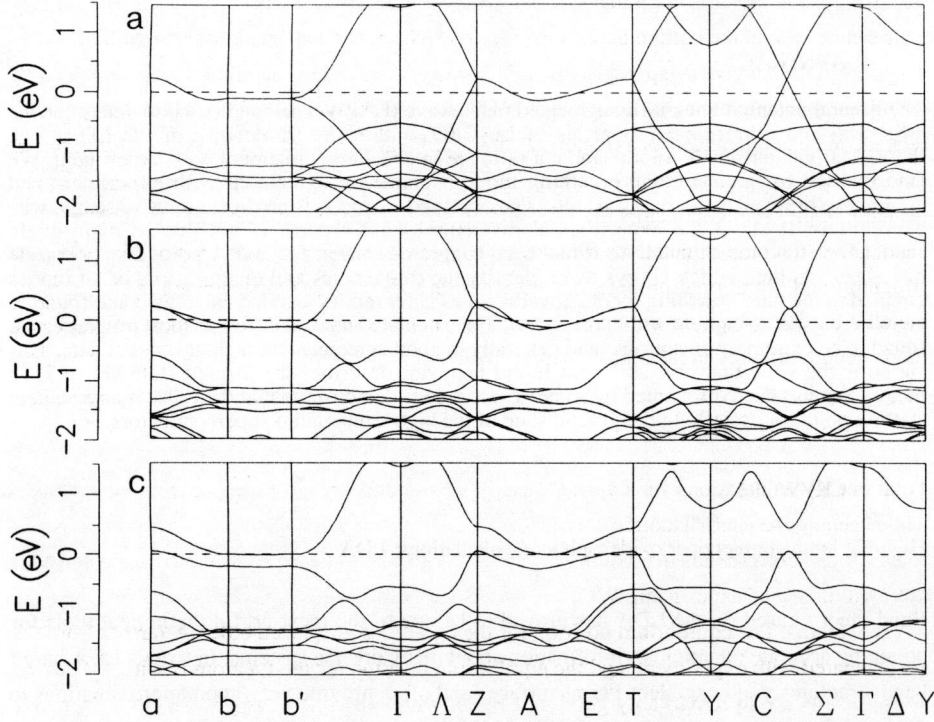

Fig. 1. Band structure for (a) the undistorted tetragonal structure, with bands folded back into the orthorhombic Abma Brillouin zone, (b) the experimental orthorhombic structure, and (c) the frozen breathing mode with 0.076 Å displacement. Points a, b, and b' are at (100), (110), and (111) in units of π/a, π/b, π/c.

If it is indeed true that the LDA cannot even qualitatively describe the ground state and low energy excitations of these materials, then one would expect that the LDA should also badly predict their structural, vibrational, and transport properties. As we shall show here, this is not at all the case. Instead, the LDA does as good a job in this regard as it does in any of the many other materials to which it has been successfully applied.

BAND STRUCTURE AND MAGNETISM

The band structure for La_2CuO_4 near the Fermi energy E_F shown in Fig. 1 is folded back into the orthorhombic Brillouin zone. This way of displaying the bands is revealing, because it underscores the fact that not much of a perturbation is required to open up a band gap near E_F. This can be seen in the bottom panel which displays the bands for the frozen in breathing mode. The LDA does not give a gap large enough to yield an insulating ground state for La_2CuO_4 or to stabilize a magnetic moment (Leung et al., 1988, Temmerman et al., 1988). These are probably related, and can be viewed as a manifestation of the "gap problem". An example of this occurs in Ge, which is calculated to have a negative gap (Pickett and Wang, 1986) instead of being an insulator. Other semiconductors like Si are predicted to have too small band gaps. On the other hand, LDA structural calculations for these semiconductors do a very good job of describing the structural and vibrational properties. Furthermore, superconductivity appears in the "doped" metallic oxides, which do not exhibit magnetic order. Thus this problem may not present a fundamental difficulty in the high T_c superconductors.

LATTICE STATICS

Total energy calculations for La_2CuO_4 were performed varying the volume and c/a rations, while keeping the internal atomic coordinates at their experimental positions. The minimum total energy found corresponds to a volume which is within 4% of the experimental volume and a c/a ratio within 1% of experiment. We then varied the axial positions of the La and O_z (the apex oxygen) atoms. The equilibrium position for the O_z atom is displaced 0.025 Å farther from the Cu compared with experiment, and the equilibrium position for the La atom is within 0.002 Å. These results are in remarkably good agreement with experiment, as good if not better than found in other systems.

FIRST PRINCIPLES PHONON CALCULATIONS

We have also performed a large number of total energy calculations to determine the frequencies and eigenvectors for zone center and zone boundary (X-point) phonons using the so-called "frozen-phonon" approximation. This approach implicitly assumes the validity of the

Born-Oppenheimer approximation and uses the LDA to determine the total energy as a function of frozen-in finite-amplitude distortions. Fitting a harmonic form to the total energy yields the phonon frequency. The method, however, yields the anharmonic contributions as well.

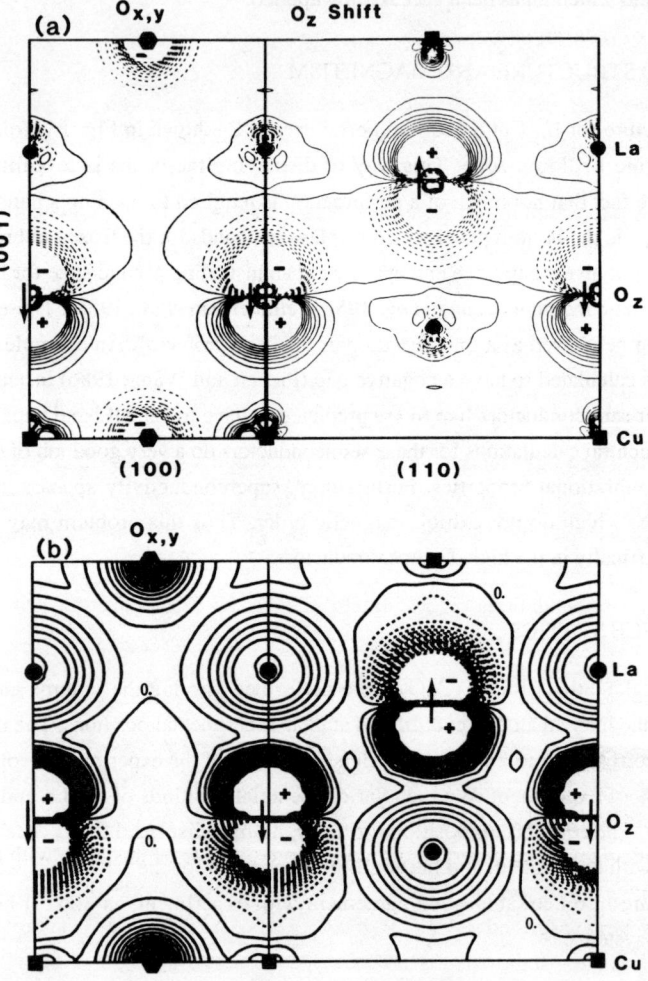

Fig. 2. Self-consistent change in (a) the valence charge density and in (b) the one-electron potential in La_2CuO_4 due to a displacement of the O_z atoms in the axial breathing normal mode.

It is crucial that these self-consistent LDA calculations adequately reproduce the electronic screening induced by the distortions in order to give sensible results. This screening can be quite long-range in the high T_c copper oxide materials due to their mixed ionic and metallic character, and can be expected to lead to unusually strong electron-phonon interactions. This point is illustrated in Fig. 2, which depicts the change in the charge density and potential induced by moving the O_z oxygen atoms axially.

In most familiar metals, the induced change is confined to the immediate vicinity of the atom that moves and is screened out rapidly as one moves away from this site. This behavior justifies the commonly used rigid-ion or rigid-muffin-tin approximations used to evaluate the electron-phonon interaction in conventional superconductors. Because of the ionic character and poor screening perpendicular to the metallic Cu-O layers, the behavior is quite different here. There are certainly large on-site changes evident in Fig. 2, but there are also very large off-site changes induced in the Cu-O layer. It is evident from this figure that one needs to go beyond rigid-ion-type approximations and that one can expect unusual electron-phonon interactions in these materials.

An early parametrized tight-binding calculation of the phonon spectrum of La_2CuO_4 (Weber, 1987) incorrectly predicted an unstable in-plane oxygen breathing mode, whereas experimentally the breathing mode is a high frequency mode and the tilt mode goes unstable (Boni et al., 1988). As mentioned, this result has been widely interpreted as indicating that conventional LDA calculations are not applicable in studying the high T_c copper oxides. By contrast, ionic calculations using an ab initio model (Cohen et al., 1988a) gave the breathing mode to be the highest frequency mode at the X-point, suggesting that ionic forces (neglected in Weber's calculation) are very important as mentioned above.

The more accurate LAPW method (Wei and Krakauer, 1985) based phonon calculations (Cohen et al., 1988b) are discussed next. Figure 3 shows the energy versus displacement for each mode that we have calculated. The observed tilt mode is predicted to be unstable, with a minimum at about the experimental distortion. Thus the tilt is not due to many-body interactions not present in LDA, as had been suggested earlier (Anderson, 1987). As mentioned, the breathing mode is also found to be stable. All the calculated zone center modes are found to be harmonic except for the O_z E_u mode, which corresponds to a sliding motion of the O_z ion in the xy plane. This infrared active mode is very broad and anharmonic. Schrodinger's equation for this anharmonic potential was solved to yield the lowest "phonon" frequencies.

Additional displacements of the O_z and La ions were studied in order to obtain the dynamical

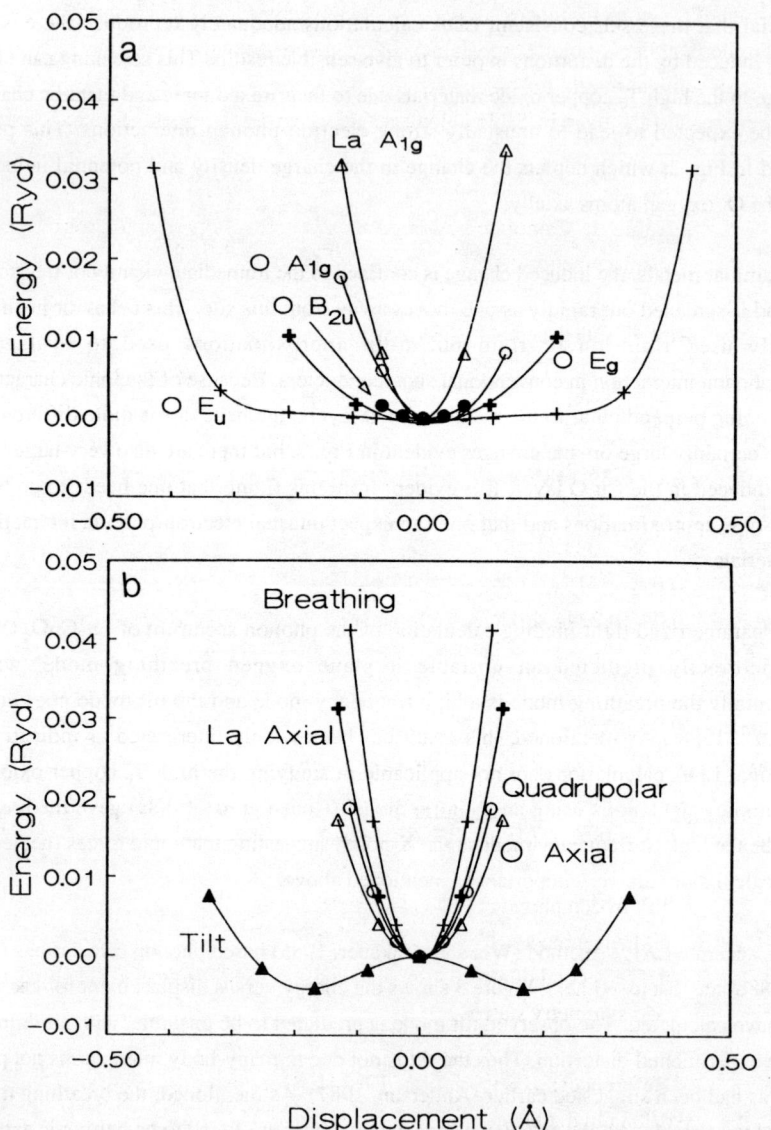

Fig. 3. Energy versus displacement for (a) zone center and (b) zone boundary (X-point) modes. The energies are per 7 atom unit cell in (a) and per 14 atom supercell in (b). The solid lines are quadratic fits to the calcualations, except for the E_u and tilt modes, where sixth order polynomials were used. For the tilt mode, the x-coordinate is the dispacement of the O_z ion in the xy plane.

matrix for the A_{1g} symmetry modes, yielding the eigenvectors and frequencies of the hybridized modes. Similarly at the X-point, combinations of the modes shown in Fig. 3b were also studied to yield the dynamical matrix for all the X-point modes having the symmetry of the breathing mode. These results are given in Table I along with the rms displacements for each first excited state.

Table I. Calculated phonon frequencies (cm^{-1}) and ground state rms displacements (Å) for La_2CuO_4.

Mode		rms	Calculated FP		Exp. EM
Γ:					
O	E_u	0.19	39,148*		110[a] 162[b]
O	B_{2u}	0.10	293		inactive
O	E_g	0.12	233		228[c]
O	A_{1g}	0.09	415	417	429[c] 424[d]
La	A_{1g}	0.04	224	220	229[c] 231[d]
X:					
La	axial	0.05	155	156	137[e]
O	axial	0.10	339	299	
Antibr.		0.09	404	475	
Breathing		0.10	609	731	710[c] 680[d]

FP: frozen phonon (symmetry mode)
EM: eigenmode
* The second number is for an excitation from the ground state to the second odd symmetry state.

References: a. Sugai *et al.*, 1987; b. Gervais *et al.*, 1988; c. Weber *et al.*, 1988a,b; d. Ohana *et al.*; e. Boni *et al.*, 1988.

The agreement with experiment is as good as that obtained in more conventional materials. This indicates that the phonons in the high T_c superconductors are not significantly dressed by magnetic or excitonic fluctuations, since this would be expected to renormalize the frequencies and give large difference between the LDA frequencies and experiment. The recent claim that the large isotope effect observed in Ba-K-Bi-O and Ba-Pb-Bi-O is due to "parasitic involvement of phonons" (Batlogg *et al.*, 1988) thus seems unlikely.

NORMAL STATE TRANSPORT PROPERTIES

Since the normal state transport properties in conventional superconductors depend on the same carriers responsible for superconductivity, it is of great interest to carry out such studies for the high T_c superconductors. Furthermore, calculation of transport coefficients requires an adequate representation of the low-energy quasiparticle excitations associated with these carriers. We have carried out detailed calculations (Allen *et al.*, 1988) to test the hypothesis that (1) the resistivity is due to quasiparticle behavior within a Fermi liquid picture, and (2) these quasiparticles are described approximately by the band electrons and holes arising from LDA calculations.

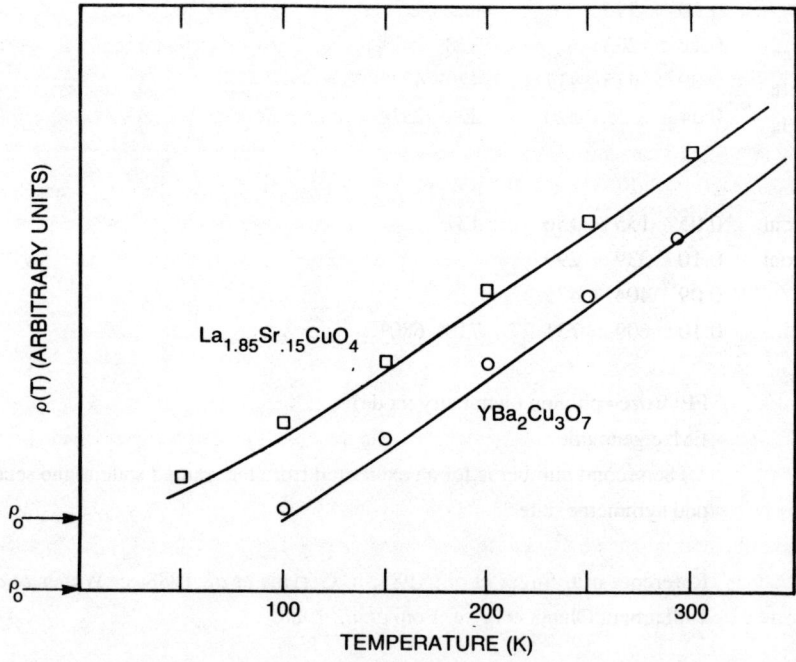

Fig. 4. Expected shape of the electron-phonon part of the resistivity according to Boltzman theory. The residual resistivity and the magnitude have been adjusted to allow a clear comparison of the shape.

The standard expression for the resistivity $\rho(T)$ using a Bloch-Boltzman approach has been calculated assuming that the scattering is due to phonons and that the electron-phonon spectral function is proportional to the measured phonon density of states. The resulting curves for metallic $La_{2-x}M_xCuO_4$ (M = Sr or Ba) and $YBa_2Cu_3O_7$ are shown in Fig. 4 together with experimental data points. It has been suggested that the measured resistivity is too linear to arise from normal quasiparticle transport. The curves in Fig. 4 are just as linear as the experimental data, indicating conventional quasiparticle behavior of the carriers.

A striking prediction was made for the Hall tensor of these materials. The Hall tensor of $La_{2-x}M_xCuO_4$ should change sign if the B-field is rotated to the xy (metallic) plane in a single crystal. The concentration dependence x for $La_{2-x}M_xCuO_4$ indicate a hole-like sign for carrier motion in the xy Cu-O layers, and a magnitude for x=0.06 within a factor of two of the experimental data (Suzuki and Murakami, 1987). This prediction also holds for $YBa_2Cu_3O_7$ and has been verified (Tozer et al., 1987). For fields in the plane, their data is T-independent and is in quantitative agreement with our predictions, and has an electron-like with magnitude similar to the calculated value. For both systems, the Hall coefficient for in-plane motion of the carriers seems to have an inverse dependence on T which is not understood. A change of sign in the Hall coefficient due to varying x was also predicted for the B-field perpendicular to the Cu-O planes in $La_{2-x}M_xCuO_4$, the sign changing to electron-like at about x=0.25. Very recently this prediction has been qualitatively confirmed (Suzuki, 1988); his data, however, show an abrupt drop in the Hall coefficient for x greater than about 0.2, and the sign does not change. Changes of sign in the Hall coefficient with respect to field orientation or doping concentration can occur naturally in a band picture, but are difficult to explain in other models.

The magnitude of the resistivity, or equivalently $d\rho/dT$, was discussed at some length by Allen et al. (1988). However, recent improvements in sample quality show that the intrinsic resistivity is a factor of two or three less than was thought to be the case several months ago, and it remains to be seen whether the intrinsic value has yet been obtained. A very satisfactory description of the normal-state transport properties is thus given by LDA-based Bloch-Boltzman transport theory. In the case of the Hall coefficient, striking predictions were made that were later confirmed by experiment.

CONCLUSIONS

LDA based total energy calculations yield excellent agreement with experiment for the structural and vibrational properties of La_2CuO_4. A stable high frequency oxygen breathing mode is found, and we correctly predict an unstable orthorhombic tilt mode in La_2CuO_4; very good agreement with experiment is also obtained for several zone center and zone boundary phonons. In this regard the LDA does as good a job for La_2CuO_4 as for any other material. This indicates that the phonons in the high T_c superconductors are not significantly dressed by magnetic or excitonic fluctuations, since this would be expected to renormalize the frequencies and give large differences between the LDA frequencies and experiment. The recent claim that the large isotope effect observed in Ba-K-Bi-O and Ba-Pb-Bi-O is due to "parasitic involvement of phonons" (Batlogg et al., 1988] thus seems unlikely. We do not find an insulating ground state for La_2CuO_4, but we suggest that this may be due to the "band gap" problem also encountered in applications of the LDA to semiconductors. The success of the transport calculations for $La_{2-x}M_xCuO_4$ and $YBa_2Cu_3O_7$ also suggests that strong correlations (beyond those included in the LDA) are not very significant around the Fermi level, since these would be expected to substantially change the low-energy quasiparticle excitations from those approximated by the LDA band structure.

Computations were carried out on the IBM 3090 at the Cornell National Supercomputer Facility, on a Multiflow Trace 7 at NRL, and on Cray X-MPs at the Pittsburgh Supercomputing Center and at NRL. This work was supported in part by the Office of Naval Research, and H.K. was supported by NSF grant DMR-87-19535.

REFERENCES

Allen, P.B., and B. Mikovic (1982). *Solid State Physics*, 32, 1.
Allen, P.B., W.E. Pickett, and H. Krakauer (1988). *Phys. Rev. B*, 37, 7482.
Anderson, P.W. (1987). In: *Novel Superconductivity* (S.A. Wolf and V.Z. Kresin, ed.), p.295. Plenum, New York.
Battlogg, B., R.J. Cava, L.W. Rupp, Jr., A.M. Mujsce, J.J. Krajewski, J.P. Remeika, W.F. Peck, Jr., A.S. Cooper, and G.P. Espinosa (1988). *Phys. Rev. Lett.*, 61, 1670.
Boni, P., J.D. Axe, G. Shirane, R.J. Birgeneau, D.R. Gabbe, H. P. Jenssen, M.A. Kastner, C.J. Peters, P.J. Picone, and T.R. Thurston (1988). *Phys. Rev. B*, 38, 185.
Cohen, R.E., W.E. Pickett, L.L. Boyer, and H. Krakauer (1988a). *Phys. Rev. Lett.*, 60, 817.
Cohen, R.E., W.E. Pickett, and H. Krakauer (1988b), to be published.
Gervais, F., P. Echegut, J.M. Bassat, and P. Odier (1988). *Phys. Rev. B*, 37, 9364.
Klein, B.M., L.L. Boyer, and D.A. Papaconstantopoulos (1979). *Phys. Rev. Lett.*, 42, 530.
Leung, T.C., X. W. Wang and B.N. Harmon (1988). *Phys. Rev. B*, 37, 384.
Ohana, I., M.S. Dresselhaus, Y.C. Liu, P.J. Picone, D.R. Gabbe, H.P. Jenssen, and G. Dresselhaus, unpublished

Sugai, S., M. Sato, S. Hosoya, S. Uchida, H. Takagi, K. Kitazawa, and S. Tanaka (1987). *Jpn. J. Appl. Phys.*, 26, 1003.

Suzuki, M. and T. Murakami (1987). *Jpn. J. Appl. Phys.*, 26, L524.

Suzuki, M. (1988). Preprint, submitted to *Phys. Rev. B*.

Temmerman, W.M., Z. Szotek, and G.Y. Guo (1988). *J. Phys. C*, 21, L867.

Tozer, S.W., A.W. Kleinsasser, T. Penney, D. Kaiser, and F. Holtzberg (1987). *Phys. Rev. Lett.*, 59, 1236.

Weber, W. (1987). *Phys. Rev. Lett.*, 58 1371.

Weber, W.H., C.R. Peters, B.M. Wanklyn, C. Chen, and B.E. Watts (1988a). *Phys. Rev. B.*, 38, 917.

Weber, W.H., C.R. Peters, B.M. Wanklyn, C. Chen, and B.E. Watts (1988b). *Sol. State Commun.*, to be published.

Wei, S.H. and H. Krakauer (1985). *Phys. Rev. Lett.*, 55, 1200.

CORRELATED ELECTRONS, A NARROW CONDUCTION BAND, AND INTERLAYER PAIRING IN YBCO

J. Ashkenazi[a] and C.G. Kuper[b]

[a] Physics Department, University of Miami, Coral Gables, Florida 33124, U.S.A.
[b] Physics Department, Technion - Israel Institute of Technology, Haifa, Israel

ABSTRACT

Starting from a generalized Hubbard model, we show that strong correlations lead to a narrow ($\simeq$ 0.1 eV) conduction band in YBCO, consisting mainly of O2-O3 p-orbitals. The Cooper pairs consist of electrons on two distinct CuO_2 planes, bound via an electronic breathing mode of charge transfer within the broken CuO_3 chains between them.

Local-density approximation (LDA) (Massidda et al., 1987; Jaejun Yu et al., 1987; Krakauer and Pickett, 1987) band structure calculations for $YBa_2Cu_3O_{7-\delta}$ ("YBCO") predict fairly wide overlapping bands (typically of width ~ 2 eV), and the Fermi surface (FS) resides in them. However, several independent experiments (Voronel et al., 1988; and also Inderhees et al., 1988; showed the existence of a λ-type specific heat singularity. Kresin and Wolf (1987) and Kresin et al. show that this is most naturally intepreted in terms of a narrow-band picture. Fisher et al., 1988; Bar-Ad et al.; Genossar et al., Takahashi et al.) suggest that the conduction band (CB) in narrow, ~ 0.1 to 0.3 eV. In particular, thermoelectric power data (Fisher et al., 1988; Bar-Ad et al.; Genossar et al., Takahashi et al.) and other transport properties are most easily interpreted (Fisher et al., 1988; Bar-Ad et al.; Genossar et al., Takahashi et al.; Kresin and Wolf, 1987; Kresin et al.; Deutscher, 1988) by assuming a CB of width $\simeq$ 0.1 eV, which is full for $\delta = 1$, and half full when $\delta = 0$. The valence (Oyanagi et al., 1987; Inoue et al., 1987) of the Cu1 atoms (our notation for the various Cu and O sites is that of Beno et al., 1987) falls from +2 to +1 as δ increase from 0 to 1. Spectroscopic evidence (Bianconi et al., 1988; Nücker et al., 1988) indicates that the carriers are mainly O2-O3 p-holes. It seems to be generally agreed that strong intra-atomic correlations are present, and modify the LDA band structure.

We are thus led to use a tight-binding model, and to explicitly include moderately large intra-atomic Coulomb integrals (~ 4 to 8 eV), and interatomic integrals (~ 1 - 2 eV) between neighbouring atoms. We model YBCO by two CuO_2 planes and the broken CuO_3 chains betwen them. (We note in passing that the chains are always broken - even when $\delta = 0$, some of the O1 atoms (Beno *et al.*, 1987) lie along the *a*-direction, with some *b*-vacancies - and that, in a one-dimensional structure such as a CuO_3 chain, missing links are very important). We define Fermi field operators thus: $a_{i\sigma}{}^{\ell\alpha}$ is an annihilation operator for a plane electrons in unit i, with spin σ, lying in the plane ℓ (= ± 1), and in the orbital α, belonging to an O2-O3 (2p) or a Cu2 (3d) shell - 11 orbitals in all. Similarly, for the chain electrons, we define operators $c_{i\sigma}{}^\alpha$, for an orbital α in a Cu1 (3d) or O1-O4 (2p) shell - 14 orbitals in all, since there are *two* O4 atoms in a unit cell.

Our model Hamiltonian, including both band structure and the Coulomb integrals which cause the strong correlations, is

$$H = H_a + H_c + H_i, \tag{1a}$$

$$H_a = \sum_{ij\ell\alpha\beta\sigma}\left[t_{i-j}^{\alpha\beta}a_{i\sigma}^{\ell\alpha\dagger}a_{j\sigma}^{\ell\beta} + \tfrac{1}{2}\sum_\tau U_{i-j}^{\alpha\beta}a_{i\sigma}^{\ell\alpha\dagger}a_{i\sigma}^{\ell\alpha}a_{j\tau}^{\ell\beta\dagger}a_{j\tau}^{\ell\beta}\right], \tag{1b}$$

$$H_c = \sum_{ij\alpha\beta\sigma}\left[s_{ij}^{\alpha\beta}c_{i\sigma}^{\alpha\dagger}c_{\beta\sigma}^{\beta} + \tfrac{1}{2}\sum_\tau V_{ij}^{\alpha\beta}c_{i\sigma}^{\alpha\dagger}c_{i\sigma}^{\alpha}c_{j\tau}^{\beta\dagger}c_{j\tau}^{\beta}\right], \tag{1c}$$

$$H_i = \sum_{ij\ell\alpha\beta\sigma\tau} W_{ij}^{\alpha\beta}a_{i\sigma}^{\ell\alpha\dagger}a_{i\sigma}^{\ell\alpha}c_{j\tau}^{\beta\dagger}c_{j\tau}^{\beta}. \tag{1d}$$

Here i-j stands for $R_i - R_j$, and U,V and W are Coulomb integrals. It is convenient to go over to a hole representation, defining

$$\tilde{t}_i^{\alpha\beta} \equiv t_i^{\alpha\beta} + \delta_{i0}\delta_{\alpha\beta}\sum_{j\gamma}(2 - \delta_{j0}\delta_{\alpha\gamma})U_j^{\alpha\gamma}. \tag{2}$$

(The $\tilde{t}$ are renormalized, to include average interatomic interactions). The dispersion of the CB is found by diagonalizing H_a to second order in $\tilde{t}_{i\text{-}j}{}^{\alpha\beta}$; explicitly, we diagonalize the matrix

$$\varepsilon_{\tilde{k}}^{\alpha\beta} = \tilde{\delta}_{\alpha\beta}\tilde{t}_0^{\alpha\beta} + \sum_j\left\{[(1-\tilde{n}_\alpha)(1-\tilde{n}_\beta)]^{\tfrac{1}{2}}(1-\delta_{j0}\tilde{\delta}_{\alpha\beta})\tilde{t}_j^{\alpha\beta}\right.$$
$$+ \tfrac{1}{4}\sum_{i\gamma}\tilde{n}_\gamma(2-n_\gamma)[(\tilde{t}_0^{\alpha\alpha} - \tilde{t}_0^{\gamma\gamma} - \sum_{\gamma'}\tilde{\delta}_{\gamma\gamma'}n_{\gamma'}U_0^{\gamma\gamma'})^{-1} \tag{3}$$
$$\left. + (\tilde{t}_0^{\beta\beta} - \tilde{t}_0^{\gamma\gamma} - \sum_{\gamma'}\tilde{\delta}_{\gamma\gamma'}n_{\gamma'}U_0^{\gamma\gamma'})^{-1}]\tilde{t}_{j-i}^{\alpha\gamma}\tilde{t}_i^{\gamma\beta}\right\}\exp{(i\vec{k}\cdot\vec{R}_j)}.$$

Here $\tilde{\delta}_{\alpha\beta} \equiv 0$ unless the orbitals α and β are on the same site, in which case $\tilde{\delta}_{\alpha\beta} \equiv 1$; n_α is the (hole) occupancy of the orbital α, and $\tilde{n}_\alpha$ is the *total* (hole) occupancy of its site. The main contributions are second-order hopping via hole-occupied Cu sites (characterized by transfer integrals ≤ 1.3 eV) and first-order hopping to hole-*vacant* O sites (transfer integrals ≤ 0.5 eV).

To perform the diagonalization of Eq. (3) explicitly, we should need the values of the parameters appearing in it. Using LDA band-structure (Massidda *et al.*, 1987; Jaejun Yu *et al.*, 1987; Krakauer and Pickett, 1987) results to estimate these parameters, we find that the highest electron band (which is mainly Cu2 ($d_{x^2-y^2}$)), of width $\simeq 0.5$ eV is empty, while, about 1.5 eV below it, there are six "minibands" consisting mainly of O2-O3 *p*-orbitals, separated by hybridization gaps, within an overall width of ~ 2 eV; the width of each miniband is ~ 0.1 to 0.3 eV. The highest miniband is the partly-filled CB. (The $Cu(d_z2)$ band is inaccessible for singly-occupied hole states). This miniband picture is confirmed by angular-resolved photoemission data (Takahashi *et al.*). However, high-energy spectroscopy may not resolve them, and may "see" only a single wide band.

When $\delta > 0.6$, neutron diffraction (Brewer *et al.*, 1988; Tranquada *et al.*, 1988) shows that these Cu orbitals are antiferromagnetically (AF) ordered. However, when δ is significantly smaller than unity, frustration (Aharony *et al.*, 1988; Birgeneau *et al.*, 1988) can suppress the intraplanar AF ordering; the remaining *inter*plane AF correlation represents interplane singlet pairing of electrons (IPSP). Since IPSP formation is a shortrange phenomenon, it can persist well above the Néel temperature since the entropy cost is small.

We define an IPSP annihilation operator $P_{ij}^{\alpha\beta}$

$$P_{ij}^{\alpha\beta} = \frac{1}{\sqrt{2}}(a_{i\uparrow}^{1\alpha} a_{j\downarrow}^{1\beta} - a_{i\downarrow}^{1\alpha} a_{j\uparrow}^{1\beta}), \tag{4a}$$

It satisfies the condition

$$(P_{ij}^{\alpha\beta})^2 = a_{i\uparrow}^{1\alpha} a_{i\downarrow}^{1\alpha} a_{j\downarrow}^{-1\beta} a_{j\uparrow}^{-1\beta}. \tag{4b}$$

In the absence of interplane hopping, IPSP's give an alternative description of the plane electrons, exactly equivalent to the usual single-electron description. The IPSP's will be uncorrelated in this case, and the band has two degenerate states (of the two planes) per unit cell - in agreement with experiment (Fisher *et al.*, 1988; Bar-Ad *et al.*; Genossar *et al.*).

Next we consider the chains. They consist of O4-Cu1-O4 ("OCO") complexes, linked by O1 atoms where they are present, and broken into quasimolecular clusters by the vacant O1 sites. We propose the following picture of the stoichometric dependence of the structure, based partly on estimates from LDA band structure (Massidda *et al.*, 1987; Jaejun Yu *et al.*, 1987; Krakauer and Pickett, 1987):

(a) For a one-OCO group, both the Cu(d) and O(p) shells are full.

(b) In a two-OCO cluster, there is a single hole in the ground state, since it is profitable for an O1-O4 p-hole orbital to hop to a Cu1 site. The hole wave function is symmetric or antisymmetric about the Cu core.

(c) for larger clusters, OCO-O-(OCO$_{n-2}$)-O-OCO, Cu1(d) holes hybridized with O4 and O1 orbitals, introduce n - 2 holes in the cluster, together with an O(p) hole at each end.

(d) The Coulomb integrals exclude any possibility of having *two* holes on a single OCO cluster. But there will be low-lying excitonic excitations consisting of a d-electron and p-hole or vice versa - these are electronic breathing modes (EBM's) of quadrupolar charge transfer within an OCO.

(e) When the stoichometric parameter $\delta = 1$, the OCO units are disjoint, and the CB is full - YBa$_2$Cu$_3$O$_6$ is an AF insulator.

(f) Initially, each additional O1 atom will link two OCO's together, and will introduce approximately one hole into the CB; this trend will continue until $\delta \simeq 0.5$, when all the OCO's are paired.

(g) In the range $0.5 > \delta > 0.2$, the two-OCO groups are progressively linked into four-OCO clusters, while no additional holes are injected into the CB.

(h) Finally, introduction of further O1's will group the OCO's into still larger clusters, while introducing two CB holes per O1 atom, until, for $\delta = 0$, the CB is half full (Fig. 1).

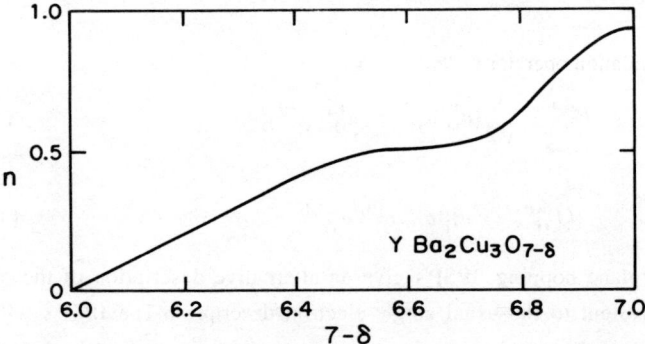

Fig. 1 Number n of conduction-band holes *vs* stoichiometry.

An optical transition (Kamarás *et al.*, 1987; Timusk *et al.*, 1988) at ~ 0.1 to 0.5 eV may possibly be an EBM. We can understand the low-temperature non-metallic nature of specimens with $\delta > 0.5$ as follows: most of the OCO clusters are small and their arrangement in the crystal is disordered - their disorder drives an Anderson transition in the CB (this picture is consistent with the experiments of Yu Mei *et al.*, 1987).

The interaction term H_i between the plane orbitals and the O4's leads to an attractive interaction between the holes on the two planes. A conduction hole in the plane $\ell = 1$ will attract electrons on the O4 atom nearest to this Cu2. But because the charge *density* in the chain complexes is always symmetrical about the Cu1's, the effect will be to excite an EBM exciton, which, in its turn, will attract a hole on the $\ell = 2$ plane (Fig. 2 is a typical diagram contributing to this interaction). When δ is sufficiently reduced from unity, this attraction may lead the IPSP's to become correlated (i.e. to become Cooper pairs); evaluation of T_c for this mechanism is in progress. Because the $\ell = +1$ and $\ell = -1$ planes are separated by ~ 9 Å, their Coulomb repulsion will be small, and the present attractive mechanism does not have to compete with the usual strong Coulomb repulsion.

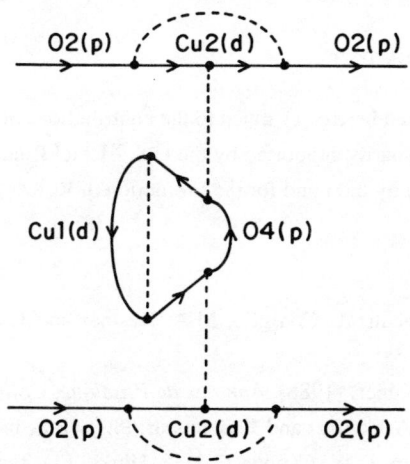

Fig. 2. A typical interplanar interaction diagram, via an EBM in an OCO complex.

An immediate consequence of this picture is that in the *c*-direction the coherence length will be rather less than a lattice vector. In the plane perpendicular to *c*, the coherence length should still be anomalously small, because the defects will weaken the coherence - although we do not have any quantitative estimate at this stage. We can understand the critical behaviour (Voronel *et al.*, 1988; Inderhees *et al.*, 1988; Kresin *et al.*) at T_c, as an immediate consequence of the short coherence lengths and the narrowness of the CB (which implies that *a large fraction of* the conduction electrons are Cooper-paired).

Both the EBM and its coupling to the CB, and hence also T_c, should be sensitive to the position of the O4 atoms. It is possible that this can even lead to an isotope effect *specific* to the O4 atoms, since the EBM's will couple to phonon modes of similar character.

The mechanism proposed here may also apply to other copper-oxide-based high-temperature superconductors. The role of the OCO complexes will be taken over by inter-plane LaO complexes in the case of doped La_2CuO_4, and by Tl-O or Bi-O clusters in the mewer materials. The higher T_c of $Tl_2Ba_2Ca_2Cu_3O_{10}$, compared with $Tl_2Ba_2CaCu_3O_8$ may arise from the fact that it has *three* adjacent CuO_2 planes; the central one will be primarily a "passenger", but it will tend to isolate the $\ell=+1$ layer of one unit cell from the $\ell=-1$ layer of the unit cell lying above it.

A fuller account of this work will appear shortly (Ashkenazi and Kuper, 1988).

ACKNOWLEDGEMENT

The band structure proposed here owes much to the contributions of B. Fisher, J. Genossar and S. Bar-Ad. The work was partly supported by the U.S. - Israel Binational Science Foundation, Grant No. 087-00195, and by the Fund for the Promotion of Research at Technion.

REFERENCES

Aharony, A., R.J. Birgeneau, A. Coniglio, M.A. Kastner and H.E. Stanley, (1988). *Phys. Rev. Lett.*, 60, 1330.

Ashkenazi, J., and C.G. Kuper, (1988). *Annales de Physique, Colloq.* 13, in press.

Bar-Ad, S., B. Fisher, J. Ashkenazi and J. Genossar, Physica C, in press.

Beno, M.A., L. Soderholm, D.W. Capone II, D.G. Hinks, J.D. Jorgensen, Ivan K. Schuller, C.U. Segre, K.Zhang and J.D. Grace, (1987). *Appl. Phys. Lett.*, 51, 57.

Bianconi, A., M. De Santis, A. Di Cicco, A. Clozza, A. Congiu Castellano, S. Della Longa, A. Gargano, P. Delogu, T. Dikonimos Makris, R. Giorgi, A.M. Flank, A. Fontaine, P. Lagarde and A. Marcelli, (1988). *Physica C*, 153-155, 115.

Birgeneau, R.J., M.A. Kastner and A. Aharony, (1988). *Z. Phys. B*, 71, 57.

Brewer, J. H., E.J. Ansaldo, J.F. Carolan, A.C.D. Chaklader, W.N. Hardy, D.R. Harshman, M.E. Hayden, M. Ishikawa, N. Kaplan, R. Keitel, J. Kempton, R.F. Kiefl, W.J. Kossler, S.R. Kreitzman, A. Kulpa, Y. Kuno, G.M. Luke, H. Miyatake, K. Nagamine, Y. Nakazawa, N. Nishida, K. Nishiyama, S. Okhuna, T.M. Riseman, G. Roehmer, P. Schleger, D. Shimada, C.E. Stronach, T. Takabatake, Y.J. Uemura, Y. Watanabe, D.Ll. Williams, T. Yamazaki and B. Yang, (1988). *Phys. Rev. Lett.*, 60, 1073.

Deutscher, G., (1988). *Physica C*, 153-155, 15.

Inderhees, S.E., M.B. Salomon, Nigel Goldenfield, J.P. Rice, B.G. Pazol, D.M. Ginsberg, J.Z. Liu and G.W. Crabtree, (1988). *Phys. Rev. Lett.*, 60, 1178.

Inoue, Z., S. Sasaki, N. Iyi and S. Takekawa, (1987). *Jpn. J. Appl. Phys*, 26, L1365.

Jaejun Yu, A.J. Freeman and S. Massidda, (1987). *Novel Superconductivity*, Ed. V.Z. Kresin and S.A. Wolf, (N.Y., Plenum Press), p. 367.

Fisher, B., J. Genossar, I.O. Lelong, A. Kessel and J. Ashkenazi, (1988). *J. Superconductivity*, 1, 53; and *Physica C*, 153-155, 1349.

Genossar, J., B. Fisher, J. Ashkenazi and L. Patlagan, *Physica C*, submitted.

Kamarás, K., C.D. Porter, M.G. Doss, S.L. Herr, D.B. Tanner, D.A. Bonn, J.E. Greedan, A.H. O'Reilly, C.V. Stager and T. Timusk, (1987). *Phys. Rev. Lett.*, 59, 919.

Krakauer, H., and W.E. Pickett, (1987). *Novel Superconductivity*, Ed. V.Z. Kresin and S.A. Wolf, (N.Y., Plenum Press), p. 501.

Kresin, V.Z., and S.A. Wolf, (1987).*"Novel Superconductivity"*, Ed. V.Z. Kresin and S.A. Wolf (N.Y., Plenum Press), p. 287.

Kresin, V.Z., G. Deutscher and S.A. Wolf, *"High T_c Superconductivity World"*, in press.

Massidda, S., Jaejun Yu, A.J. Freeman and D.D. Koelling, (1987). *Phys. Lett. A*, 122, 198.

Nücker, N., J. Fink. J.C. Fuggle, P.J. Durham and W.M. Temmerman, (1988). *Physica C*, 153-155, 119.

Oyanagi, H., H. Ihara, T. Matsubara, M. Tokumoto, T. Matsushita, M. Hirabayashi, K. Murata, N. Terada, T. Yao, H. Iwasaki and Y. Kimura, (1987). *Jpn. J. Appl. Phys.*, 26, L1561.

Takahashi, T., H. Matsuyama, H. Katayama-Yoshida, Y. Okabe, S. Hosoya, K. Seki, H. Fujimoto, M. Sato and H. Inokuchi, (1988) *Proc. of 1st Int. Conf. on Superconductivity*, Nagoya, in press.

Timusk, T., S.L. Herr, K. Kamarás, C.D. Porter, D.B. Tanner, D.A. Bonn, J.D. Garret, C.V. Stager, J.E. Greedan and M. Reedyk, (1988). *Phys. Rev. B*, 38, 6683.

Tranquada, J.M., H. Moudden, A.I. Goldman, P. Zolliker, D.E. Cox, G. Shirane, S.K. Sinha, D. Vaknin, D.C. Johnston, M.S. Alvarez and A.J. Jacobson, (1988). *Phys. Rev. B*. in press.

Voronel, A.V., D. Linsky, A. Kisliuk, S. Drislikh, B. Fisher and A. Kessel, (1988). *Physica C*, 153-155, 1086.

Yu Mei, C. Jiang, S.M. Green, H.L. Luo and C. Politis, (1987). *Z. Phys. B*, 69, 11.

THERMODYNAMIC AND ELECTRODYNAMIC PROPERTIES IN SYSTEMS WITH LOCAL ELECTRON PAIRING

J. Ranninger*, R. Micnas** and S. Robaszkiewicz**

*Centre de Recherches sur les Très Basses Températures
Centre National de la Recherche Scientifique
B.P. 166X, 38042 Grenoble Cédex, FRANCE
**Institute of Physics, A. Mickiewicz University
60-769 Poznan, POLAND

ABSTRACT

We show that the picture of a charged hard core Bose gas on a lattice and its ultimate superfluid state may be an explanation for numerous physical features in the superconducting and normal state of high T_c materials. This description is independent on a particular mechanism leading to real space electron bound pairs which form the bosons.

The high T_c materials La_2CuO_4, $YBa_2Cu_3O_{7-\delta}$ and their derivatives as well as $BaBi_xPb_{1-x}O_3$ and its latest high T_c descendant $Ba_{1-x}K_xBiO_3$ show clearly tendencies toward lattice instabilities. These are either of the nature of tetragonal-orthorhombic phase transitions in the Cu based samples or cubic-monoclinic transitions in the Cu free materials. As a rule the highest values for the critical temperatures T_c are observed for concentrations near where the structural transitions occur. Moreover, these materials exhibit anomalously large amplitude vibrations of certain O atoms (Capponi et al., 1987; J. Röhler). Upon decreasing the temperature below T_c one observes unusually large shifts of certain Raman active modes (Mihailovich et al., 1987; Mac Farlaine et al., 1987; Thomsen et al., 1988; Ruf et al.). On the contrary, the fact that one is close to a lattice instability does not seem to show up in the characteristic shift in the phonon density of states (Renker et al., 1987, 1988). An anomalous increase of the sound velocity upon lowering the temperature below T_c (associated with a hardening of the bulk modulus) has been observed (Bhattarcharya et al., 1988; Saint-Paul). This goes in hand with an abrupt change of (a-b)/(a+b) where a,b are the lattice constants in the a-b plane (Horn et al., 1988). The stiffening of the lattice in the superconducting state is opposite to what happens in classical superconductors. Moreover significant differences in the phonon density of states occur between superconducting and non-superconducting materials (i.e. $YBa_2CuO_{7-\delta}$, $\delta=0$, $\delta=1$) which cannot be accounted by structural differences only but would require large changes in the interatomic force constants which could be a signature of strong electron-lattice coupling (Renker et al., 1988). Normal state resistivity of the superconducting samples on the contrary can be easily explained in terms of weak electron (hole) phonon scattering processes (Micnas et al., 1987; Xing et al., 1988). As far as the isotope effect is concerned in these materials, it is either totally absent or weak.

This confusing and seemingly, contradictory picture which evolves upon regarding lattice effects has been partially responsible to direct the bulk of the theoretical research toward non-lattice mediated superconductivity mechanisms. Moreover as the La_2CuO_4 and $YBa_2Cu_3O_7$ based materials are close to an antiferromagnetic phase transition one was easily led to believe that superconductivity had its origin in some sort of magnetically correlated spin liquid, the RVB state (Anderson, 1987).

With the arrival of the 30K superconductor $Ba_{0.6}K_{0.4}BiO_3$ (Cava et al., 1988) and its generalized version $Ba_{1-x}K_xBiO_3$ (which up to now has shown no trace of magnetism) one should now seriously reconsider the possibility of a lattice mediated high T_c superconductivity. Photoinduced polaronic (bi-polaronic) carriers with a mass of $24m_e$ (m_e free electron mass) were recently observed in La_2CuO_4 (Kim et al., 1988; Taliani et al., 1988) together with local lattice deformations surrounding these carriers and having tetragonal instead of orthorhombic symmetry. A further indication for polaronic carriers comes from the reflectivity data (Orenstein; Thomas) which show a frequency dependent conductivity which can be interpreted by an increase in the mass m^* of the carriers from $m^*=7m_e$ as one from $\hbar\omega \gtrsim 0.3$ eV to $\omega \sim 0$.

The picture of coexisting strong electron-lattice coupling and weak electron-phonon coupling is by no means contradictory. Strong electron-lattice coupling leads to the formation of small polarons which are very stable entities, consisting of a correlated motion of the electrons and the lattice deformations which surround them. Phonons -in such a picture- are defined as the vibrational modes with respect to the deformed lattice. They couple weakly to the polarons, which is the real reason for the stability of the latter. As far as the high T_c materials are concerned, this picture is consistent with the normal state resistivity and the quasi-temperature independent phonon density of states -reminiscent of a weak phonon-electron (hole) coupling. The strong coupling between the electrons and local lattice deformations, on the contrary, is born out by the observed local lattice deformations surrounding the electrons together with the large amplitude fluctuations of these deformations (i.e. a tilting motion of the oxygen octahedra around their tetragonal configuration in La_2CuO_4 and an analogous breathing type motion around the cubic configurations in the $BaBi_xPb_{1-x}O_3$ based compounds).

The possibility of having large lattice deformations in these materials induces strong variations in the crystal field which in case of strong electron-electron correlations leads to unusually large electron-lattice coupling (Baeriswyl and Bishop, 1987; Barisic *et al.*, 1970, 1987). This effect is usually not taken into account in derivations of phonon spectra and lattice dynamics. In order to include such strong electron-lattice coupling one must resort to a representation which involves the concept of small polarons. The deformation induced crystal field effect manifests itself in the polaronic level shift (the polaron binding energy), a sizeable reduction of the band width of the carriers and finally an induced short range attraction interaction between polarons (Alexandrov and Ranninger, 1981).

In the high T_c materials the chemistry is very complex for which the simple polaronic picture can however serve as a useful qualitative guide in describing them. According to the chemical approach (Pouchard *et al.*, 1987; Wilson, 1988; Sleight, 1988) the particularity of these materials consists in containing the proper elements which can exist in an average (generally unusual) oxidation state. This, together with a suitable ligand environment leads to localization-delocalization processes with limiting oxidation states differing by two oxidation steps. Such processes are also referred to as charge disproportionation or double valence fluctuations. A ligand environment is considered to be suitable if the structures permit local deformations which help to stabilize the two oxidations states. It is believed that the strong polarizibility of the oxygens is a relevant factor in this mechanism.

Contrary to the purely polaronic mechanism where the charge disproportionation is exclusively due to strong local lattice deformations, in the chemical picture the lattice deformations are not the primary cause for charge disproportionation but do play an important role in stabilizing double valence fluctuation processes. Concerning the localization-delocalization processes proposed in the chemical picture, it is qualitatively equivalent to that of a bi-polaronic system (Alexandrov and Ranninger, 1981).

On the basis of this discussion we propose that high T_c materials should be described (to within a first rough attempt) by an effective Hamiltonian which is of the form of a generalized Hubbard model with short range attractive interaction:

$$H = \sum_{ij\sigma} t_{ij} c^+_{i\sigma} c_{j\sigma} + U \sum_i n_{i\uparrow} n_{i\downarrow} + \sum_{ij\sigma\sigma'} W_{ij} n_{i\sigma} n_{j\sigma'} - \mu \sum_{i\sigma} n_{i\sigma} \qquad (1)$$

This Hamiltonian can be derived (Alexandrov and Ranninger, 1981) from the Fröhlich Hamiltonian in the limit of very strong electron-lattice coupling. In this case the operators $c^{(+)}_{i\sigma}$ denote annihilation (creation) operators for small polarons at sites i, having spin σ and $n_{i\sigma}=c^{(+)}_{i\sigma}c_{i\sigma}$ is the number operator, t_{ij} represents an effective hopping integral and U and W_{ij} effective on-site respectively intersite electron-electron interaction. The site index i denotes either single atomic sites, diatomic units or small clusters which electronically are well separated from each other. We should point out models introducing purely electronic mechanisms of local attraction can also be described by the Hamiltonian (Equ.1).

If $U < 0$, $W_{ij} > 0$ we talk about on-site pairing, while if $U > 0$ $W_{ij} < 0$ for nearest neighbours and $W_{ij} > 0$ otherwise, we talk about intersite pairing. Electron pairing in real space (Equ. 1) gives rise to superconductivity which is qualitatively different from that of classical B.C.S. superconductors. The latter originates from an exchange coupling between pairs of electrons $c^+_{k\uparrow}c^+_{-k\downarrow}$ acting in a thin layer around the Fermi surface. Upon going from the normal state to the superconducting one this exchange mechanism leads to a macroscopic coherent quantum state which at T=0 is the well known B.C.S. ground state:

$$|\psi>_{B.C.S.} = \prod_k (u_k e^{i\phi_k/2} + v_k e^{-i\phi_k/2} c^+_{k\uparrow} c^+_{-k\downarrow}) |0> \qquad (2)$$

with phase locked Cooper pairs i.e. $<c^+_{k\uparrow} c^+_{-k\downarrow}> \neq 0$ (ϕ_k, the phase of the pair with momenta (k,-k) being independent of k). It is important to remember that the formation of Cooper pairs occurs simultaneously with the onset of superconductivity. v^2_k measures the probability of finding a Cooper pair in the state $(k\uparrow,-k\downarrow)$; $u^2_k + v^2_k = 1$.

In the case of real space pairing (Equ. 1) we have to distinguish two qualitatively different regimes. The first one is rather similar to B.C.S. and occurs when the electron pairing potential is weak enough in order not to form bound pairs. The conditions for that strongly depend on the dimensionality and strength of the pairing potential (Micnas et al., 1988). Contrary to B.C.S., real space pairing being a static interaction (rather than retarded !) involves all the electrons of the Fermi sea in the pairing mechanism. This is manifest in the mean field critical temperature which for real space pairing in the weak attraction limit is given by (Robaszkiewicz et al., 1982; Alexandrov and Russ, 1983):

$$T_c = 1.14 \, D \, \sqrt{n(2-n)} \exp(-2D/|U|) \qquad (3)$$

where the band half width D (times the n dependent factor) replaces the characteristic phonon frequency in the B.C.S. expression for T_c and the effective coupling constant $\lambda = |U|/2D$. The ground state wave function for the Hamiltonian Equ. (1) is still given by expression Equ. (2) provided that |U|/D is small enough so that no real space pairs form.

If on the contrary we have the right conditions for the formation of real space bound pairs (Micnas *et al.*, 1988) the physics of the superconducting and normal state is qualitatively different from what we just discussed above. In that case we are faced with a problem of an interacting charged hard core Bose gas on a lattice which has properties which are similar to ^{4}He showing superfluidity in a solid. In the remainder of this lecture we shall discuss the thermodynamic and electrodynamic properties of such a system in the case of on-site pairing and confront our results with the experimental ones on high T_c materials.

If the condition for on-site pair formation is satisfied the Hamiltonian Equ. (1) can be rewritten as: (Alexandrov and Ranninger, 1981; Pouchard *et al.*, 1987; Wilson, 1988; Sleight, 1988)

$$H = \sum_{ij} J_{ij} (\rho_i^+ \rho_j^- + \rho_j^+ \rho_i^-) + \sum_{ij} K_{ij} \rho_i^z \rho_j^z - \bar{\mu} \sum_i (2\rho_i^z + 1)$$

$$J_{ij} = \frac{2t_{ij}^2}{|U|}, \qquad K_{ij} = J_{ij} + 2 W_{ij}, \qquad \bar{\mu} = \mu + \frac{1}{2}|U| - zW_0 \qquad (4)$$

z being the coordination number and W_0 zero momentum Fourier component of W_{ij}. The charge operators for the on-side electron pairs are:

$$\rho_i^+ = c_{i\uparrow}^+ c_{i\downarrow}^+, \qquad \rho_i^- = c_{i\downarrow} c_{i\uparrow}, \qquad \rho_i^z = (n_{i\uparrow} + n_{i\downarrow} - 1)/2 \qquad (5)$$

and the ground state of the Hamiltonian Equ. (4) is given by:

$$|\psi\rangle_{LP} = \prod_i (u_i e^{i\phi_i/2} + v_i e^{-i\phi_i/2} c_{i\uparrow}^+ c_{i\downarrow}^+)|0\rangle \qquad (6)$$

which is the real space analogue of the B.C.S. ground state, Equ. (2) and v_i^2 measures the probability of finding a local pair (LP) at site i, $u_i^2 + v_i^2 = 1$.

Several configurations of the ground state Equ. (6) are feasable depending on the coupling constant and the concentration of LP's. For concentrations below some critical n_c (depending on the coupling constants of the Hamiltonian Equ. (4)), the ground state is the homogeneous superconducting state with u_i; v_i and ϕ_i all independent on i. For $n>n_c$ the ground state is a state in which superconductivity and charge order coexist. The finite temperature properties of the LP system are quite different from that of weak local pairing. Upon increasing the temperature in a

LP system the superconductivity order breaks down because of the breaking up of the phase locked state by thermally excited fluctuations of the local phase ϕ_i of the LP's. The pairs continue to exist above T_c and eventually break up at some higher temperature T_p comparable to the binding energy of the pairs. Above T_p the system is metallic while for $T_c < T < T_p$ it has the properties of a system of doubly charged bosons. The full phase diagram (Robaszkiewicz et al., 1981) as a function of LP concentration is reproduced in Fig.(1). It should be noted that T_c increases as :

$$T^{(3)}{}_c \sim 3.31 n^{2/3}/m^*a^2, \qquad T^{(3)}{}_c \sim \pi n/2m^*a^2 \qquad (7)$$

for small concentration (n<<1) and d=3 and d=2+ε respectively. This behaviour is characteristic of a condensation of a free Bose gas, $m^*(= (3\Sigma_j J_{ij}/a^2)^{-1})$ denoting the effective mass of the bosons and a the intersite spacing. Note that the maximum of T_c is attained at the transition to the charge ordered state when the concentration of bosons has become large enough in order to bring into play the repulsive interaction between them.

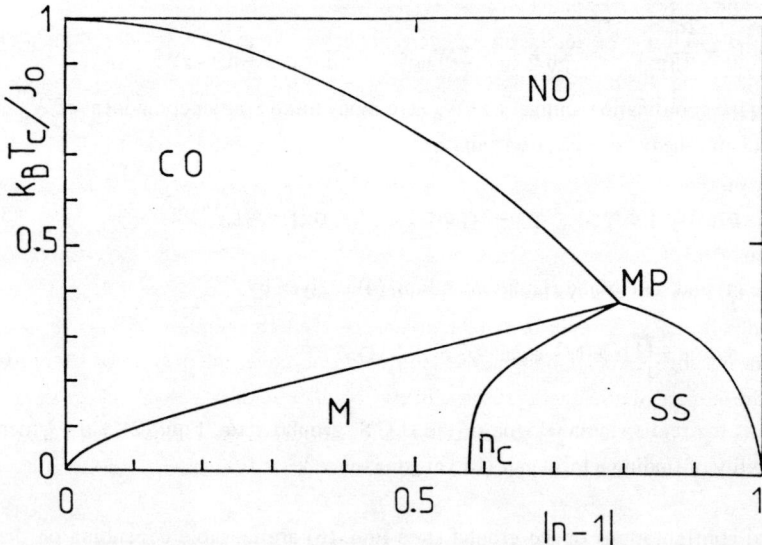

Fig. 1   Phase diagram of a local pair system as a function of concentration n. SS denotes the homogeneous superconducting phase, analogous to ^{4}He II. M represents a mixed phase where superconductivity and charge order coexist. CO is the charge ordered state and NO denotes the phase of non-ordered local pairs. The maximum of T_c is obtained at the multicritical point MP.

Let us now focus on the homogeneous superconducting state SS. The low lying excitations are collective modes of phase fluctuations which in the superconducting state are coupled to the density fluctuations. For this reason the excitation spectrum is sound like with a dispersion $\omega_k(T) = s(T) k + \alpha k^2$ in the long wavelength limit. $s(T) \sim (J_0(J_0+K_0))^{1/2} <\rho^+>_T$. J_0 and K_0 denoting the zero momentum Fourier components of J_{ij} and K_{ij}. As one increases the concentration of LP's (going from 0 to 1) at zero temperature, $s(0)$ increases while ω_{kB} (the frequency at the Brillouin zone) decreases and finally tends to zero at n_c. In the M phase there are two branches of collective oscillations, a sound wave like one relevant for superfluidity and an optical one characteristic for CO. As the temperature increases the acoustic branch tends to zero at $T=T_c$ where the superconductivity breaks down. Such a drastic effect onto the sound velocity arises from the fact that the LP's are on a rigid lattice and Umklapp-processes destroy the momentum of charge fluctuations. The linearity of the excitation spectrum depends on the range of the repulsive interaction between LP's and must be short ranged. For long range interaction we have $\omega_k(T) = \omega_{p\ell}(T) + \alpha k^2$ where $\omega_{p\ell} = 2e <\rho^+>_T (4\pi/m^*)^{1/2}$ is the plasma like frequency of the LP's which tends to zero as one approaches T_c. On the basis of this we can conjecture about the specific heat C as a function of temperature. In general we distinguish three regimes $T<T_c$, $T_c<T<T_p$ and $T>T_p$. For $T<T_c$, C will vary either like T^d (d dimensionality) in the limit $T\rightarrow 0$, if the LP interaction is short ranged or like $\exp(-(\omega_{p\ell}(0)/T)$ if this interaction is long ranged. As we approach T_c, C should show a λ-like transition as in ^{4}He II. For $T_c<T<T_p$, the specific heat is governed by a bosonic band. In the dilute limit when interactions between LP's can be neglected the specific heat should be essentially that of Bosons with a square density of states for which superfluidity has been artificially supressed (Alexandrov et al., 1986a). Finally for $T>T_p$, C_v should show the usual linear in T behaviour arising from Fermions. In Fig.2 we give a schematic plot of C_v versus T. The region of disordered LP's ($T_c<T<T_p$) may not always be accessible since its existence strongly depends on the binding conditions for LP's. It might be rendered visible by suitably suppressing the superconducting state, for instance upon applying a magnetic field. Moreover the disordered LP phase ought to show the normal diamagnetic susceptibility together with a Van Vleck contribution. The resistivity of this phase ought to be that of itinerant bosons on a lattice which are diffused by impurities and phonons. It is expected to vary linearly in T (Xing et al., 1988).

Let us now discuss the electrodynamic properties of the superconducting state of LP's. The Meissner effect of charged Bosons shows a penetration depth :

$$\lambda_H = (m^*c^2/16\pi n_s e^2)^{1/2} \qquad (8)$$

where $n_s = <\rho^+> 2/T$

denotes the concentration of condensed bosons (Alexandrov and Ranninger, 1981a). The upper critical field Hc_2 for a dilute weakly interacting Bose gas or bosons being scattered by impurities shows a characteristics temperature variation (Alexandrov *et al.*, 1986; Alexandrov *et al.*, 1988)

$$Hc_2(T) \sim 0.64 \, (\phi_0/2\pi) \, (n/\ell)^{1/2} \, [T_c/T(1-(T/T_c)^{3/2}]^{3/2} \tag{9}$$

where ℓ is the mean free path of the bosons. Notice in particular the positive curvature of $Hc_2(T)$ which is distinctely different from B.C.S.

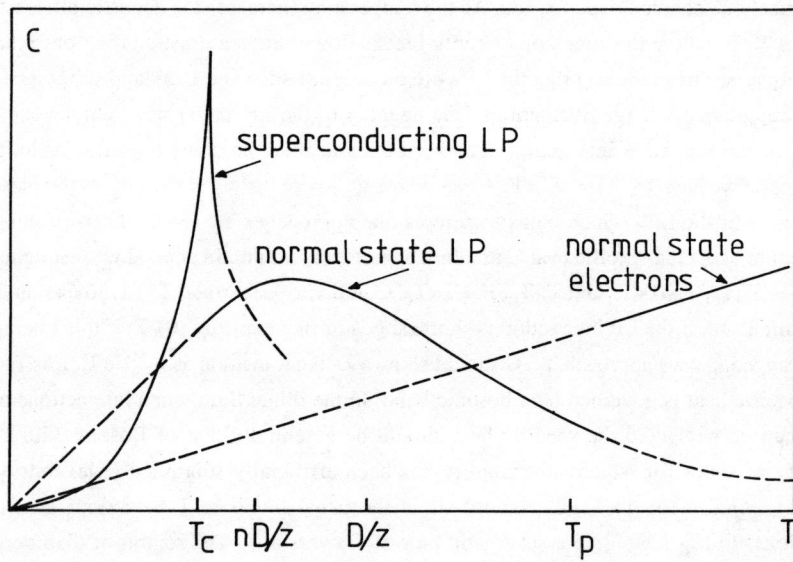

Fig.2 Schematic plot of the specific heat of the strong correlation negative U Hubbard problem. Below the superconducting transition temperature T_c, C is given by the LP's in the superfluid phase. Between T_c and T_p (pair breaking temperature) C is determined by that of a weakly interacting Bose gas on a lattice. Above T_p, C follows the linear T behaviour of electrons. It is assumed in this plot that the mass of the LP's is bigger than the electron mass by a factor 3.

The generalization (Alexandrov *et al.*, 1986) of the Ginzburg-Pitaevskii equation for a neutral Bose gas to a charged one permits one to estimate the zero temperature coherence length (Alexandrov *et al.*, 1988):

$$\xi(0) = (2m^* \, N_s K_0)^{1/2} / \hbar \tag{10}$$

where K_o is the zero wave vector Fourier component of the Boson-Boson interaction and N_s denotes the density of condensed Bosons. This leads to the Ginzburg-Landau parameter:

$$\kappa = \frac{\lambda_H}{\xi(0)} = \frac{m^*c}{\hbar}\left(\frac{K_o a^3}{8\pi e^2}\right)^{1/2} \tag{11}$$

Taking the effective radius of the interaction $(K_o a^3/8\pi e^2)^{1/2}$ to be of the order of a few lattice constants and $m^* \gg m_e$, then $\kappa \gg 1$ and we have an extreme type II superconductor. The lower critical field can be derived upon using the classical result:

$$H_{c1} = (\phi_o/4\pi \lambda^2_H)\, \ln(\lambda_H/\xi_o) \tag{12}$$

The thermodynamical critical field is given by (Alexandrov et al., 1988):

$$H_c = n\, 4\pi\, K_o \tag{13}$$

Let us now in the remainder of this talk demonstrate why the idea of LP's might be relevant to high T_c materials. One of the few things which have been established beyond doubt until now is that in these materials the coherence length is very small (a few tens of Ås') and comparable to the interparticle distance. This rules out any type of B.C.S. mean field formalism whatever the mechanism for pairing might be or however big the cutoff frequency for the exchange of pairs might be.

We are in a regime which is fluctuation dominated; the critical region being comparable to T_c itself. Therefore the usual Ginzburg Landau theory with Gaussian fluctuations does not apply. The experimentally measured $\Delta C/k_B n \sim 1$ (where ΔC is the jump of the specific heat at T_c) would give $T_c \sim T_F$ if we applied the B.C.S. relation $\Delta C/k_B n \sim 7 T_c/T_F$. It is known that T_c must be $\ll T_F$ in order that B.C.S. theory holds. The small number of carriers together with the short coherence length leads one there fore naturally to consider the picture of LP's and ultimately their Bose condensation.

What experimental evidence is then there to support such a picture? In the Cu free materials $BaBi_{1-x}Pb_xO_3$ and $Ba_{1-x}K_xBiO_3$ we know that in the regime of x where we have a charge ordered state, existence of LP's is established. It is highly probable (though not proven at the moment) that in the superconducting state these local pairs exist as well. The negative magnetic susceptibility (Batlogg et al., 1988; Kondoh et al.) observed in the normal state of the superconducting samples are a strong indication for that.

Concerning high T_c materials there is now a general concensus that the superconductivity originates from pairing of holes which are predominantly situated on the oxygen ions in the CuO_2 planes. For these materials one finds (Tajima *et al.*, 1988; Uemura *et al.*, 1988) over a wide regime of either Sr doping (in La_2CuO_4) or O depletion (in $YBaCuO_{7-\delta}$):

$$T_c \sim \omega^2_{p\ell} \sim 1/\lambda^2_H \tag{14}$$

This is an indication for Bose condensation; see Equ. (7) and remember that the plasma frequency $\omega_{p\ell} = (4\pi e^2/\epsilon(\infty))^{1/2}(n/m)^{1/2}$. Another indication for a charged superfluid state may lie in the positive curvature of Hc_2 (Equ. (9)) observed experimentally for both, fields perpendicular and parallel to the CuO_2 planes.

The experimentally observed thermal power shows a temperature behaviour which resembles in nothing that of a Fermionic system and moreover is independent on an external magnetic field up to 30 T (Yu *et al.*, 1988). The latter suggests spin less particles which could be diamagnetic LP's.

Recent specific heat measurements (Butera *et al.*, 1988; Ishikawa *et al.*, 1988; Voronel *et al.*, 1988) show a pronounced λ-like peak some times followed, only a few degrees above, by a (first) order phase transition. Such a behaviour is compatible with the picture developed in this lecture and the schematic plot of C_v in Fig.(2).

If the superconductivity in high T_c materials is due to a Bose condensation of a dilute lattice gas of LP's one should expect a gap in the single electron excitation spectrum which corresponds to the binding energy of the LP's. It is given by the energy difference between the narrow LP (bi-polaron) band and the bottom of the wider electron (polaron) band and hence this gap would be temperature independent. Tunneling experiments (Geerk *et al.*) seem to confirm this.

With the availability of better and better single crystals it should be possible in the future to examine the critical regime and determine the critical exponents. The critical behaviour ought to be quite different from the classical Ginzburg Landau one and should correspond to that of a quantum $S = 1/2$, X-Y model.

The proposal presented here that high T_c materials are essentially bosonic systems which undergo a superfluid transition does not depend on any particular mechanism. We showed that qualitatively such a picture fits rather well to the present experimental situation. But also quantitatively it turns out that this picture is rather good. Taking an effective mass of the Bosons of $\sim 20 m_e$ and a carrier concentration of $\sim 5.10^{21}/cm^3$ we are able to obtain values for T_c, the normal state specific heat and the penetration depth which are in good agreement with experiments. $m^* \sim 20\ m_e$ is also roughly equal to the measured mass (Kim *et al.*, 1988; Taliani *et al.*, 1988) of the photo induced carriers in these materials. For a very detailed comparison of the LP picture and experiments on high T_c materials we refer the reader to a recent review by de Jongh (De Jongh, 1988).

ACKNOWLEDGEMENT

We are indebted to A.S. Alexandrov for many valuable discussions. R.M. and S.R. acknowledge financial support from the Polish Academy of Sciences within the projects RPB 01.09 and CRBP 01.12. J.R. and RM. are grateful for the hospitality at the workshop on "High temperature superconductions, Concepts, Models and Methods" organized by the Institute of Scientific interchange at TORINO, where part of this manuscript was prepared.

REFERENCES

Alexandrov, A.S. and J. Ranninger, (1981). *Phys. Rev. B*, 23, 1796.
Alexandrov, A.S., and J. Ranninger, (1981a). *Phys. Rev. B*, 24, 1164.
Alexandrov, A.S., J. Russ., (1983). Phys. Chemistry, 57, 167.
Alexandrov, A.S., J. Ranninger, and S. Robaszkiewicz, (1986). *Phys. Rev. B*, 33, 4526.
Alexandrov, A.S., J. Ranninger, and S. Robaszkiewicz, (1986a). *Phys. Rev. Lett.*, 56, 949.
Alexandrov, A.S., D.A. Samarchenko and S.V. Tavern, (1988). *Sov. Phys. JETP*, 66, 567.
Anderson, P., (1987). *Science*, 235, 1196.
Baeriswyl, D. and A.R. Bishop, (1987). *Physica Scripta T*, 19, 239
Barisic, S., J. Labbé, and J. Friedel, (1970). *Phys. Rev. Lett.*, 25, 919.
Barisic, S., I. Batistic and J. Friedel, (1987). *Europhys. Lett.*, 3, 1231.
Batlogg, B., R.J. Cava, L.W. Rupp Jr, A.M. Mujsce, J.J. Krajewski, J.P. Remeika, W.F. Peck Jr, A.S. Cooper and G.P. Espinosa, (1988). *Phys. Rev. Lett.*, 61, 1670.
Bhattacharya, S., M.J. Higgins, D.C. Johnston, A.J. Jacobson, J.P. Stokes, J.T. Lewandowski, and D.P. Gashorn, (1988). *Phys. Rev. B*, 37, 5901.
Butera, R.,(1988). *Phys. Rev. B*, 37, 5909.
Cava, R.J., B. Batlogg, J.J. Krajewski, R. Farrow, L.W. Rupp Jr., A.E. White, K. Short, W.F. Peck, and T. Komentani, (1988). *Nature*, 332, 814.
Capponi, J.J., C. Chaillout, A.W. Hewat, P. Lejay, M. Marezio, N. Nguyen, B. Raveau, J.L. Souberoux, J.L. Tholence and R. Tournier, (1987). *Europhys. Lett.*, 3, 1301.
De Jongh, L.J., (1988). *Physica C*, 152, 171.

Geerk, J., X.X. Xi, and G. Linker, preprint.

Horn, P.M., D.T. Keane, G.H. Held, J.L. Jordan-Sweet, D.L. Kalser, F. Holtzberg and T.M. Rice , (1988). *Phys. Rev. Lett.*, 59, 2772.

Ishikawa, M., Y. Nakazawa, T. Takabatake, A. Kishi, R. Kato, and A. Maesono, (1988). *Solid State Comm.*, 66, 201.

Kim, Y.H., C.M. Foster, A.J. Heeger , S. Cox, G. Stucky, (1988). *Phys. Rev. B*, 38, 6478.

Kondoh, S., M. Sera, F. Fukuda, Y. Ando and M. Sato, to be published.

Mac Farlaine, R.M., H. Rosen, H. Seki, (1987). *Solid State Comm.*, 63, 831.

Micnas, R., J. Ranninger, and S. Robaszkiewicz, (1987). *Phys. Rev. B*, 36, 4051.

Micnas, R., J. Ranninger, S. Robaszkiewicz, and S. Tabor, (1988). *Phys. Rev. B*, 37, 9410.

Mihailovic', D., M. Zgonik, M. Copic, M. Morvat, (1987). *Phys. Rev. B*, 36, 3997.

Orenstein, J., G.A. Thomas, D.H. Rapkine, A.J. Millis, L.F. Schneemeyer and Waszczak, Preprint.

Pouchard, M., J.C. Grenier, and J.P. Doumec, (1987). Compte Rendu Ac. Sc. (Paris), 305, 571.

Renker, B., F. Gompf, E. Gering, N. Nücker, D. Ewert, W. Reichardt, and H. Rietschel, (1987). *Z. Phys. B*, 67, 15.

Renker, B., F. Gompf, E. Gering, G. Roth, W. Reichardt, D. Ewert, H. Rietschel and H. Mutka, (1988). *Z. Phys. B*, 71, 437.

Robaszkiewicz, S., R. Micnas and K.A. Chao, (1981). *Phys. Rev. B*, 23, 1447.

Robaszkiewicz, S., R. Micnas and K.A. Chao, (1982). *Phys. Rev. B*, 36, 3915.

Röhler, J., these proceedings.

Ruf, T., C. Thomsen, R. Liu and M. Cardona, Preprint.

Saint-Paul, M., Private Communication.

Sleight, A.W., (1988). Workshop "*Superconductivity, new models, new applications*", Les Houches; Böttcher Meeting of the German Physical Society, Bad Honeff unpublished.

Tajima, S., T. Nakahashi , S. Uchida and S. Tanaka, (1988). *Physica C*, 156, 90.

Taliani, C., R. Zamboni, G. Ruani, F.C. Matacotta and K.I. Pokhodnya, (1988). *Solid State Commun.*, 66, 487.

Thomas, G. A. in these proceedings.

Thomsen, C., R. Lin, A. Wittlin, L. Gensel, M. Cardona, W. König, M.V. Cabanas, E. Garcia, (1988). *Solid State Comm.*, 65, 219.

Uemura, Y.J., V.J.Emery, A.R. Moodenbaugh, M. Suenaga, D.C. Johnston, A.J. Jacobson, J.T. Lewandowski, J.H. Biewer, R.F. Kiefl, S.R. Kreitzman, G.M.Luke, T. Riseman, C.E. Stronach, W.J. Kossler, J.R. Kempton, X.H. Yu, D. Opieand and H.E.Schone, (1988). *Phys. Rev. B*, 39, 909.

Voronel, A.V., O. Linsky, A.Kislink, and S. Drislikh, (1988). *Physica C*, 153, 1086.

Wilson, J.A., (1988). *J. Phys. C*, 21, 2067.
Xing, D.Y., M. Liu , and C.S. Ting, (1988). *Phys. Rev. B*, 37, 9769.
Xing, D.Y., W.Y. Lai, W.P. Su and C.S. Ting, (1988a). *Solid State Commun.*, 65, 1319.
Yu, R.C., M.J. Naughton, X. Yan , P.M. Chaikin, F. Holtzberg, R.L. Greene, J. Stuart, and P. Davis, (1988). *Phys. Rev. B*, 37, 7963.

OPTICAL PROPERTIES

THE ENERGY GAP AND TWO-COMPONENT ABSORPTION IN A HIGH T_c SUPERCONDUCTOR

G.A. Thomas, M. Capizzi*, J. Orenstein, D.H. Rapkine, A.J. Millis,
P. Gammel, L.F. Schneemeyer and J.V. Waszczak

AT&T Bell Laboratories, Murray Hill, NJ 07974 USA

ABSTRACT

Evidence is discussed favoring an energy gap of conventional size and a frequency dependent conductivity with two components.

KEYWORDS

Optical properties of metals; superconductivity; energy gap; conductivity.

INTRODUCTION

We have engaged in a systematic search for the intrinsic optical conductivity of $Ba_2YCu_3O_{7-\delta}$ in the a,b plane as a function of frequency, temperature and O-concentration (Orenstein et al., 1988; Thomas et al., 1988; Thomas et al.) The experiments involved strong interactions between the optical results and the crystal growth procedures to provide feedback for improved preparation procedures. More than two dozen crystals with varying surface preparation, annealing history and age were screened with reflectivity measurements. A compendium of many of these results (Thomas et al.) has been published to emphasize the substantial variation from sample to sample and the danger of basing conclusions on the results from one crystal. Progress came when two samples were grown which had low frequency reflectivities that were within experimental error of 1. These samples were measured immediately after growth and had

---

*Permanent address: Department of Physics, University of Rome, Rome, Italy.

reduced T_c's because they were not annealed. We speculate that such a procedure gives relatively uniform O concentration and clean surfaces. Results for one of these samples are shown in Figures 1-4, and have been discussed previously. (Orenstein *et al.*, 1988; Thomas *et al.*, 1988)

The reduced T_c values of these two samples turned out to provide the advantage of being fairly different (50K and 70K) so that when we found apparent superconducting energy gaps that were also different, the variation supported the interpretation of our region of perfect reflectivity (within error) as being a relatively convincing measure of 2Δ. The ratio of $2\Delta/T_c$ for both of these crystals was near 3.5. While these crystals appear to be ralatively good, we have no definitive evidence that they show intrinsic behavior and we are unable (because of the microtwinning) to investigate the possible anisotropy in the a,b plane.

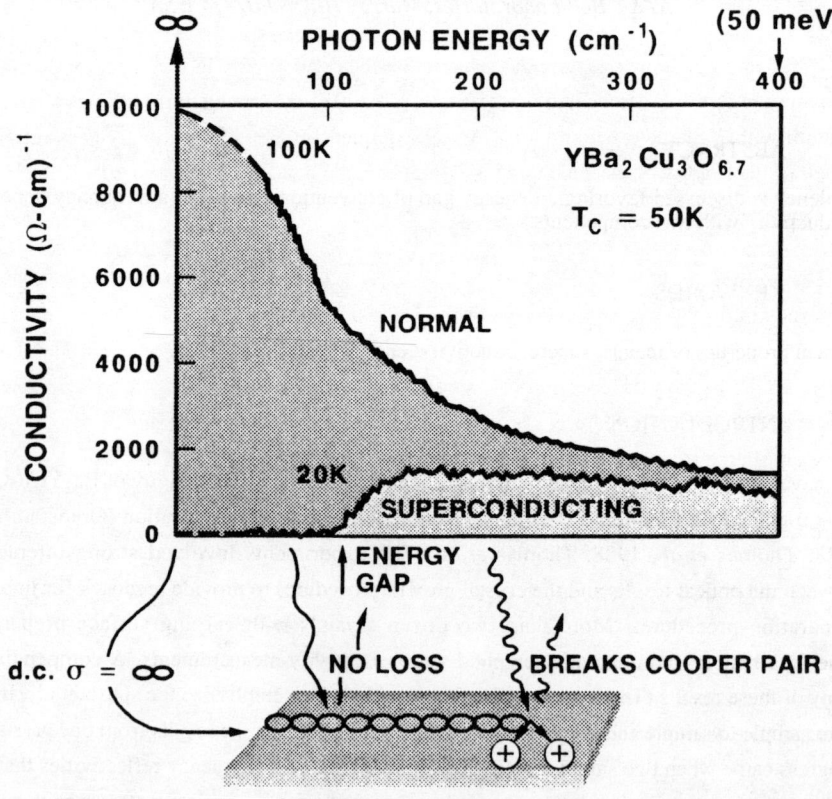

Fig. 1   Conductivity as a function of frequency in the region of the superconducting energy gap.

THE ENERGY GAP

Figure 1 shows a plot of the conductivity as a function of frequency in the region that may be the superconducting energy gap. This crystal had a T_c of 50K as measured by a relatively sharp anomaly in magnetization, and an O concentration that we estimate from T_c. The ratio $2\Delta/T_c$ is about 3.4 (with about 20% random uncertainty) for this data, so the energy gap is of a normal size expected for a BCS superconductor in the weak coupling approximation.

The upper curve is the normal state (100K) and the lower curve is the superconducting state (20K). There are substantial uncertainties in the measured reflectivity in this region[1], but within these limits we have taken a smoothed spectrum with an extrapolation to zero frequency (indicated by a dashed curve in the Figure) and performed a Kramers-Kronig transformation (using higher frequency data as well) to obtain the conductivity and dielectric constant.

We shall attempt some pedagogical remarks on the conductivity in the superconducting state shown in Figure 1. As indicated by the heavy line (20K) the conductivity follows a remarkable pattern with 3 characteristic regions. At zero frequency (d.c.) the conductivity is infinite as a result of the rigidity of the superconducting macroscopic wave function which is impervious to ordinary scattering.

The wavy line in the crystal at bottom represents a Cooper pair in the condensed state as a superposition of 2 waves traveling with wave-vectors $\pm k_F$. At finite, low frequencies, the conductivity drops to zero, i.e. to a state with no loss, because the photon energy is too small to break up the pairs in the condensed state. In this frequency region below 2Δ the reflectivity is perfect because, while there is no loss, the light is reflected by the large negative dielectric constant. In the frequency region above 2Δ the photon energy is sufficient to excite the carriers (holes in this case) out of the condensed state and the conductivity returns toward that of the normal state.

THE DIELECTRIC CONSTANT

The dielectric constant for this sample is shown in Figure 2. At lower frequencies than those shown, the curve continues smoothly and monotonically to lower values, and appears to diverge. The zero-crossing indicates the plasma frequency.

TWO COMPONENTS OF THE CONDUCTIVITY

The high values of reflectivity at low frequency observed in the two samples discussed above encouraged us to consider that the spectra at higher frequency might be predominantly intrinsic. The results for the conductivity at somewhat higher frequency are shown in Figure 3. The main point that we wish to illustrate with this Figure is the presence of two distinguishable types of absorption. First, a low frequency, T-dependent component (unshaded) and second, a higher frequency, T-independent component (shaded). The low frequency part, which we shall call "Drude", probably does not strictly follow the Drude form. In other words, it probably does not have a frequency independent scattering rate, Γ. However, it is similar to the Drude form (given as "simple theory" in the Figure), as shown by the reasonable fit (below 200 cm^{-1}) to the data taken at T=100K.

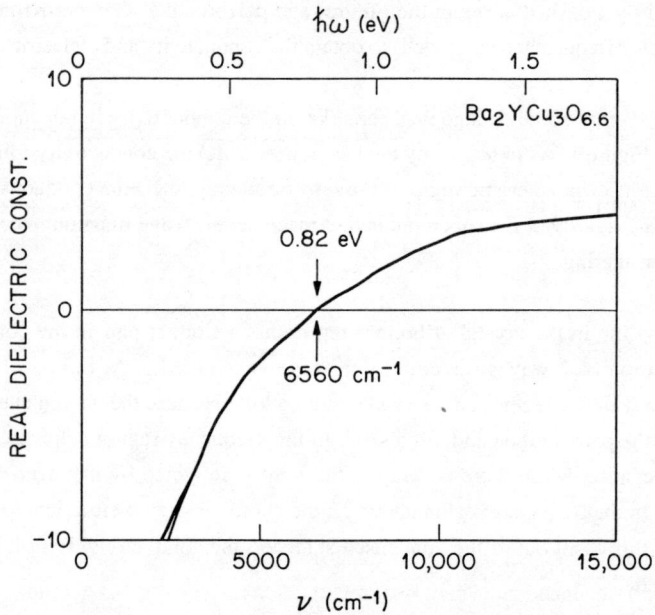

Fig. 2. Dielectric constant as function of frequency as it passes through the plasma frequency.

One possible interpretation of the two absorption regimes is shown in Figure 3 where the charge carriers are surrounded by a cloud of excitations. Qualitatively, we might expect the low frequency photons to move the carrier and cloud together, with the carrier subject to a variety of possible T-dependent scattering processes.

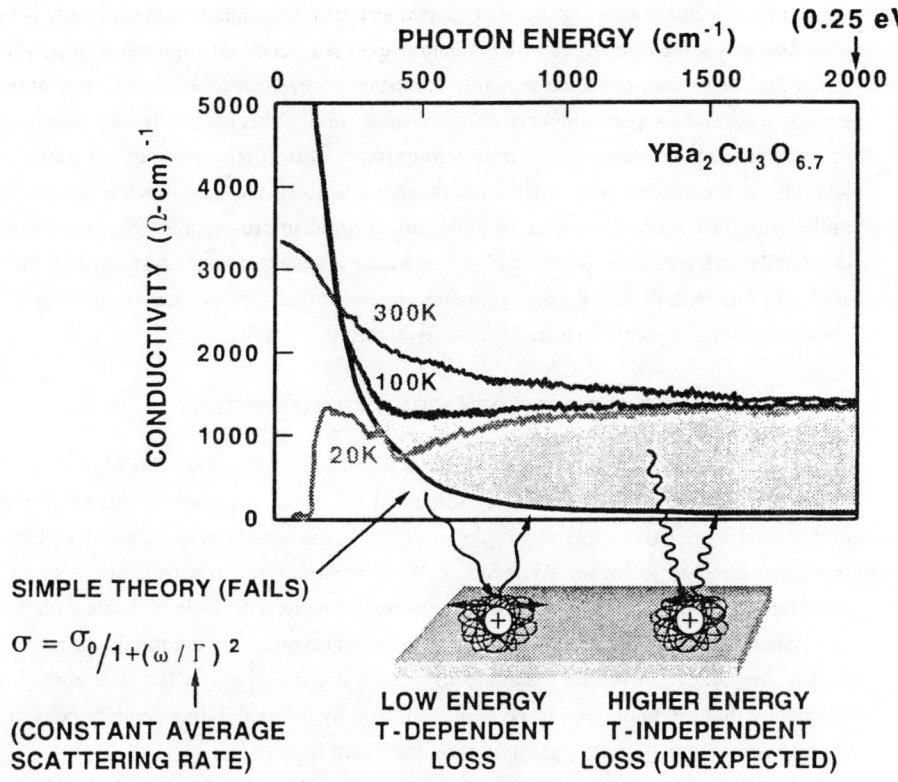

Fig. 3. Conductivity as a function of frequency showing the two components (shaded and unshaded), and the feature near 50 meV (430cm^{-1}).

At higher frequencies, one might expect that shorter range excitations could occur and be relatively T-independent. This class of excitations is unexpected and not included within the simple theory. These qualitative comments on the spectrum are intended to be as nearly generic as possible because we do not understand the origin of the absorption processes at this time.

FEATURE IN THE CONDUCTIVITY NEAR 50 meV

The shaded area in Figure 3 is shown as extending down in frequency to where the simple theory deviates from the 100K data. In fact, we do not know how to separate the two components of the conductivity discussed above. One possible demarkation point is the feature that appears as a minimum in the conductivity at about 430 cm^{-1} (or about 50 meV). This feature is most distinct at the lowest T, but is present at 100K. Because of the T-dependence of the reflectivity this feature has been suggested as the superconducting energy gap, 2Δ. We do not think this is the case because the feature occurs at a relatively high value of conductivity, rather than near zero, as shown in Figure 3. Furthermore, we see the same feature at the same energy in several samples with very different values of T_c. This feature has also been attributed to a plasma edge resulting from a near-zero-crossing of the dielectric constant (rather than an anomaly in the conductivity). Although this appears to be the case in some data on pressed pellet samples, such ensembles of small anisotropic micro-crystals give results that are drastically different from what we observed in the a,b plane of our crystals. In particular, as shown in Figure 2, there is no near-zero-crossing in the dielectric constant near 50 meV, and as shown in Figure 3, there is a distinct anomaly in the conductivity.

SENSITIVITY OF THE CONDUCTIVITY TO O CONTENT

A third consistent observation in our studies is that the conductivity of this high T_c materials is extremely sensitive to O content. One example of this result is shown in Figure 4. In the upper part of the Figure the conductivity is plotted for the same sample as in the previous Figures but is shown here up to higher frequencies. Also shown is the spectrum for a sample with substantially reduced O. In the $O_{6.6}$ sample the minimum near 50 meV is visible with the Drude component at lower frequency and the T-independent component comprising a broad peak at higher frequency. In the $O_{6.1}$ sample the electronic contribution is small enough at low frequency that the phonons can be seen, followed by a broad T-independent peak at higher frequencis and finally, a stronger peak near 15000 cm^{-1}.

DISTRIBUTION OF SPECTRAL WEIGHT

The bottom half of Figure 4 shows the integrated area under the two curves in the upper Figure. This integration is carried out up to some frequency and the area is plotted as a function of this frequency. This integral has the units of a number and is shown in the Figure as the number per unit volume, n, times the volume of the unit cell, V_{CELL} (170 Å^3), times the ratio of the free electron mass, m_e, to the carrier effective mass, m*. For m* = m_e, the curves measure the number of holes peer cell. We have indicated on the curves the point where this number reaches 0.2. This point is of interest because for the $O_{6.1}$ sample, the 0.1 O beyond 6 may contribute two holes to the non-conducting states. This part of the spectrum is shaded in the upper Figure. For the $O_{6.6}$ sample, the 0.1 O beyond 6.5 may contribute two holes to the conducting states. Again this part of the spectrum is shaded. In some qualitative sense these two regions may represent the non-conducting and conducting parts of the spectrum. As we indicated above we cannot clearly define the proper separation between these two components of the absorption, but the qualitative indication seems clear.

DETERIORATION OF THE CRYSTALS

We have observed a deterioration of the crystals through a decrease in the reflectivity with time.

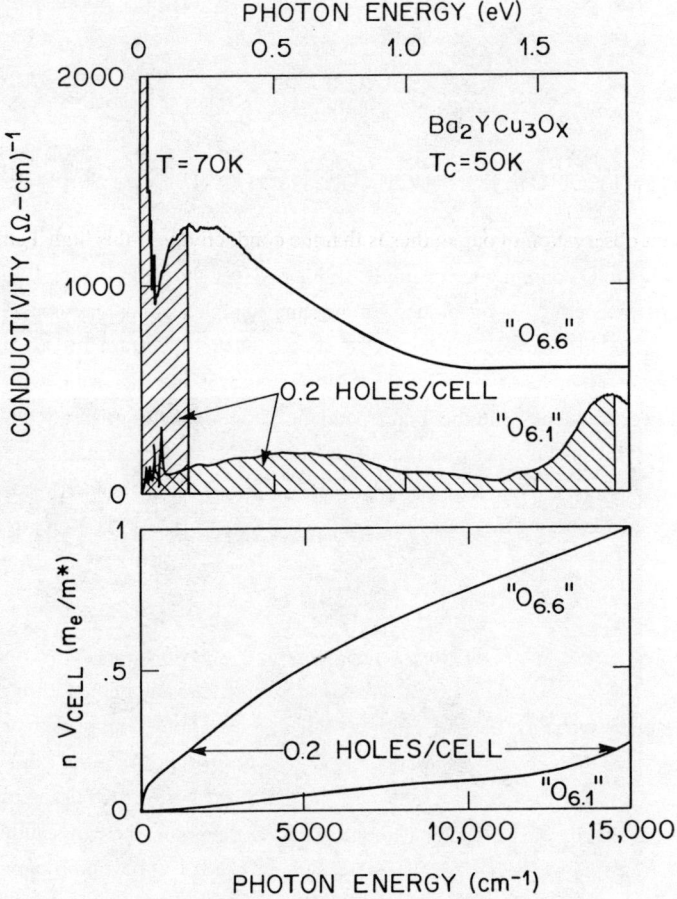

Fig. 4. Conductivity as a function of frequency for two samples with different O concentrations. The shaded area indicates the spectral weight from 0.2 holes/cell (with mass ratio 1) derived from the integrated area under the curves plotted in the lower part of the Figure.

The samples were kept in a N_2 gas atmosphere except for brief periods of careful sample handling in the room air. The drop in R observed corresponds to a drop in conductivity, and has generally been much smaller than that intentionally produced in the sample shown in Figure 4. In samples with slight degradation we have found spots appearing in electron microscope pictures of the sample surfaces as shown in Figure 5.

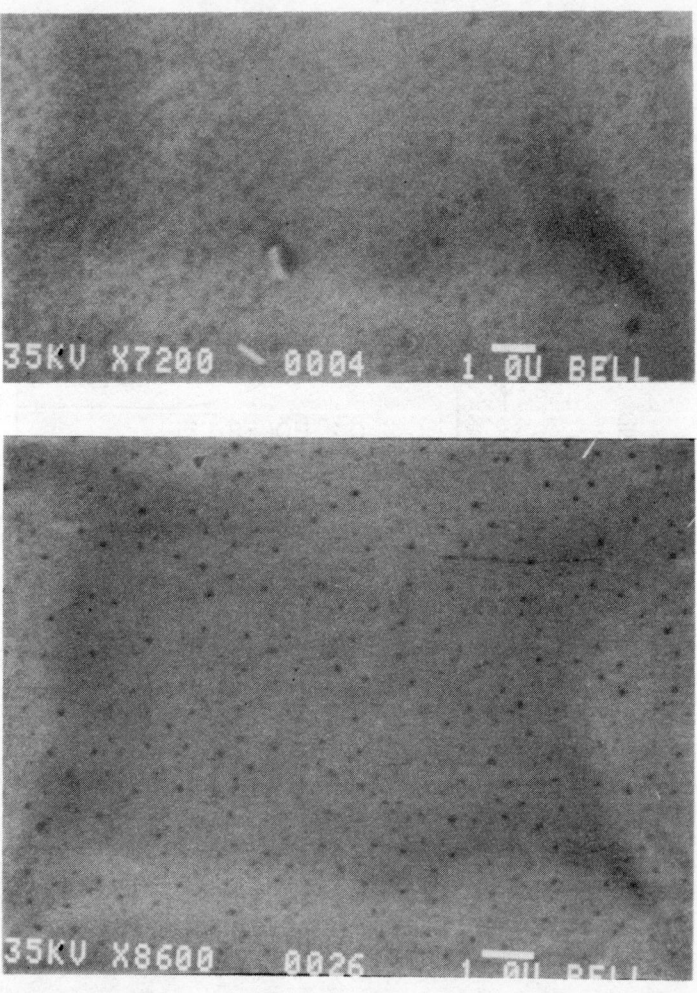

Fig. 5. Electron micrograph of a slightly degraded crystal (different from the sample discussed above).

After substantial degradation the spots have been found to cover the surface. The region inside the spots is of lower conductivity than the surrounding surface, but we do not know the chemical composition of the degraded regions as yet.

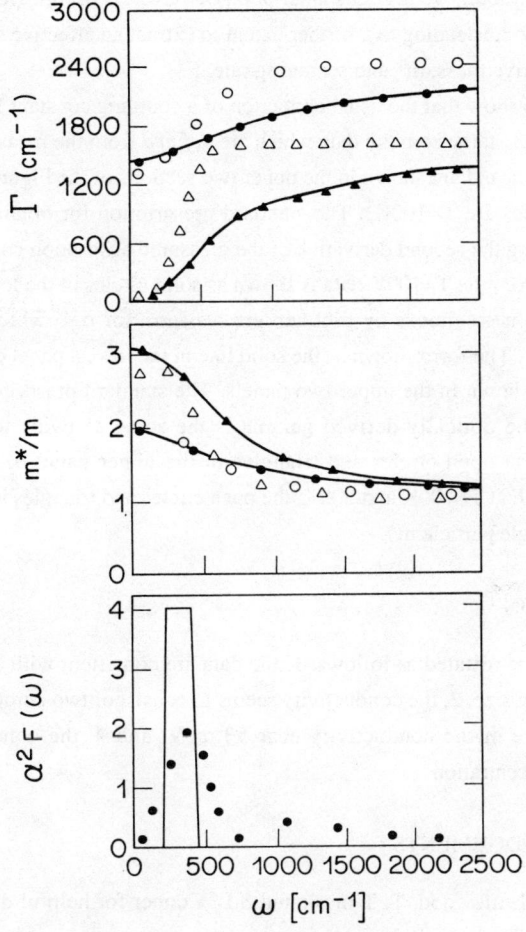

Fig. 6. The scattering rate (upper), effective mass (middle), and coupling constant (lower) as a function of frequency, with data (T=100K, solid triangles; T=300K, solid circles) and calculated values (open circles, lines) obtained as described in text.

THE FREQUENCY DEPENDENT SCATTERING AND MASS VIEWPOINT

An alternative inerpretation of the spectra to the two-component picture discussed above can be made in terms of a frequency dependent mass and scattering rate. Based on the variation of the spectra with O concentration, we find this latter point of view less useful. However we wish to comment that it may be misleading to a further extent to extract an effective coupling constant, $\alpha^2 F(\omega)$ from the effective mass, m* and scattering rate, Γ.

Figure 6 is intended to show that the usual extraction of a coupling constant from Γ is not self-consistent. The analysis starts from the data which are *defined* from the measured conductivity using the Drude formula and are shown in the upper two sections of the Figure (solid circles are T=300K, solid triangles are T=100K). The standard prescription for obtaining the coupling function involves taking the second derivative of the measured absorption coefficient. The $\alpha^2 F$ obtained in this way from the T=100K data is shown as solid circles in the lowest panel of Fig. 6. A better procedure is to choose by trial and error a form for $\alpha^2 F$ which reproduces the measured conductivity. The form shown as the solid line in the lowest panel of Fig. 6 yields the curves for Γ and m* shown in the upper two panels. The standard prescription is inaccurate because it assumes the optically derived gamma is the same as twice the single particle scattering rate Γ_s. The open circles and triangles in the upper panel of Fig. 6 show $2\Gamma_s$ calculated from the $\alpha^2 F$ at T=100K and 300K; the open circles and triangles in the central panel similarly show the single particle m*.

CONCLUSION

Our main points can be restated as follows: 1. the data are consistent with the presence of an energy gap of ordinary size, 2. the conductivity seems to consist of two components, 3. there is a reproducible feature in the conductivity near 50 meV, and 4. the conductivity changes drastically with O concentration.

ACKNOWLEDGEMENTS

We wish to thank P. Littlewood, T. Timusk and S.L. Cooper for helpful discussions, and L. W. Rupp and R.E. Miller for technical assistance.

REFERENCES

Orenstein, J., G.A. Thomas, D.H. Rapkine, A.J. Millis, L.F. Schneemeyer, and J.V. Waszczak, (1988). "Reflectivity of $Ba_2YCu_3O_{7-\delta}$: Normal State Dynamics", *Physica C*, 153-155, 1740.

Thomas, G.A., J. Orenstein, D.H. Rapkine, M. Capizzi, A.J. Millis, R.N. Bhatt, L.F. Schneemeyer and J.V. Waszczak, (1988) "$Ba_2YCu_3O_{7-\delta}$: Electrodynamics of Crystals with High Reflectivity", *Phys. Rev. Lett.*, 61, 1313.

Thomas, G.A., M. Capizzi, T. Timusk, S.L. Cooper, J. Orenstein, D.H. Rapkine, S. Martin, L.F. Schneemeyer and J.V. Waszczak, (in press) "Variations in the Far-Infrared Reflectivity of $Ba_2YCu_3O_x$," *Journal of the Optical Society of America*.

VIBRATIONAL MODES IN THE CuO_2 PLANES OF $YBa_2Cu_3O_{7-\delta}$

M. Cardona and C. Thomsen

*Max-Planck-Institut für Festkörperforschung, Heisenbergstr. 1
D-7000 Stuttgart 80, Federal Republic of Germany*

ABSTRACT

We have studied the vibrational properties of the CuO_2 planes in $YBa_2Cu_3O_{7-\delta}$ using Raman scattering. The out-of-phase vertical vibration of OII, III has features linking it directly to the superconductivity properties of these materials. The onset of the anomalous frequency softening of this mode at T_c is lowered in temperature when a magnetic field (H ≤ 12 T) is applied. The anomaly is absent in non-superconducting $YBa_2Cu_3O_{7-\delta}$ ($\delta \simeq 1$) and $PrBa_2Cu_3O_{7-\delta}$ ($\delta \simeq O$).

KEYWORDS

CuO_2 planes; lattice vibrations; phonon softening; superconducting energy gap; Fano line shape.

INTRODUCTION

The CuO_2 planes seem to play a crucial role in the high-temperature superconductivity mechanism. They are present in all materials with transition temperature $T_c \geq 35K$ and it is likely that they are, at least partially, responsible for the high transition temperatures observed. We have investigated the vibrational porperties of these planes using Raman scattering in order to find a clue for a possible involvement of optical phonons in the pairing mechanism. One of the three Raman-active modes related to vibrations of copper or oxygen in the planes has indeed two properties that link it to superconductivity: an anomalous softening of the phonon when the temperature is lowered below T_c and an asymmetric Raman line shape. We show Raman scattering results and discuss both of these properties which are absent in non-superconducting samples.

EXPERIMENTAL RESULTS

All spectra presented were taken on single crystals grown by a slow-cooling method (Schneemeyer *et al.*, 1988). The superconducting $YBa_2Cu_3O_{7-\delta}$ crystals had a T_c of 89K and a sharp transition (3K) as measured by the shielding effect. The crystal with $\delta \simeq 1$ was tetragonal and did not have a transition to superconductivity. The $PrBa_2Cu_3O_{7-\delta}$ crystals had $\delta \simeq 0$, were tetragonal (a = 3.892 Å, c = 11.648 Å), and non-superconducting as determined by Morán *et al.* (1988). The Raman spectra were excited with the 5145 Å line of an Ar^+-ion laser (< 10 mW), dispersed by a triplemate monochromator (SPEX, 1988) and recorded with a Mepsicron multichannel detector (ITT, 1988). Due to the weakness of the scattered intensities each spectrum had to be accumulated for several hours.

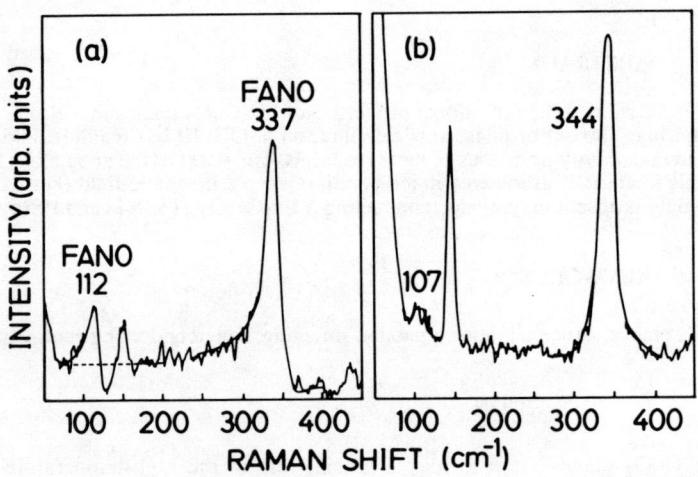

Fig. 1. (a) Raman spectra showing a Fano asymmetry in single crystals of $YBa_2Cu_3O_{7-\delta}$ ($\delta \approx 0$). The dashed line indicates the background due to scattering off of the electronic continuum. The dip on the high-energy side of the phonon peak originates from an interference of the discrete phonon frequency with the continuous background. (b) In semiconducting $YBa_2Cu_3O_{7-\delta}$ ($\delta \approx 1$) no asimmetry is observed. (The phonon line shapes are emphasized by smooth lines near the peaks)

In Figs. 1a and 1b we show the spectra of the superconducting ($\delta \approx 0$) and the semiconducting ($\delta \approx 1$) $YBa_2Cu_3O_{7-\delta}$ crystals, respectively, recorded at 15K. We show here only the spectral range between 50 and 450 cm^{-1}; for a complete spectrum see, for example, Thomsen and Cardona (1988). Clearly visible are the asymmetric line shapes of the phonons at 112 cm^{-1} and 337 cm^{-1} in the superconductor (Fig. 1a) (peak frequencies given; the actual oscillator frequencies are shifted slightly to lower frequencies due to the asymmetry). In the semiconductor (Fig. 1b) the corresponding two peaks are symmetric and have a Lorentzian line shape as is normally expected.

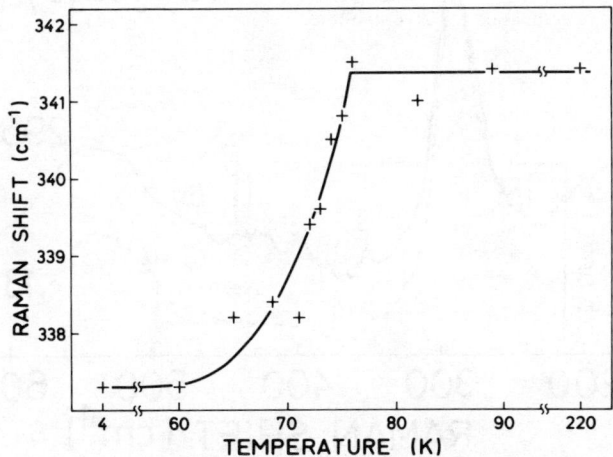

Fig. 2. Anomalous softening at T_c of the Raman-active phonon involving vibrations in the CuO_2 planes of $YBa_2Cu_3O_7$. The discrepancy between T_c (89 K) and the onset of softening (T_s = 75 K) is due to laser heating of the sample.

In Fig. 2 [taken from Thomsen and Cardona (1988)] we show the temperature dependence of one of the peaks. Between room temperature and T_c its frequency does not vary appreciably, but near T_c it suddenly drops (within ~10 K) by about 1 to 2% to remain again roughly constant down to 4 K. This softening, first observed by Macfarlane *et al.* (1987) in ceramic samples as a broad transition, is particularly sharp in Fig. 2; the deviation of the onset of softening (T_s = 75 K) from T_c is due to heating of the illuminated spot by the laser. Measuring at various power levels Thomsen and Cardona (1988) were able to show that T_s extrapolates to T_c for zero incident laser power. The frequency of the corresponding phonon in the semiconducting phase does not soften upon cooling the sample.

Fig. 3 shows Raman spectra of $PrBa_2Cu_3O_{7-\delta}$ ($\delta = O \pm 0.1$) at room temperature and at 13 K. It is seen that the CuO_2-plane phonon (B_{1g} symmetry) under discussion (here at 292 cm^{-1}) does not shift anomalously with temperature (Thomsen et al., 1988b). Rather, it remains constant in frequency. Similarly, the line shape is Lorentzian and not asymmetric as in the superconductors [Fig. 1a]. It should be noted that while two phonons in the superconductor have an asymmetry only one shows a softening at T_c.

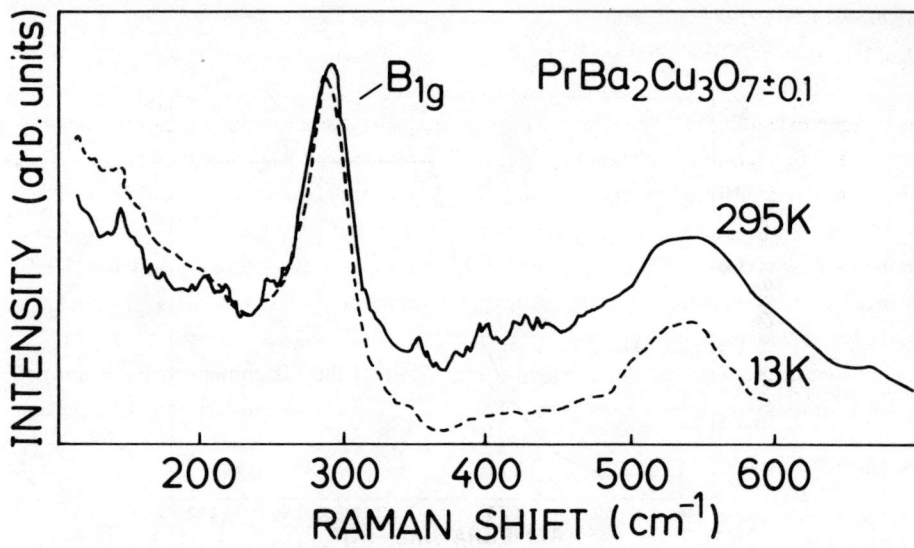

Fig. 3. Raman spectra of single crystalline $PrBa_2Cu_3O_{7\pm0.1}$. The line shape of the B_{1g} phonon is symmetric and a softening between room temperature and 13K is not observed.

DISCUSSION

First it is necessary to establish the origin of the Raman peaks, i.e. to know what the eigenvectors of the vibrations are. Using the symmetry properties of the near-tetragonal crystal and the polarization selection rules in Raman scattering, it was possible to prove that the peak at 335 cm^{-1} corresponds to an out-of-phase vertical vibration (with near B_{1g} symmetry) of the two oxygen atoms in each CuO_2 plane. Corresponding oxygen atoms in two different planes in the unit cell are also out of phase as a consequence of the even parity of Raman modes with respect to the inversion center (Thomsen and Cardona, 1988). The mode at 112 cm^{-1}, according to a lattice dynamical calculation, corresponds to a Ba (A_g-symmetry) vertical vibration (Liu et al., 1988). It has been suggested that for reasons related to the asymmetric line shape of the Raman peak this mode should originate from the CuO_2 planes, i.e. it should be the Cu vertical vibration (Cooper et al., 1988). We would like to point out, however, that the strong anisotropy reported for the 112 cm^{-1} mode in untwinned single crystals strongly suggests that Ba has the largest displacement in the eigenvector of this mode (Cooper et al., 1988b; Thomsen et al., 1988).

In discussing the results, the following questions must be addressed: What is the physical origin of 1) the softening 2) the asymmetric line shape and 3) what, if any, is the connection with superconductivity? A possible origin for the softening is the opening of a superconducting energy gap extending to or beyond the phonon frequency, and implying a gap parameter of $2\Delta/kT_c \geq 5$. In a recent calculation Fujimori (1988) has reported that a small softening may be consistent with a charge transfer from the CuO_2 planes into oxygen of the BaO planes. The asymmetric line shape may originate from a Fano resonance, i.e. the interaction of the (discrete) phonon frequency with a background of continuous electronic excitations (Fano, 1961). Interference causes a reduction of the scattering due to the electronic background, best seen for the 112 cm^{-1} phonon in Fig. 1a where the electronic background has been indicated by a dashed line (Thomsen et al., 1988 and Cooper et al., 1988). If the assignment to Ba is correct, however, it remains a question, how the phonon at 112 cm^{-1} can interact with an electronic continuum.

The absence of both the softening of the 335 cm^{-1} phonon and the Fano line shapes in $YBa_2Cu_3O_{7-\delta}$ ($\delta \approx 1$) is, of course, consistent with the above explanation of the effects. In the semiconductor the superconducting gap is absent and so is the electronic continuum. $YBa_2Cu_3O_6$, however, has a somewhat different structure and it should be more appropriate to study these effects in a non-superconducting material even more closely related to $YBa_2Cu_3O_7$. We have confirmed in two ways that the anomalous softening is indeed related to the occurrence of superconductivity.

First we studied the Raman spectra of $PrBa_2Cu_3O_7$ which has the same number of oxygen atoms per unit cell as does $YBa_2Cu_3O_7$ but is not superconducting. The Pr-crystals had tetragonal symmetry, a fact which appears not to be related to the absence of superconductivity: $PrBa_2Cu_3O_7$ has also been prepared with orthorhombic symmetry (in ceramic form) and is not superconducting either (Tarascon et al., 1987). We have also performed experiments on ceramic, orthorhombic $PrBa_2Cu_3O_7$ and not found the softening. In Fig. 3 we see that the anomalous softening is not present in the crystal either. Therefore, we conclude that mere structural effects of the seventh oxygen are not sufficient to explain the softening.

The most stringent test of the role of superconductivity in this context is however to study a sample at a particular temperature and then "switch on" of "off" superconductivity with a magnetic field H > H_{c2} of that temperature. Alternatively, one may study the onset of softening, T_s, for various fixed applied magnetic fields. It is expected that T_s is lowered for a field H from T_c (H = O) to T^1_c (H = H_{c2}), i.e. to a temperature where H does just not exceed the critical magnetic field H_{c2}. Should the phonon softening, however, originate from e.g. a structural change in the lattice at or near T_c (H = O), an applied magnetic field should have no influence on T_s. These experiments have been performed by Ruf et al., (1988), and show that T_s is indeed a function of applied magnetic field. The slope dT_s/dH is found to be roughly coincident with dT_c/dH_{c2} for a magnetic field (O ≤ H ≤ 12.5 T) parallel to the c-axis of the crystal. Measurements with H perpendicular to the c-axis have not yet been reported due to the larger magnetic fields required in that geometry.

Hence, superconductivity appears to be a necessary condition for the softening of the oxygen vibrational mode to occur in the perovskite-like superconductors. This anomaly should then also be observed in the Bi, Tl, and Pb superconductors (with 2 or more CuO_2 planes) all of which have a Raman-active mode similar to that discussed here. The oxygen mode has been identified in $Bi_2(Sr_{1-x}Ca_x)_3Cu_2O_{8+\delta}$ and, recently, in $Pb_2Sr_2EuCu_3O_{8+\delta}$ but a temperature anomaly has not yet been observed (Thomsen and Cardona, 1988 and Cardona et al., 1988). This mode is quite broad in the Bi compound and the shift may be too small to be observed if it exists. In the far infrared (fir) the softening has been seen most clearly for two phonons at 275 and 315 cm^{-1} in ceramic $YBa_2Cu_3O_7$ samples. The relative shift is somewhat larger (-3%) between temperatures slightly above T_c and 10K than in the Raman spectra. Anomalous line shapes have also been observed in the fir, but their origin has not yet been analyzed in detail (Genzel et al., 1988).

The asymmetric Raman line shape indicative of the interaction between phonons and the electronic continuum is absent for the $PrBa_2Cu_3O_7$ crystal. Figure 3 shows this clearly for the data taken at room temperature; in the spectrum taken at 13K a varying background makes this more difficult to see. The temperature dependence of the asymmetry in superconducting samples has been reported by Feile et al. (1988). They show a weak temperature dependence together with a small discontinuity at T_c. There is, however, no reason to expect the low-temperature Raman peak in Fig. 3 to develop a significant asymmetry not at all present at 273K. The peak near 112 cm^{-1} could not be resolved in the $PrBa_2Cu_3O_7$ crystal due to the increasing background at low frequencies.

CONCLUSION

Two of the Raman-active vibrational modes of the $YBa_2Cu_3O_{7-\delta}$ superconductor have been shown to interact with the electronic system. In particular, interferences between two phonons and an electronic background resulted in asymmetric line shapes. One of these phonons is known to originate from the CuO_2 planes. The other, lower frequency one, probably stems from vibrations of the Ba atoms. The frequency softening of the CuO_2- plane mode has been shown to be directly related to superconductivity, since the onset depends on the application of a magnetic field in the same way as does T_c.

These results, of course, do not imply that the optical phonons discussed are responsible for the formation of Cooper pairs; they merely show an interaction of electrons with phonons. Theoretical calculations have been reported, however, showing that one must go beyond weak-coupling BCS theory for a possible mechanism. Assuming a strong-coupling theory to be valid, a coupling parameter of $\lambda^* \geq 1.7$ has been calculated by Zeyher and Zwicknagl (1988) from the phonon softening reported here. Weak coupling, on the other hand, predicts a hardening for all phonons when cooling below T_c.

ACKNOWLEDGEMENT

We thank B. Gegenheimer and E. Morán for providing us with the crystals used in this work and R. Liu for experimental assistance.

REFERENCES

Cardona, M., C. Thomsen, R. Liu, M.A. Alario-Franco, Hj. Mattausch, and W. König (1988), *to be published*.

Cooper, S.L., M.V. Klein, B.G. Pazol, J.P. Rice, and D.M. Ginsberg (1988). Raman Scattering from Superconducting Gap Excitations in Single Crystal $YBa_2Cu_3O_{7-\delta}$. *Phys. Rev. B*, 37, 5920-5923.

Cooper, S.L., F. Slakey, M.V. Klein, J.P. Rice, E.D. Bukowski and D.M. Ginsberg (1988b). Raman Scattering Studies of Coupled-Modes and Superconducting Gap Excitations in Single Crystal $YBa_2Cu_3O_{7-\delta}$. *JOSA B*, special issue, to be published.

Fano, U. (1961). Effects of Configuration Interaction on Intensities and Phase Shifts. *Phys. Rev.*, 127, 1866-1878.

Feile, R., P. Leiderer, J. Kowalewski, W. Assmus, J. Schubert and U. Poppe (1988). Temperature Effects on the Phonon Spectrum in $YBa_2Cu_3O_7$ Single Crystals and Thin Films. *Z. Phys. B*, to be published.

Fujimori, A. (1988). Character of Doped Holes and Low-Energy Excitations in High-T_c Superconductors: Roles of the Apex Oxygen Atoms, *this volume*.

Genzel, L., A. Wittlin, M. Bauer, M. Cardona, E. Schönherr and A. Simon (1988). Phonon Anomalies and Superconducting Energy Gaps from Infrared Studies of $YBa_2Cu_3O_{7-\delta}$. *To be published*.

ITT, 1988. Fort Wayne, IN, USA

Liu, R., C. Thomsen, W. Kress, M. Cardona, B. Gegenheimer, F.W. de Wette, J. Prade, A.D. Kulkarni and U. Schröder (1988). Frequencies of k = O Phonons in $YBa_2Cu_3O_{7-\delta}$: Theory and Experiment. *Phys. Rev. B*, 37, 7971-7974.

Macfarlane, R.M., H. Rosen and H. Seki (1987). Temperature Dependence of the Raman Spectrum of the High-T_c Superconductor $YBa_2Cu_3O_7$. *Solid State Commun.*, 63, 831-834.

Morán, E., U. Amador, M. Barahona, M.A. Alario-Franco, A. Vegas and J. Rodriguez-Carvajal (1988). $Ba_2PrCu_3O_7$: Crystal Growth, Structure and Magnetic Properties. *Solid State Commun.*, 67, 369-372.

Ruf, T., C. Thomsen, R. Liu and M. Cardona (1988). Raman Study of the Phonon Anomaly in Single-Crystal $YBa_2Cu_3O_{7-\delta}$ in the Presence of a Magnetic Field. *Phys. Rev. B*, 38, 11985.

Schneemeyer, L.F., J. V. Waszczak, T. Siegrist, R.B. van Dover, L.W. Rupp, B. Batlogg, R.J. Cava and D.V. Murphy (1988). Superconductivity in $YBa_2Cu_3O_7$ Single Crystals. *Nature*, 328, 601-603.

SPEX, 1988. Spex Industries, Edison, NJ, USA.

Tarascon, J.M., W.R. McKinnon, L.H. Greene, G.W. Hull and E.M. Vogel (1987). Oxygen and Rare-Earth Doping of the 90-K Superconducting Perovskite $YBa_2Cu_3O_{7-x}$. *Phys. Rev. B*, 36, 226-234.

Thomsen, C. and M. Cardona (1988). Raman Scattering in High-T_c Superconductors. In: *The Properties of High-Temperature Superconductors* (D.M. Ginsberg, Ed.) (World Scientific Publishers, Singapore) to be published.

Thomsen, C., M. Cardona, B. Gegenheimer, R. Liu and A. Simon (1988). Untwinned Single Crystals of $YBa_2Cu_3O_{7-\delta}$: an Optical Investigation of the a-b Anisotropy. *Phys. Rev. B*, 37, 9860-9863.

Thomsen, C., R. Liu, M. Cardona, U. Amador and E. Morán (1988b). CuO_2-plane Vibrational Modes in Single Crystals of $PrBa_2Cu_3O_{7-\delta}$.*Solid State Commun.*, 67, 271-274.

Zeyher, R. and G. Zwicknagl (1988). Phonon Self-Energy Effects Due to Superconductivity: Evidence of the Strong-Coupling Limit in $YBa_2Cu_3O_{7-\delta}$. *Solid State Commun.*, 66, 617-622.

ABSORPTION AND PHOTOINDUCED ABSORPTION STUDIES OF La_2CuO_4 and $YBa_2Cu_3O_{6+x}$: ROLE OF DEFECT STATES

A.J. Epstein, J.M. Ginder, J.M. Leng, R.P. McCall, M.G. Roe[+], and H.J. Ye

The Ohio State University Columbus, Ohio 43210-1106

W.E. Farneth, E.M. McCarron III, and S.I. Shah

E.I. du Pont de Nemours and Company, Inc. Wilmington, DE 19880

ABSTRACT

Linear and photoinduced optical absorption studies of $YBa_2Cu_3O_{6+x}$ are described and compared with those of La_2CuO_4. Absorption studies of thin films of $YBa_2Cu_3O_{6+x}$ show a series of optical absorptions at 0.5, 1.6, 3.0, 4.3, and 4.8 eV for x=0. As x increases, the energies of each of the peaks are unchanged except for the highest-energy peak, which shifts considerably with increasing oxygen concetration. Photoinduced optical absorptions studies of $YBa_2Cu_3O_{6+x}$ reveal similar results for x=0 and x=0.3. These spectra differ from those of La_2CuO_4 in that the crossover from photoinduced absorption to photoinduced bleaching is shifted down in energy from 2.0 eV to 1.5 eV. There remain two photoinduced electronic absorptions, though their energies and character differ substantially from those observed in La_2CuO_4. For x=0.0 samples, the high energy photoinduced absorption is very broad, with a weak maximum at ~ 1.2 eV. The low energy peak is much sharper, with a maximum at ~0.13 eV. The infrared spectrum of x=0.0 samples is similar to that previously reported. Photoinduced infrared studies demonstrate the important role of electron-phonon coupling. The phonon modes shift as expected with substitution of ^{18}O for ^{16}O. A new photoinduced mode at 315 cm^{-1} is observed for $YBa_2Cu_3{}^{16}O_{6.0}$. This mode is associated with the breaking of the tetragonal symmetry to orthorhombic or lower symmetry upon photoexcitation of charge carriers. It is likely associated with the formation of O^{1-} sites upon photoexcited charge transfer.

KEYWORDS

High T_c superconductivity; photoinduced absorption; defect states; electron-phonon coupling; symmetry breaking.

INTRODUCTION

Spectroscopic studies of the $La_{2-x}Sr_xCuO_4$ and $YBa_2Cu_3O_{6+x}$ systems contribute to an understanding of the insulator-to-metal transitions that occur in these systems upon strontium and oxygen doping, and of the mechanism for pairing and superconductivity (Halley, 1988). There have been extensive attempts to correlate these results with the outcome of detailed electronic structure calculations for these materials. Photoinduced optical absorption provides an additional probe that enables a sensitive measure of the nature of the local charge states in the semiconducting form of these materials without perturbing the lattice through doping. Photoinduced UV-visible spectroscopy gives direct information on the presence of energy gaps and the electronic structure of the photoinduced defects, while photoinduced infrared spectroscopy gives insight into the nature and strength of the electron-phonon coupling and the role of symmetry breaking in these materials. Both of these photoinduced spectroscopies give insight into the time dynamics of the photoexcited charge states.

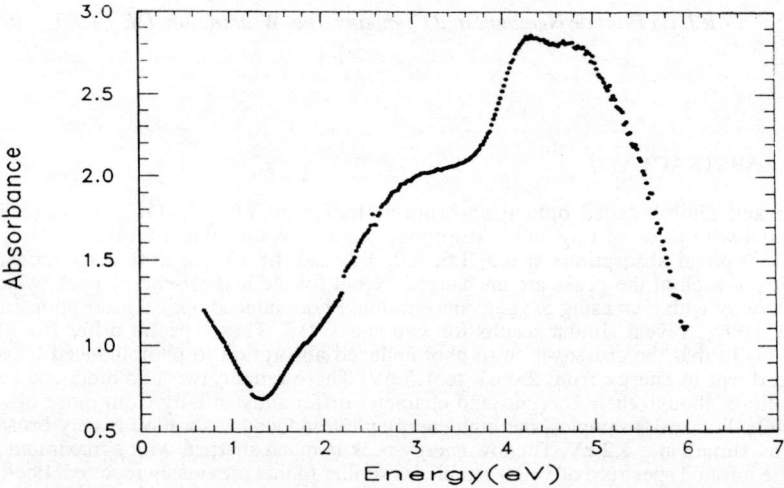

Fig.1. Absorbance spectrum of a 2500 Å thick film of $YBa_2Cu_3O_{6.0}$.

We present here a summary of our recent results in the application of absorption and photoinduced absorption spectroscopies to the $YBa_2Cu_3O_{6+x}$ system (Ye et al., and Leng et al.) and compare these results to those previously obtained for the La_2CuO_4 material (Ginder et al., 1988; Epstein et al., 1988). We conclude that there is a substantial reduction in the (charge-transfer) gap in going from the La_2CuO_4 system to the $YBa_2Cu_3O_{6+x}$ system. In addition, the details of the photoinduced optical absorption created by the photoexcited charge carriers are changed substantially. The origin of this change is important in an understanding of the origins of the metallic state in these systems, as well as the origin of superconducting coupling. Photoinduced infrared results demonstrate the important role of symmetry breaking upon photoexcitation, and reveal a significant, though still weak, electron-phonon coupling.

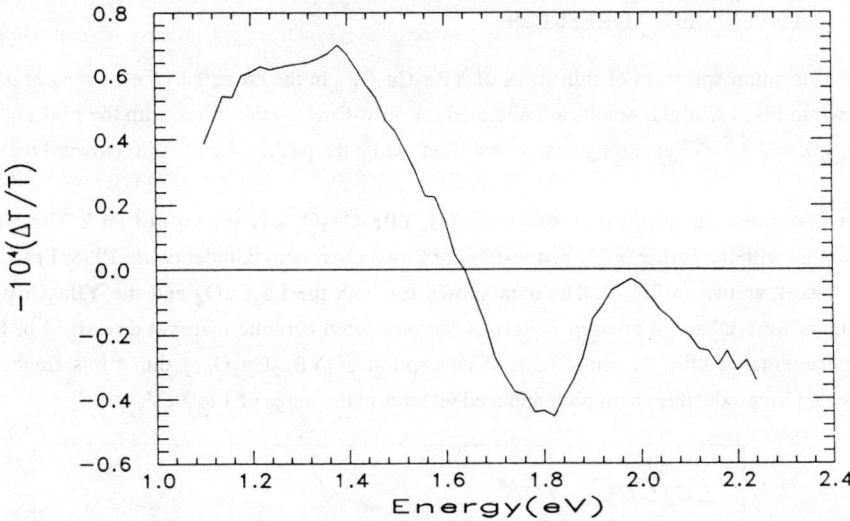

Fig. 2. The photoinduced UV-visible absorption spectrum of $YBa_2Cu_3O_{6.0}$ at T=10K.

EXPERIMENTAL

Powders of $YBa_2Cu_3O_{6+x}$ with $0.0 \leq x \leq 0.3$ were prepared as described previously (Farneth et al., 1988). To summarize briefly, samples of variable oxygen content were prepared by both low temperature vacuum annealing and rapid quenching techniques. Pressed, unsintered pellets were equilibrated for two hours in flowing air, nitrogen, or oxygen, at temperatures for which the equivalent oxygen content was known. Samples were then quenched by immersion in liquid N2. Oxygen contents were subsequently measured by thermogravimetric analysis and shown to be consistent with values expected from the equilibrium temperature reached prior to quenching. Thin films (2500-5000Å) of $YBa_2Cu_3O_{6+x}$ on MgO substrates were produced by annealing films of co-sputtered reactants (Shah). Near-steady-state photoinduced UV-visible photoinduced absorption measurements were carried out by exciting the sample with 2.71 eV photons from a chopped argon ion laser beam; the sample transmission T is probed by light from an incandescent lamp filtered through a monochromator and detected by the appropriate photodiodes. The induced change in sample transmission, ΔT, was measured with a lock-in amplifier. The quantitaty ΔT/T is proportional to the change in the absorption coefficient of the sample. Infrared measurements were made with a Nicolet 60SX FTIR spectrometer. For these photoinduced measurements, the fractional transmission change, - ΔT/T, was determined from the difference in absorption before and after excitation with an argon ion laser.

EXPERIMENTAL RESULTS

The absorption spectrum of thin films of $YBa_2Cu_3O_{6.0}$ in the range 1 to 6 eV (Leng *et al.*) is shown in Fig. 1. Similar results are obtained for $YBa_2Cu_3O_{6.3}$ thin films with the peaks at 0.5, 1.6, 3.0, and 4.3 eV remaining nearly unshifted, while the peak at 4.8 eV shifts toward 6 eV.

The photoinduced absorption spectrum for $YBa_2Cu_3O_{6.0}$ is shown in Fig. 2 These data contranst with the earlier published results for La_2CuO_4 system (Ginder *et al.*, 1988; Epstein *et al.*, 1988), shown in Fig. 3. The data shown for both the La_2CuO_4 and the $YBa_2Cu_3O_{6+x}$ systems were taken on pressed pellets of the powdered ceramic material despersed in KBr. Experimental studies of photoinduced absorption of $YBa_2Cu_3O_{6+x}$ thin films frequently revealed irreproducible sharp photoinduced features in the range of 1 to 3 eV.

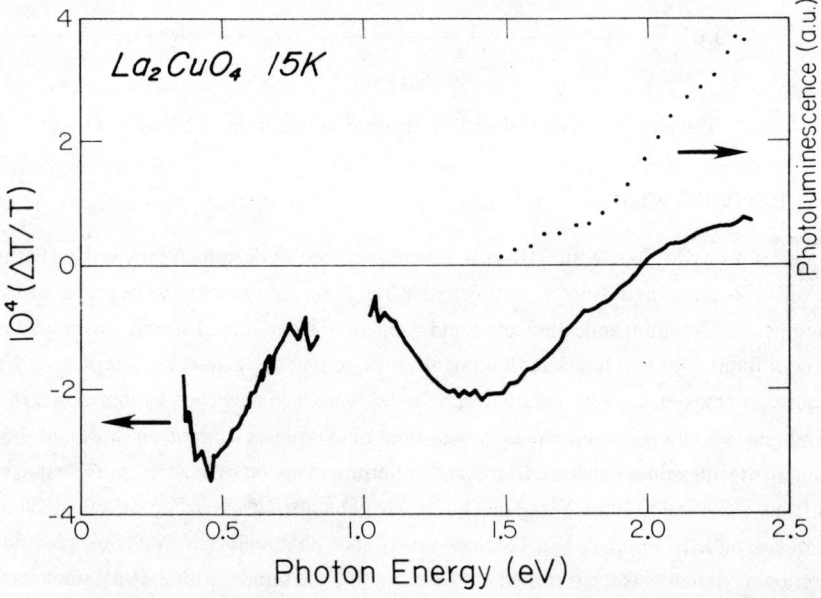

Fig.3. Solid line: photonduced UV-visible absorption spectrum of La_2CuO_4 at T=10K. Dotted line: photoluminescence spectrum of La_2CuO_4 at T=10K (Ginder *et al.*, 1988).

Photoinduced absorption studies were extended to the mid- and far-infrared region, as shown in Fig. 4. The spectrum is similar to that reported earlier by Taliani *et al.*, (1988) and by Kim *et al.*, (1988) with the exception that one new feature at 315 cm^{-1} is now clearly observed.

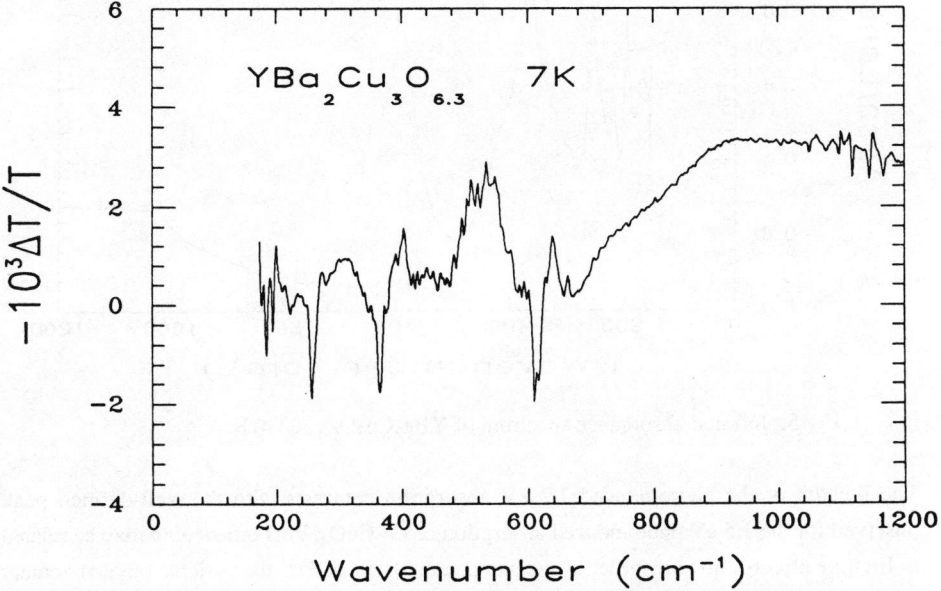

Fig.4. The photoinduced IR absorption spectrum of $YBa_2Cu_3O_{6.3}$ at T= 7K.

The photoinduced IR spectra can be contrasted with the direct IR absorption for $YBa_2Cu_3O_{6.0}$, shown in Fig. 5. Both the direct absorption modes and the photoinduced modes shift in frequency as expected for the substitution of ^{16}O with ^{18}O.

DISCUSSION

The origins of the photoinduced absorption are important both in the context of the models for the insulator-to-metal transition as well as models for pairing in the system. It was suggested that the 2 eV photoinduced bleaching observed in La_2CuO_4 (Ginder *et al.*, 1988; Epstein *et al.*, 1988) corresponds to an energy gap photoexcitation while the two photoinduced absorptions correspond to defect levels in the gap formed by the positive and negative photoinduced charges. The relevance of this model for the $YBa_2Cu_3O_{6+x}$ system is less clear. The photoinduced bleaching is likely related to the bleaching of a charge-transfer band caused by exciting an electron from an O^{2-} to a Cu^{2+} ion, both of which lie in the CuO_2 plane.

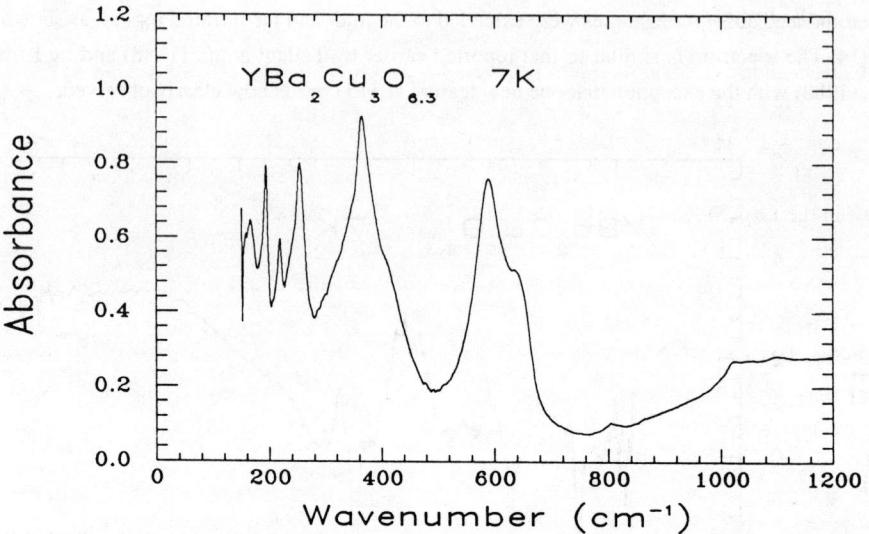

Fig.5. Infrared absorbance spectrum of $YBa_2Cu_3O_{6.3}$ at T=7K.

The breadth of the photoinduced 1.2 eV absorption contrasts with the well-defined peak observed for the 1.5 eV photoinduced absorption of La_2CuO_4. This broadening may be related to lifetime effects due to disorder of the site potentials caused by the variable oxygen content within the chain-containing Cu-O layers. Similarly, the low-energy peak at 0.13 eV is distinctly different in character from the low energy peak at 0.13 eV is distinctly different in character from the low energy peak at 0.5 eV in La_2CuO_4. The presence of a photoinduced infrared mode at 315 cm^{-1} suggests the important role of symmetry breaking in this system. The photoinduced negative defect state, D, is likely a Cu^{+1} site in the CuO_2 layers formed upon photoinduced charge transfer from a nearby 0^{2-}. The net negative charge is expected not to change the symmetry about the copper lattice site, hence it is unlikely to be associated with the 315 cm^{-1} photoinduced infrared mode. In contrast, the creation of a photoexcited positive defect state, D^+, whose origin is associated with the formation of an O^{1-} site, will distort the symmetry around the copper site, making an otherwise symmetric vibrational mode become infrared active given the nonzero electron-phonon coupling.

ACKNOWLEDGEMENT

This work is supported in part by the Defense Advanced Research Projects Agency through a contract monitored by the U.S. Office of Naval Research.

REFERENCES

+ Current address: Naval Research Laboratories, Washington, D.C. 20375

Epstein, A.J., M.G. Roe, J.M. Ginder, F. Zuo, R.P. McCall, Y. Song, X.D. Chen, J.P. Golden, J.R. Gaines, P.E. Wigen, and E. Ehrenfreund, (1988). Photoinduced absorption in the $La_{2-x}Sr_xCuO_{4-\delta}$. *Mat. Res. Soc. Symp. Proc.*, 99, 857.

Farneth, W.E., R.K. Bordia, E.M. McCarron III, M.K. Crawford, and R.B. Flippen, (1988). Influence of oxygen stoichiometry on the structure and transition temperatures of $YBa_2Cu_3O_{6+x}$ *Solid State Commun.*, 66, 953.

Ginder, J.M., M.G. Roe, Y. Song, R.P. McCall, J.R. Gaines, E. Ehrenfreund, and A.J. Epstein, (1988). Photoexcitations in La_2CuO_4 : 2 eV energy gap and long-lived defects. *Phys. Rev. B*, 37, 7506.

Halley, J.W. ed., *Theories of High Temperature Superconductivity*, (1988). (Addison-Wesley, Redwood City).

Kim, Y.H., C.M. Foster, A.J. Heeger, S. Cox, and G. Stucky, (1988). Photoinduced self-localized structural distortions in $YBa_2Cu_3O_{7-\delta}$ *Phys. Rev. B*, 38, 6478.

Leng, J. M., J.M. Ginder, M.G. Roe, A.J. Epstein, W.E. Farneth, S.I. Shah and E.M. McCarron III, (1989). Photoinduced excitations in semiconducting $YBa_2Cu_3O_{7-\delta}$ *Bull. Am. Phys. Soc.* 34, 1037 *and to be published.*

Shah, S.I. and P.F. Carcia, (1987). Superconductivity and the resputtering effects in RF sputtered $YBa_2Cu_3O_{7-\delta}$ thin films. *Appl. Phys. Lett.* 51, 2146.

Taliani, C.R., Zamboni, G. Ruani, F.C. Matacotta, and K.I. Pokhodnya, (1988). *Solid State Commun.*, 66, 487.

Ye, H.J., R.P. McCall, W.E. Farneth, E.M. McCarron III, and A.J. Epstein. (1989). Photoinduced IR absorption in $YBa_2Cu_3{}^AO_{6.3}$ (A=16,18) *Bull. Am. Phys. Soc.* 34, 1038 *and to be published.*

PHOTOINDUCED OPTICAL EXCITATIONS IN THE YBa$_2$CU$_3$O$_{7-y}$ HIGH T$_c$ SUPERCONDUCTING SYSTEM

C. Taliani, R. Zamboni, G. Ruani and A. J. Pal
Istituto di Spettroscopia Molecolare, CNR, Via de' Castagnoli 1, I-40126 Bologna, Italy

F. C. Matacotta
Istituto ITM, CNR, Via Induno 10, I-20092 Cinisello Balsamo, Italy

Z. Vardeny and X. Wea
Physics Department, University of Utah, Salt Lake City, 84112 USA

ABSTRACT

The IR and near-IR photoinduced absorption (PA) spectra of the semiconducting tetragonal phase of YBa$_2$Cu$_3$O$_{7-y}$ (y = 0.79 ÷ 0.85) are investigated in the 0.045-2 eV spectral range at low temperature.
Two electronic PA bands are observed at 0.13 and 1 eV and are attributed to midgap states generated or populated by photoexcitation of e-h pairs. Several hypothesis on the origin of these bands are proposed.
PA phonon bands observed at 435 and 510 cm^{-1} correspond to totally symmetric vibrations of YBa$_2$Cu$_3$O$_{7-y}$ involving the Cu-O bonds perpendicular to the Cu-O planes. A vibronic activation mechanism is proposed in order to account for their IR activity and the polarization of the .13 eV midgap state is discussed in this framework.

INTRODUCTION

Many investigations on the optical properties of the ceramic superconductors have been performed in order to study the electron-phonon and the electron-electronic excitation coupling mechanisms in the high temperature superconductors.

The YBa$_2$Cu$_3$O$_{7-y}$ high T$_c$ superconducting system is particularly interesting in this regards because it allows to change the electronic structure from a semiconductor to a superconductor by varying the oxygen stoichiometry from y = 1 to y = 0. This may be neatly achieved using the gettered annealing technique.

Recent experiments performed on single crystals (Tanaka *et al.*, 1988) and ceramics (Garriga *et al.*, 1988) confirmed the presence of a low-lying electronic transition in the superconductor at about .3-.5 eV, but its nature is still controversial. Several Raman studies on single crystals (Krol *et al.*, 1987; 1988; Denisov*et al.*, 1988) as well as the study of the variation of phonon modes upon the oxygen vacancy ordering (Thomsen *et al.*, 1987;Ruani *et al.*), have contributed to the assignment of the phonon spectrum.

A direct evidence of the coupling between optical excitations and charges may derive from the measurement of the variation of the infrared spectrum of the semiconducting phase $YBa_2Cu_3O_6$ upon photogeneration of charge carriers. In view of this the IR photoinduced absorption (PA) spectrum of the low temperature gettered annealed $YBa_2Cu_3O_{7-y}$ (y = 0.79 ÷ 0.85) has been previously reported by our group (Taliani *et al.*, 1988).

EXPERIMENTAL

Samples of $YBa_2Cu_3O_{7-y}$ with an oxygen content corresponding to y = 0.79 and 0.85 were prepared by means of the oxygen annealing technique according to (Cava et.al., 1987). Characterization of the structure was performed by x-ray diffraction and two separate methods were used for the determination of the oxygen content. Pressed pellets, prepared by mixing and grinding the ceramic with CsI in an agata mortar, were used for the spectroscopic experiments; the dimensions of the ceramic particles were in the range between 10 and .1 microns. Details are reported in Ruani et.al., in press.

The spectra in the 45-400 meV range were obtained by using a modified FT-IR interferometer; a c.w. 15 mW He-Ne laser was used for excitation at a photon energy of 1.96 eV. The power incident on the sample was kept lower than 50 mW/cm^2 in order to minimize the thermal effects. In the 0.1-2 eV range the PA spectra were obtained in a dispersive instrument by means of phase sensitive technique (PST) modulating the Ar$^+$ ion laser pump beam.

The PA spectrum was derived by calculating -$\Delta T/T$.

PHOTOINDUCED ABSORPTION

The PA spectra of the $YBa_2Cu_3O_{7-y}$ (y-=0.79) in the spectral range of 45-400 meV and .1-2 eV are shown in Fig. 1 and 2 respectively. Comparison of the PA spectrum with the thermal modulation spectra insured that the PA spectrum does not show any artifact due to thermal modulation. Evidence for this is reported in our previous paper (Taliani *et al.*, 1988) on the PA in $YBa_2Cu_3O_{7-y}$. The observation of a PA effect by exciting at 1.96 eV is an indirect evidence that the energy gap in $YBa_2Cu_3O_{7-y}$ (y=0.79) is at lower energy in accordance with (Garriga *et al.*, 1988).

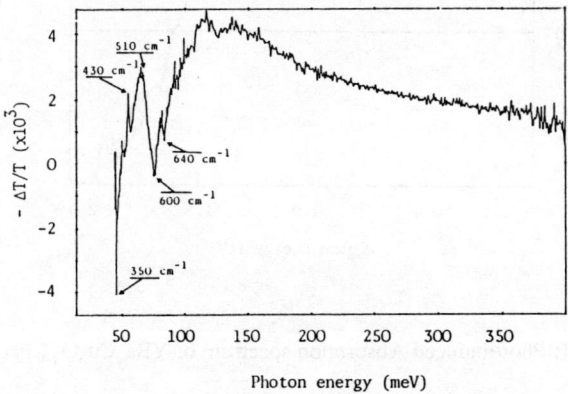

Fig. 1. FT-IR Photoinduced Absorption spectrum of $YBa_2Cu_3O_{7-y}$ (y=0.79) at 20K.

ELECTRONIC SPECTRA

Two PA bands are observed on the high energy side of the PA spectrum at .13 and 1 eV (see Fig. 2). The remarkable intensities of these bands suggest that they are dipole allowed electronic transitions; this is consistent with a transition to midgap states which are generated or populated in the *normal* semiconducting state of $YBa_2Cu_3O_{7-y}$, by photogeneration of e-h pairs.

Similar transitions have been observed at .5 and 1.4 eV in La_2CuO_4 by J. M. Ginder *et al.* (1988). The absence of the onset of the 1 eV band in the FT-IR PA spectrum (Fig. 1), may be related to the different lifetimes of the species investigated in this experiment compared to the PTS experiment (Fig. 2).

The observation of two PA electronic transitions in the gap may be explained in several ways:
a) It may correspond to the optical transitions to two distinct trapped hole and electron states in

the gap as it was suggested in the PA of the La_2CuO_4 ceramic. (Ginder et al., 1988)

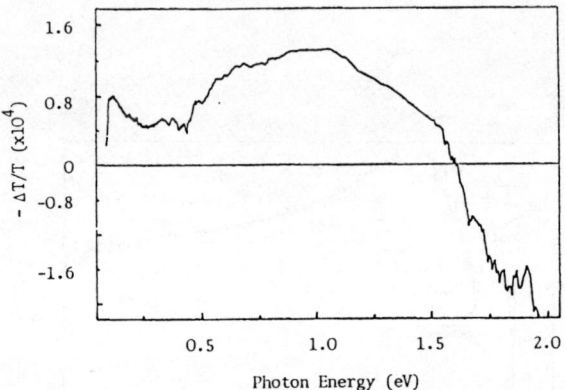

Fig. 2. Near IR Photoinduced Absorption spectrum of $YBa_2Cu_3O_{7-y}$ (y=0.79) at 80K.

Alternatively the two optical transitions may origin from the same species. This may be generally the case of bosons which may be envisaged either as local bosons with singlet ground and first excited states in the gap or by polaron or bipolaron states.

b) In the first case the .13 eV band may correspond to the dipole allowed transition from the ground to the first singlet excited state and the 1 eV band to the transition from the ground state to the continuum. This may account for the relatively large difference of the width of the two electronic PA bands.

c) In the second case we assume a strong electron-phonon coupling which allows the formation of hole pairs via strong lattice relaxation (bipolarons). The two PA bands may correspond to transitions from the valence band to the first and second excited levels of bipolarons.

d) An alternative explanation for the low energy .13 eV PA band may be referred to the charge fluctuation mechanism proposed by Tesanovich et al., (1989). The non-bonding $Op(\pi)$ level may be populated upon photoexcitation and the $Op(\pi)$-- $Cud(y^2-z^2)$ transition becomes possible.

This transition has been calculated to occur at .1 eV (Tesanovic et al.), it may then correspond to the .13 eV band. It is impossible at this point to establish which of these hypothesis is correct.

For a better understanding of the nature of these PA bands we are presently carrying out studies of the lifetime, temperature and pump intensity dependence of the PA bands in order to establish if the .13 and the 1 eV bands belong or not to the same species.

PHONON SPECTRUM

On the low energy side two PA bands at 510 and 435 cm^{-1} (63 and 54 meV respectively) are observed. There is a perfect coincidence of the 435 cm^{-1} band with a Raman mode of $YBa_2Cu_3O_7$, and the 510 cm^{-1} band is very close to the 502 cm^{-1} Raman band (Thomsen *et al.*., 1987). Both bands have been assigned to totally symmetric A_g modes of the $YBa_2Cu_3O_7$ phase. The 435 cm^{-1} Raman band has been assigned to the totally symmetric bending mode Cu(2)-O(2),O(3)[3,12]; the 510 cm^{-1} band has been on the other hand assigned to the Cu(2)-O(4) axial stretching vibration (Krol *et al.*, 1987; Cardona *et al.*, 1987). The numbering of the atoms follows the convention reported in Jorgensen *et.al.* (1987). The coincidence of these two PA bands with Raman modes indicates that the photogenerated carriers couple to the symmetric deformation of the Cu-O_5 pyramids; the breaking of the local symmetry around the photogenerated charges makes the Raman modes formally allowed in the IR.

Moreover the PA activity of these modes constitutes a direct evidence of the coupling of the photogenerated carriers with the sub-lattice of the Cu-O planes as well as the axial deformation of the CuO_5 pyramids.

Furthermore their IR activity requires not only the breaking of the local symmetry around the defect charges, which would only make them formally IR allowed, but requires also a plausible mechanism that would impart some intensity to these modes. This mechanism may be provided by an intensity transfer from an allowed electronic state.

The IR activation of totally symmetric vibrations, which are formally IR forbidden in centrosymmetric structures, has been already observed in doped or photoinduced conjugated polymers and is attributed to a vibronic coupling between the low energy bipolaron states and charged defect phonon modes.

A similar mechanism may be invoked for the appearance of A_g modes in the PA spectrum of $YBa_2Cu_3O_{7-y}$. The excited electronic state from which the intensity is *borrowed* may be the .13 eV state. The energy difference between this electronic state and the phonons is of the order of .06 eV and hence the vibronic activation process could be very efficient because of the energy denominator effect. In other words the phonon states acquire some mixed character.

If this is the case then we may derive some considerations on the nature of the excited electronic state at .13 eV. In fact, in this view, the symmetry and the polarization of phonons should be such that the atomic displacements occurring in the vibrations should give rise to the same distortion that occurs in the transition from the ground to the excited electronic state. The phonon modes, in order to give rise to a vibronic coupling, should have the same polarization of the electronic state.

This implies that the defect state at .13 eV should be polarized perpendicular to the Cu-O planes.

The photoinduced bleaching at about 350 cm^{-1} (43 meV) corresponds to the IR active in plane vibration of O(4) atoms (Thomsen et al.., 1988). The photoinduced bleachings at 640 and 590 cm^{-1} (79 and 73 meV) have been assigned by us (Ruani et al..) to the Cu(1)-O(1) IR active stretching at k = 0 and k ≠ 0 (defect) modes respectively. Hence all the bleachings correspond to strong IR active vibrations that have a dipole component in the *ab* plane. The intensity decrease of such infrared bands upon photoexcitation may imply that the photoexcited carriers move freely on the *ab* planes giving rise to the screening of the modes which are plane polarized. Alternatively the bleaching of the strong defect mode at 590 cm^{-1} suggests that photogenerated carriers induce some lattice distortion from the tetragonal to the orthorhombic structure which consequently reduces the disorder that is intrinsic of the tetragonal phase with 6.2 oxygen stoichiometry.

ACKNOWLEDGEMENTS

Financial support from C.N.R. "Progetto strategico Superconduttività" is acknowledged.
One of the authors (A.J. Pal) has carried out this work with the support of the "ICTP Programme for Training and Research in Italian Laboratories, Trieste, Italy."

REFERENCES

Cardona, M., L. Genzel, R. Liu, A. Wittlin, Hj. Mattausch, F. Garcia-Alvarado and E. Garcia-Gonzales, (1987). Infrared and Raman spectra of the MBa$_2$Cu$_3$O$_7$-type high T$_c$ superconductors. *Solid State Commun.*, 64, 727.

Cava, R. J., B. Batlogg, C. H. Chen, E. A. Rietman, S. M. Zahurak, and D. Werder, (1987). Single-phase 60K bulk superconductor in annealed Ba$_2$YCu$_3$O$_{7-\delta}$ (0.3<δ<0.4) with correlated oxygen vacancies in the Cu-O chains. *Phys. Rev. B*, 36, 5719.

Denisov, V. N.,B. N. Mavrin, V. B. Podobedov, I. V. Alexandrov, A. B. Bikov, A. F.

Goncharov and O. K. Mel'nikov, (1988). Raman spectra of the $YBa_2Cu_3O_x$ single crystals. *Phys. Lett. A*, 130, 411.

Garriga, M. , U. Venkateswaran, K. Syassen, J. Humlicek, M. Cardona, Hj. Mattausch and E. Schonherr, (1988). Optical response of $MBa_2Cu_3O_{7-\delta}$ -type materials. *Physica C*, 153-155, 643.

Ginder, J. M., M.G. Roe, Y. Song, R. P. McCall, J. R. Gaines, E. Ehrenfreund and A. Epstein, (1988). Photoexcitations in La_2CuO_4: 2eV energy gap and long-lived defect states. *Phys. Rev. B*, 37, 7506.

Jorgensen, J. D., M. A. Beno, D. G. Hinks, L. Soderholm, K. J. Volin, R. L. Hitterman, J. D. Grace, I. K. Schuller, C. U. Segre, K. Zhang and M. S. Kleefisch, (1987). Oxygen ordering and the orthorhombic to tetragonal phase transition in $YBa_2Cu_3O_{7-x}$. *Phys. Rev..B*, 36, 3608.

Krol, D. M. , M. Stavola, W. Weber, L. F. Schneemeyer, J. V. Waszczak, S. M. Zahurak and S. G. Kosinski, (1987). Raman spectroscopy and normal-mode assignments for $Ba_2MCu_3O_x$ (M = Gd,Y) single crystals. *Phys. Rev. B*, 36, 8325.

Krol, D. M., M. Stavola, L. F. Schneemeyer, J. V. Waszczak and W. Weber, (1988). Vibronic Raman scattering from HT_c single crystals. *Mat. Res. Soc. Symp. Proceedings*, 99, 781.

Ruani, G., C. Taliani, R. Zamboni, D. Cittone and F. C. Matacotta. Dependence of the IR absorption on oxygen deficiency in the $YBa_2Cu_3O_{7-y}$ superconducting system. *J. Opt. Soc. Am B*, in press.

Taliani, C., R. Zamboni, G. Ruani, F. C. Matacotta and K. I. Pokhodnya, (1988). Infrared photoinduced absorption in the $YBa_2Cu_3O_{7-y}$ high T_c superconducting system. *Solid State Commun.*, 66, 487.

Tanaka, J. , K. Kamiya, M. Shimitzu, M. Simada, C. Tanaka, H. Ozeki, K. Adachi, K. Iwahashi, F. Sato, A. Sawada, S. Iwata, H. Sakuma and S. Uchiyama, (1988). Optical spectra and electronic structures of high T_c oxide superconductors. *Physica C*, 153-155, 1752.

Tesanovic, Z. , A. R. Bishop, R. L. Martin and C. Harris. Charge fluctuations, dynamic polarizability and High T_c in layered oxide superconductors. *Phys. Rev. Lett*, in press.

Thomsen, C., R. Liu, M. Bauer, A. Wittlin, L. Genzel, M. Cardona, E. Schönherr, W. Bauhofer and W. König, (1987). Systematic Raman and infrared studies of the superconductors $YBa_2Cu_3O_{7-x}$ as a function of oxygen concentration (0<x<1). *Solid State Commun.*, 65, 55.

Thomsen, C., M. Cardona, W. Kress, R. Liu, L. Genzel, M.Bauer, E. Shonherr and U. Shoder, (1988). Raman and IR studies of the oxygen deficient semiconducting phase of the superconducting cuprate perovskites. *Solid State Commun.*, 65, 1139.

ANOMALOUS HEATING OF OXYGEN VIBRATIONS IN RAMAN SPECTRA OF $YBa_2Cu_3O_{7-\delta}$

D. Mihailović and J. Solmajer

J. Stefan Institute, Jamova 39, 61111 Ljubljana, Yugoslavia

ABSTRACT

Phonon temperature determined from Raman Stokes/Antistokes ratios as a function of laser intensity for the 152, 340 and 505 cm^{-1} modes of $YBa_2Cu_3O_7$ show that the higher frequency modes have temperatures significantly above ambient. This suggests that photoexcited carriers lose energy by scattering preferentially from high-energy phonon vibrations involving oxygen atoms. Our measurements suggest that the assignments of Thomsen *et. al.*, (1988) for the 116 cm^{-1} modes should be reversed. Appearance of additional modes in the spectra at higher excitation intensities indicate anharmonic coupling to infrared modes. The method, could be used to investigate the electron-phonon interactions in these materials.

KEYWORDS

Raman scattering; high-T_c superconductors; hot phonons

INTRODUCTION

Measurements of phonon Raman spectra (Liu *et al.*, 1988, Thomsen *et al.*, 1987, and Macfarlane *et al.*, 1987, 1988 and many others) have been very important in characterizing the atomic vibrations in the high-T_c superconducting oxides. Due to the large number of ions per unit cell and relatively low symmetry of $YBa_2Cu_3O_{7-\delta}$, for the tetragonal ($1<\delta<0.8$) and D_{2h} for the orthorhombic ($0.5<\delta<0$) structure, we observe a large number of Raman active modes. In order to investigate the interaction between phonons and photoexcited (PE) carriers, we have measured the stokes/antistokes scattering ratios of the resolvable modes as a function of laser intensity. We will assume for the moment the assignments for the observed modes as given by Thomsen *et al.* (1988). The behavior of the 152 cm^{-1} and 340 cm^{-1} modes, involving the out-of phase z-axis motion of the O(2) and O(1) atoms and the Cu(2) z-axis motion respectively (both in the Cu-O planes) are compared to the behavior of the bridging O(4) z-axis vibration (at 505 cm^{-1}). The polarizability components of the 340 cm^{-1} and 505 cm^{-1} modes reflect the directionally of their respective Cu-O bonds: the polarizability in the z direction of the 505 cm^{-1} mode is more than an order of magnitude greater than in x or y ($\alpha_{zz} \gg \alpha_{xx}, \alpha_{yy}$), whereas for the (also totally symmetric in O_7) 340 cm^{-1} mode $\alpha_{zz} \ll \alpha_{xx}, \alpha_{yy}$. Different effective temperatures of the modes may indicate differences in the interaction of PE carriers with in-plane and out-of-plane atoms.

The effect of laser heating on the Raman spectra for both tetragonal and orthorhombic $YBa_2Cu_3O_{7-\delta}$ has been measured, and the laser damage threshold identified by Denisov *et al.* (1988). The Raman spectrum has been carefully studied *in situ* during oxygen treatment by McCarthy *et al.* (1988), and changes of the spectral features associated with changes in oxygen content known (Macfarlane *et al.*, 1988).

Assuming the spectrometer and detector response have been normalized, the S/A scattering ratio is given by the Boltzman factor exp (h ν/k_BT).This holds for phonon Raman scattering far from resonance. (The scattering intensity and hence S/A ratio can be significantly disturbed on either side of the laser line by proximity of the scattered photon with a resonance in the material.) In tetragonal, semiconducting $YBa_2Cu_3O_6$, a 1.7 eV resonance has been observed, but in the orthorhombic, superconducting material this resonance is smeared out (see for example Garriga *et al.* 1988). Using laser photon energies far from 1.7 eV for $\delta<0.1$, we should be clear from resonant effects and the S/A should reflect the effective phonon temperature in the scattering volume. In any case we should not expect resonant effects on the S/A ratio to change with laser intensity.

MEASUREMENTS

In the present experiments, two lasers were used: a pulsed frequency-doubled mode-locked Nd:YAG (Quantronix 416) laser and a CW Ar+ (Omnichrome) laser with photon energies 2.34 eV (532 nm) and 2.42 eV (514.5 nm) respectively. The repetition rate of the pulsed Nd:YAG laser was 100 MHz, with pulses approximately 90 ps long. No significant differences were observed in the Raman spectra, as long as the CW intensity incident on the sample was the same. The spectral response of the spectrometer/detector combination was calibrated using quartz and undoped saphire phonon lines.

From the Meissner signal (onset 90K, midpoint 80K) and the room temperature frequency (505 cm^{-1}) of the O(4) Raman-active z-axis vibration we conclude that the sample (size (0.5 mm)^3) is orthorhombic with $\delta<0.1$. The absence of other lines in the spectra, especially around 570 cm^{-1} where disorder induced IR-active bands appear (McCarthy *et al.*,1988, and Denisov *et al.*, 1988), indicate that the surface is of good quality concerning oxygen disorder. X-ray measuremets show that the major part of the sample is untwinned, but we cannot exclude the possibility that in the laser spot (approx. 50 μm diameter) includes twinned regions. The incident k-vector is along a or b and the light is polarized either along the c axis or in the a-b plane. In this (strictly backscattering geometry), we observe, as already mentioned, the 116, 152 and 505 cm^{-1} modes in zz polarization (weakly in xx,yy) and the 340 cm^{-1} mode in the xx (or yy) polarization (but not in zz).

RESULTS

The phonon spectra as a function of intensity are shown in Figure 1a) for laser intensies range up to 4kW/cm^2. Figure 1b) shows the change in the spectrum around the 505 cm^{-1} line for laser intensities up to 8kW/cm^2. In figure 1c) we present the spectrum of the same spot on the sample with low intensity laser excitation, but after it has been exposed to intensities above 8 kW/cm^2. We observe in Figure 1c) that the spectrum has changed dramatically, and the sharp 640 cm^{-1} line, which is possibly from $BaCuO_2$ or more likely an infrared-active vibration involving Cu(1)-O(2) which has become Raman active. In either case, damage has occured on the sample surface. We conclude: no change is observed in the spectral features up 4kW/cm^2, reversible changes are observed for laser intensities up to 8kW/cm^2, and damage occurs above this.

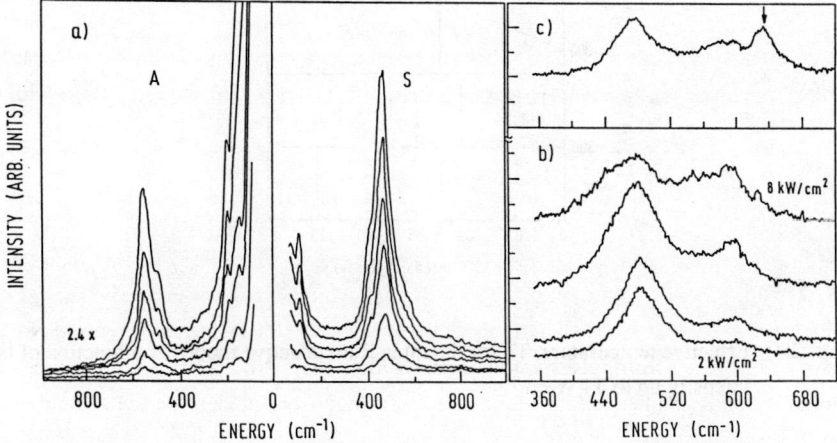

Fig. 1. Stokes and antistokes Raman spectra for laser intensities up to a) 4kW/cm^2, b) 8kW/cm^2 and c) at 0.5 kW/cm^2, but after exposure above 8 kW/cm^2.

Intensities for stokes and antistokes scattering are obtained by fitting the phonon lines to Lorentzian lineshapes. Phonon temperatures for the three clearly resolved lines at 152, 340 and 505 cm^{-1} calculated from the Boltzman factors are shown in Fig. 2. At the lowest laser intensity the phonon "temperature" is near 300 K for all modes, as expected. At higher laser intensity however, we find that the phonon temperature does not change in the same way for all three modes. Whereas the 505 cm^{-1} phonon temperature increases and reaches a temperature near 500 K, the 340 cm^{-1} line temperature reaches less than 400K, and the 152 cm^{-1} line heats hardly at all. Since the measurements of line intensities were done for all the modes on the same measurement (but different polarization for the 340 cm^{-1} mode) the effects could not be introduced by different excitation intensity.

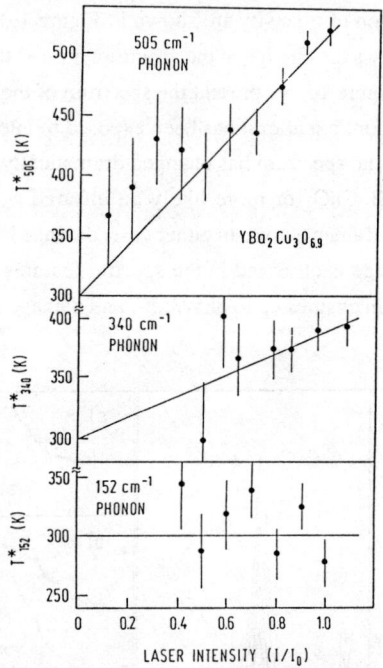

Fig. 2. Effective temperature, T* for the three Raman active modes as a function of laser intensity up to 4 kW/cm^2.

DISCUSSION

We describe the situation in terms of a dynamic equilibrium, whereby PE carriers (both electrons and holes) lose their excess energy by scattering selectively with phonons, while the lattice as a whole is near ambient temperature. The interaction of PE carriers with phonons, and associated hot phonon effects have been studied in detail in GaAs for example (for example see Shah, 1970, Esipov, 1987). However, a distribution of phonon temperatures such as observed here, has to our knowledge not been hitherto reported, probably because these effects were never investigated in materials of low symmetry and a large number of Raman active modes.

Performing an analysis analogous to that done on GaAs by Shah et al. (1970) we obtain a 505 cm^{-1} phonon lifetime τ_p < 3 ps, which would imply an intrinsic width of this mode greater than 15 cm^{-1}, not in disagreement with observations. The analysis assumes the Fröhlich interaction of electrons with (IR-active and Raman inactive in YBaCuO) polar optic modes, and cannot be applied with confidence without better knowledge of the e-p interaction. The same analysis performed with a deformation potential type interaction of PE carriers with non-polar modes will give τ_p >1µs, so the question then arises what interaction is responsible for the heating of the Raman active (non-polar) modes. We suggest that this analysis could be used to deduce the form of the electron-phonon interaction, provided the homogeneous linewidth ($1/\tau_p$) can be measured another way, for example by direct measurement of τ_p. The details of the analysis and assumptions made, the possible effects of Fano interference on the analysis, as well as results of further measurements on different samples will be given in a forthcoming publication. A change in infrared photoinduced absorption (PA) at Raman-active phonon frequencies have already been observed (Kim et al., 1988, Ruani et al., 1988), and we suggest that anharmonic coupling of the IR modes with Raman-active modes could be important. The converse effect (IR modes appearing in Raman spectra) which is observed under conditions of high laser excitation densities, or raised ambient temperature (McCarthy et al., 1988) suggests anharmonic coupling in the vicinity of the O(4) atom.

If we reverse the assignments of Thomsen et al., (1988) for the 116 cm^{-1} and 152 cm^{-1} modes we find that a weak interaction of PE carriers with Ba vibrations is a much more comfortable hypothesis than in-plane Cu vibrations. The observation of Fano interference for the 116 cm^{-1} mode and not for the 152 cm^{-1} mode (Cooper et. al., 1988) is further suggestion that the 152 cm^{-1} vibration belongs to a Ba vibration.

We conclude that both the in-plane O and out-of-plane bridging oxygen interact with PE carriers much more strongly than the Ba vibrations. Resonant effects cannot explain the change in phonon temperature with laser power.

REFERENCES

Denisov, V.N., B.N. Marvin, V.B. Podobedov, I.V. Alexandrov, A.B. Bykov, A.F. Goncharov and O.K. Mel'nikov (1988). Raman spectra of YBa$_2$Cu$_3$O$_x$ single crystals. *Phys. Lett. A*, 130, 411.

Garriga M., U.Venkateswaran, K.Syassen, J. Humlicek, M. Cardona, Hj. Mattausch and E. Schönherr (1988). Optical response of MBa$_2$Cu$_3$O$_{7-\delta}$ - type materials. *Physica C*, 153-155, 643.

Kim Y.H., C.M. Foster, A.J. Heeger, S. Cox, L. Acedo and G. Stucky (1988). Photogeneration of Self-localized Polarons in YBa$_2$Cu$_3$O$_{7-\delta}$ and La$_2$CuO$_4$. To be published.

Esipov S. and Y.B. Levinson (1987). The temperature and energy distribution of photoexcited hot electrons. *Adv.Phys.*, 36, 331.

Liu R., C. Thomsen, W. Kress, M. Cardona, B. Gegenheimer, F.W. de Wetet, J. Prade, A.D. Kulkarni and U. Schröder (1988). Frequencies, eigenvectors, and single-crystal selection rules for K=0 phonons in $YBa_2Cu_3O_{7-\delta}$: Theory and experiment. *Phys. Rev. B*, 37, 7971.

Macfarlane R.M., Hal Rosen and H. Seki (1987). Temperature dependence of the Raman spectrum of the high-T_c superconductor $YBa_2Cu_3O_{7-\delta}$. *Sol. Stat. Comm*, 63, 831.

Macfarlane R.M., H.J. Rosen, E.M. Engler, R.D. Jacowitz and V.Y. Lee, (1988). Raman study of the effect of oxygen stoichiometry on the phonon spectrum of the high-T_c superconductor $YBa_2Cu_3O_x$. *Phys. Rev. B*, 38, 284.

McCarty K.F., J. C. Hamilton, R.N. Shelton, and D.S. Ginley, (1988). High-temperature Raman measurements of single-crystal $YBa_2Cu_3O_{7-\delta}$. *Phys. Rev. B*, 38, 2914.

Ruani G., C. Taliani, R. Zamboni, D. Cittone and F.C. Matacotta (1988). Dependence of the infrared absorption on oxygen deficiency in the $YBa_2Cu_3O_{7-y}$ superconducting system. To be published.

Shah J., R.C.C. Leite and J.F. Scott (1970). PE hot LO phonons in GaAs. *Sol.Stat. Comm.*, 8, 1089.

Thomsen C., M. Cardona, B. Gegenheimer, R. Liu and A. Simon, (1988). Untwinned single crystals of $YBa_2Cu_3O_{7-\delta}$: An optical investigation of the a-b anisotropy. *Phys. Rev. B*, 37, 9860.

Cooper S.L., M.V. Klein, B.G. Pazol, J.P. Rice and D.M. Ginsberg, Raman scattering from superconducting gap excitations in single crystal $YBa_2Cu_3O_{7-\delta}$, *Phys. Rev. B*, 37, 5920 (1988)

MAGNETIC INTERACTIONS

NUCLEAR MAGNETIC RESONANCE IN HIGH TEMPERATURE SUPERCONDUCTOR

Y. Kitaoka, K. Ishida, K. Asayama,
H. Katayama - Yoshida*, Y. Okabe* and T. Takahashi*

*Department of Material Physics, Faculty of Engineering Science, Osaka University,
Toyonaka, Osaka 560, Japan*
**Department of Physics, Tohoku University, Sendai 980, Japan*

ABSTRACT

Results of Cu NQR (nuclear quadrupole resonance) and ^{17}O NMR (nuclear magnetic resonance) are reviewed on $La_{1.85}Sr_{0.15}CuO_4$ (T_c=38K) and $YBa_2Cu_3O_7$ (T_c=92K). The Cu nuclear spin-lattice relaxation (T_1) behaviors for both compounds above and below T_c were found to be unusual with no signatures expected for a non-magnetic metal and a BCS superconductor, respectively. It is pointed out that the behavior above T_c is dominated by an antiferromagnetic spin fluctuation. The further striking feature is that the T_1 behavior of ^{17}O is quite disparate from that of Cu in spite of their close location. The observation of a relatively large enhancement of $1/T_1$ just below T_c is of particular importance, giving a key signature that the superconductivity is of dominant s-wave type. Combined above findings with the recent result by the photoemission study that doped oxygen p holes yield a substantial density of states at the Fermi level (Takahashi et al., 1988), it is extracted that the superconductivity is realized by p-holes pairing with a dominant s-wave symmetry.

KEYWORDS

High Temperature Superconductivity, Nuclear Magnetic Resonance, Antiferromagnetic Spin Fluctuation, S-Wave Superconductivity.

INTRODUCTION

A most intriguing subject of the solid state physics is to elucidate the origin of the high temperature superconductivity. Four classes of cuparates, La-Sr(Ba)-Cu O, Y-Ba-Cu-O, Bi-Sr-Ca-Cu-O, and Tl-Ba-Ca-Cu-O (see the proceeding of the Interlaken Conference, 1988) have been well established as high T_c superconductors. As for the overall electronic structure which gives an important base for a theoretical starting point, the photoemission studies (Fujimori et al.,1987; Takahashi et al.,1987), which are also extensively presented in this volume, have revealed that doped holes are predominantly oxygen 2p-like rather than Cu-3d like due to strong on-site Coulomb interaction among Cu 3d electrons. The nuclear magnetic (NMR) and quadrupole (NQR) resonance studies are powerful tools which allow us to discriminate the microscopic properties between oxygen and copper sites in both normal and superconducting states. As well known, the nuclear spin-lattice relaxation study has been one of the crucial experiments to establish BCS theory (Hebel et al., 1959).

So far, La NQR and NMR for $La_{2-x}Sr_x(Ba)CuO_4$ and Cu NQR and NMR for $YBa_2Cu_3O_{7-y}$ were reported by many groups (see the proceeding of the Interlaken Conference, 1988). However, the results seem to be not always consistent except for the result of CuO_2 plane in $YBa_2Cu_3O_7$. The nuclear spin-lattice relaxation time T_1 of Cu in CuO_2 plane was measured in zero magnetic field by using Cu NQR observed at 31.5 MHz and the unusual behavior was found by several groups (Warren et al., 1987; Mali et al., 1987; Kitaoka et al., 1988a; Imai et al., 1988). Namely, $1/T_1$ above T_c shows a weak temperature dependence, which can not be explained by a Fermi liquid excitation giving rise to T_1T=constant law. Below T_c, $1/T_1$ decreases markedly with no enhancement just below T_c and approaches to a T^3 behavior at low temperature, which are not in the framework of BCS prediction (Kitaoka et al., 1988b). Thus the nuclear spin-lattice relaxation behavior of ^{63}Cu in CuO_2 plane has demonstrated that $YBa_2Cu_3O_7$ possesses an unconventional superconducting nature compared to a BCS superconductor (Kitaoka et al., 1988b; Imai et al., 1988).

As for the La related compounds, La NQR and NMR were only available thus far (see the poceeding of the Interlaken Conference, 1988). La NQR study on $La_{2-x}Ba_xCuO_4$ clarified that there appears a spin glass like disordered magnetic phase with an antiferromagnetic (A.F.) coherency over a long distance only within CuO_2 plane between 3D A.F. state and the superconducting state (Kitaoka et al., 1988c). The superconducting nature was, however, not yet clear because the Cu NQR and NMR were not successfully observed for the La related compounds. Quite recently, our group has discovered the Cu NQR for $La_{1.85}Sr_{0.15}CuO_4$ with T_c=38K for the first time (Ishida et al.,1988a). Here, we will compare the nuclear spin-lattice relaxation behavior af ^{63}Cu in CuO_2 plane between La and Y compounds. We point out that $1/T_1$ of ^{63}Cu is strongly enhanced by the spin fluctuation characteristic for a highly correlated system.

If one is reminded of the electronic structure of the cuparate proposed by the photoemission studies (see this volume), i.e. doped holes are predominantly oxygen p-like, one would be aware that the microscopic information of the oxygen sites is of substantial importance. So we have carried out the ^{17}O NMR investigation of $YBa_2Cu_3O_7$ by a substitution of ^{17}O in place of ^{16}O (Kitaoka et al., 1988d; Ishida et al., 1988b). Here we will compare both nuclear relaxation behaviors of ^{17}O and ^{63}Cu of $YBa_2Cu_3O_7$.

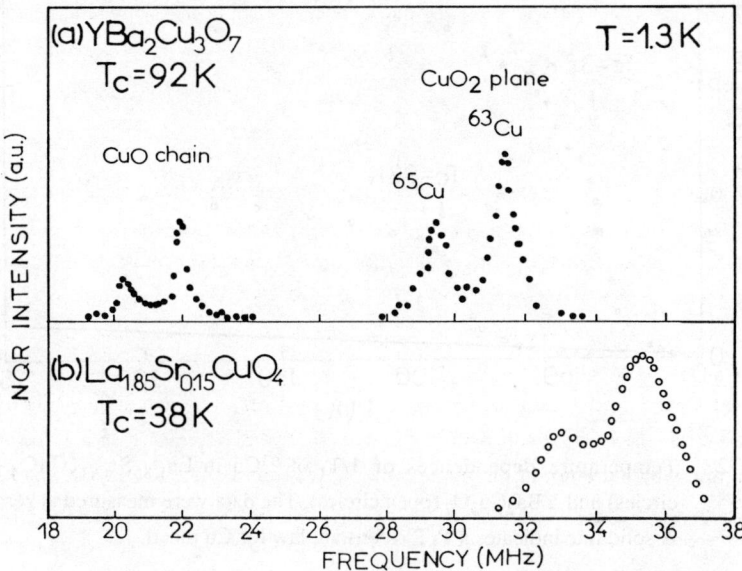

Fig. 1. Cu NQR spectra of (a) $YBa_2Cu_3O_7$ and (b) $La_{1.85}Sr_{0.15}CuO_4$ at 1.4K and zero magnetic field.

RESULTS

^{63}Cu Nuclear Relaxation Time T_1 of $La_{1.85}Sr_{0.15}CuO_4$ and $YBa_2Cu_3O_7$

Figures 1 (a) and (b) show the Cu NQR spectra for Y and La compounds, respectively. The spectra were obtained at 1.3K and zero magnetic field by changing frequency. Each pair of spectra correspond to two ^{63}Cu and ^{65}Cu isotopes. For Y compound, it is well known that there are two Cu sites, i.e. CuO_2 plane and CuO chain, respectively. For La system including one CuO_2 sheet, ^{63}Cu NQR has been discovered around 35.3 MHz (Ishida et al.,1988a). As seen in Fig 1(b), the resonance frequency for La compound is close to that for Y compound.

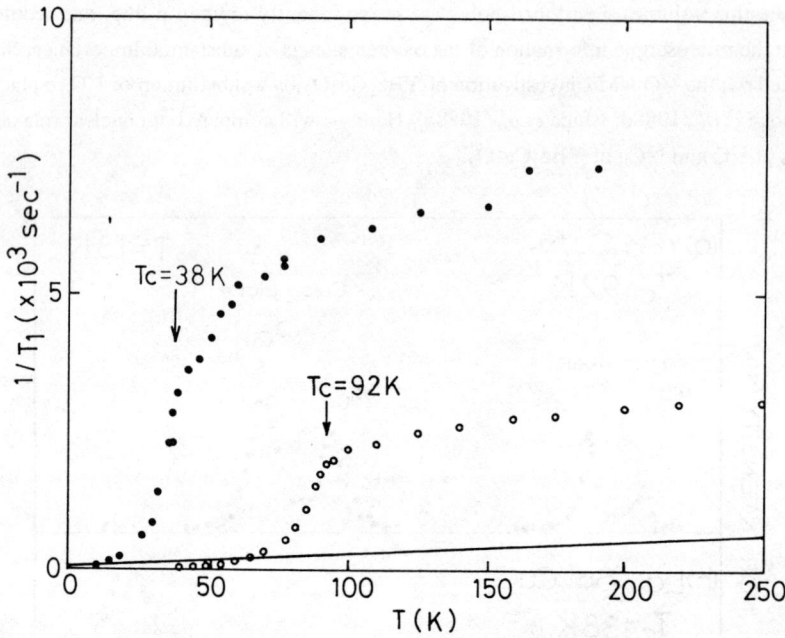

Fig. 2. Temperature dependences of $1/T_1$ of ^{63}Cu in $La_{1.85}Sr_{0.15}CuO_4$ (solid circles) and $YBa_2Cu_3O_7$ (open circles). The data were measured at zero field. A solid line indicates a T_1T = constant law for Cu metal.

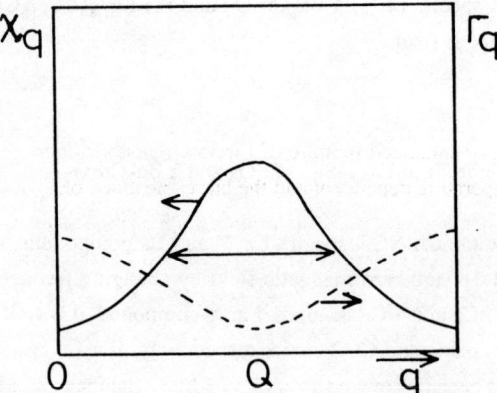

Fig. 3. The schematical wave-number (q) dependences of the susceptibility (χ_q) and the magnetic relaxation rate Γ_q of spin system for the reference.

In Fig. 2, we show the temperature dependences of $1/T_1$ of ^{63}Cu for $La_{1.85}Sr_{0.15}CuO_4$ (Ishida et al., 1988a) and $YBa_2Cu_3O_7$ (Kitaoka et al.,1988b) indicated by solid and open circles, respectively. Here we will skip the result of CuO chain for Y compound because the data are strongly sample dependent due to some difference of the oxygen deficiency in CuO chain (Warren et al., 1987; Mali et al., 1987; Kitaoka et al., 1988a; Imai et al., 1988). As seen in the figure, both $1/T_1$ are similar to each other. Thus, the relaxation of ^{63}Cu in the La system is considered to arise from the same mechanism as that of CuO_2 plane in Y compound. $1/T_1$ above T_c shows a weak temperature dependence being saturated at higher temperature. The difference between both behaviors is that $1/T_1$ for La system seems to be proportional to the temperature in the narrow temperature range between T_c and 70K and is by about three times larger than that of Y compound.

Below T_c $1/T_1$ reduces dramatically without the enhancement just below T_c characteristic for a BCS superconductor. At low temperature, both $1/T_1$ saturate, implying that the intrinsic magnetic relaxation is masked. Actually, the quadrupole relaxation process was confirmed to come up at low temperature for both cases. This origin of the quadrupole relaxation process is not yet clear. As shown already (Kitaoka et al., 1988b), we can, however, separate an intrinsic magnetic relaxation rate from the observed T_1 by measuring precisely the ratio of T_1 of two isotopes ^{63}Cu and ^{65}Cu with the different nuclear magnetic moment and electric quadrupole moment. Then $1/T_1$ for Y compound was found to approach to a T^3 behavior (Kitaoka et al., 1988b), wich is not of an exponential type expected for BCS case. Thus, the unconventional nuclear relaxation behaviors of ^{63}Cu in CuO_2 plane seem to signify that the superconducting nature of these materials is quite unusual. The further important feature is that the behavior of T_1 above T_c is not of a type expected for a non-magnetic metal. Accordingly, first of all, we should discuss about an origin of the relaxation process above T_c.

Compared with the value of $1/T_1$ for Cu metal shown by a solid line in Fig. 2, both $1/T_1$ of La and Y compounds are strongly enhanced in spite of the low density of carrier concentration. Combined both the weak temperature dependent and the large enhanced behaviors of $1/T_1$ above T_c, we suppose that $1/T_1$ of ^{63}Cu in CuO_2 plane is dominated by a spin fluctuation characteristic for a highly correlated system. In general, $1/T_1$ is described by the imaginary part of a dynamical susceptibility $\chi(q,\omega)$ as

$$1/T_1 = k\, T \sum_q A_q^2 \, \text{Im}\, \chi(q,\omega_n) \tag{1}$$

where A_q is the hyperfine coupling constant and ω_n is the resonance frequency. If we express $\text{Im}\, \chi(q,\omega_n)$ by using the magnetic relaxation rate Γ_q of the electron spin system as

$$\text{Im } \chi(q,\omega) = \chi(q)\,\omega\,\Gamma_q / (\omega^2 + \Gamma^2_q) \tag{2}$$

taking a relation of $\omega_n \to 0$ we have an expression of

$$1/T_1 = k_B T \sum_q A^2_q \chi(q) / \Gamma_q = k_B T \sum_q f(q,T) \tag{3}$$

where $\chi(q)$ is a wave number dependent spin susceptibility. The point is that $1/T_1$ reflects the q-dependent magnetic response averaged over a whole reciprocal space. This means that even if the uniform susceptibility $\chi(0)$ is temperature-independent like Pauli paramagnets (just corresponding to the high-T_c materials), $1/T_1$ is dominated by an antiferromagnetic (A.F.) spin fluctuation. To see this situation, we show schematically the q-dependences of $\chi(q)$ and Γ_q in Fig. 3. The half-width of $\chi(q)$ curve corresponds to the inverse of the magnetic coherence length ξ_M of the A.F. spin fluctuation. As ξ_M is increasing, f(q,T) has larger contribution from around q=Q (Q: antiferromagnetic q-vector), which yields the enhancement of $1/T_1$. In order to gain a rough insight into the dependences of $1/T_1$ on the temperature and the magnetic coherence length, we tentatively may have a relation of $\Sigma_q f(q,T) \sim \chi(Q,T)$ by approximating $\chi(q,T)$ curve as a rectangular shape with a height of $\chi(Q,T)$ and a width of ξ_M^{-1} and relating Γ_q to ξ^{-1}_M. In this case we have $1/T_1 \sim k_B T \chi(Q,T)$. When the magnetic coherence length increases, $\chi(Q,T)$ is expected to become more temperature dependent and larger compared to an usual metallic system, having a behavior like $\chi(Q,T) \sim C/(T+\theta)$. Thus we predict $1/T_1 \sim T/(T+\theta)$ where θ is a characteristic temperature below which the A.F. spin correlation starts to appear. We expect that $k_B \theta$ is smaller than the superexchange constant J_{ex} between Cu spins. At lower temperature than θ, it is expected that $1/T_1$ is approaching to a linear temperature dependence, which has been confirmed for La compound where $1/T_1$ of ^{63}Cu behaves like $T_1 T$ = constant below 70K. Thus, the nuclear relaxation behavior of ^{63}Cu in CuO_2 plane is considered to be well described in terms of the A.F. spin fluctuation. This type of discussion has been also presented by several groups (Imai et al., 1988; Horvatic et al., 1988).

With above scheme, the marked decrease of $1/T_1$ below T_c may be also associated with the strong depression of the A.F. spin fluctuation due to the appearance of the superconductivity. Namely, when the superconducting energy gap appears in the excitation spectrum of the mobile hole carriers (quasiparticles with spin) below T_c, it is likely that the frequency spectrum of the Cu-d spin fluctuation may have also some gap in the low frequency part because the spins are strongly coupled with each other and then $1/T_1$ enhanced above T_c is depressed below T_c as pointed out by the theory (Koyama et al., 1988). Thus the unconventional behavior of $1/T_1$ of ^{63}Cu below T_c may be associated with a depression of the Cu-d spin fluctuation on CuO_2 plane. Under such a circumstance, in order to further characterize the superconducting nature of these materials, it is of substantial importance to investigate the electronic state at oxygen sites by ^{17}O NMR technique. Below we will review the ^{17}O NMR results.

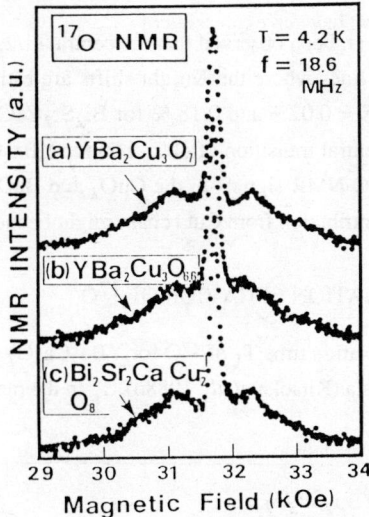

Fig. 4. ^{17}O NMR spectra at 4.2K and 18.6 MHz obtained by sweeping the magnetic field. (a)YBa$_2$Cu$_3$O$_7$ (b)YBa$_2$Cu$_3$O$_{6.65}$ (c)Bi$_2$Sr$_2$CaCu$_2$O$_8$

^{17}O NMR SPECTRA

In a gas exchange procedure of ^{17}O where the sample is prepared at high temperature of 950°C, all oxygen sites in CuO$_2$ plane, BaO layer and CuO chain for YBa$_2$Cu$_3$O$_7$ are equally exchanged by ^{17}O (Cardona et al., 1988). Since we focus on the result that T_1 of ^{17}O for YBa$_2$Cu$_3$O$_7$ exhibits a quite different feature from that of Cu (see Fig. 5), it is important to know which oxygen sites the spectrum of YBa$_2$Cu$_3$O$_7$ comes from. First, to check the contribution from CuO chain, we compare the ^{17}O spectra among various types of compounds such as YBa$_2$Cu$_3$O$_7$ (T_c=92K), YBa$_2$Cu$_3$O$_{6.65}$ (T_c=61K) and Bi$_2$Sr$_2$CaCu$_2$O$_8$ (T_c=75K) as shown in Fig. 4 (a), (b) and (c), respectively. The spectra were obtained at 4.2K by changing the external field at 18.6 MHz. All spectra indicate a typical powder pattern influenced by a first order effect of the nuclear electric quadrupole interaction. Although the oxygen content in CuO chain decreases for YBa$_2$Cu$_3$O$_{6.65}$ and Bi compound does not include CuO chain due to the replacement of Bi$_2$O$_2$ layers, any change of ^{17}O spectra does not have been observed in going from (a) to (c) in Fig. 4 except for the broadening of two satellite peaks in the spectrum of YBa$_2$Cu$_3$O$_{6.65}$. In particular, one should notice that the relative intensity of the central peak to the satellite peaks unchanges with almost the same electric quadrupole frequency of 0.72 MHz for all compounds. These features of ^{17}O NMR spectra seem to tell us that the contribution to the ^{17}O spectrum from CuO chain is small, if it was.

Quite recently, it has, however, been observed that the central (1/2,-1/2) transition in the spectra of Fig.4 is spilled into two lines where the Knight shifts are estimated to be K = 0.03% and 0.13% for YBa$_2$Cu$_3$O$_{6.65}$, K = 0.02% and 0.18 % for Bi$_2$Sr$_2$CaCu$_2$O$_8$ (Butaud et al., 1988). Thus it is evident that the central transition of (1/2,-1/2) for YBa$_2$Cu$_3$O$_7$ shown in Fig. 4 (a) is mainly composed of two ^{17}O NMR signals in the CuO$_2$ and BaO layers, though both are not well resolved and a small contribution from CuO chain might be included in the spectrum.

NUCLEAR RELAXATION BEHAVIOR OF ^{17}O

The nuclear spin-lattice relaxation time T_1 of ^{17}O for YBa$_2$Cu$_3$O$_7$ was measured in an external magnetic field of about 3 Tesla (Kitaoka et al., 1988d). T_c in the magnetic field is expected to be

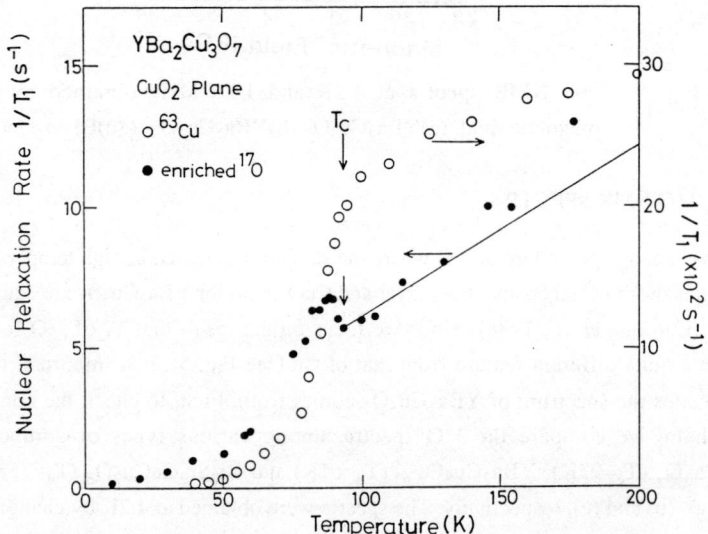

Fig. 5. Temperature dependences of $1/T_1$ of ^{17}O (solid circles) and of ^{63}Cu (open circles) in YBa$_2$Cu$_3$O$_7$. The T_1 data of ^{17}O and ^{63}Cu were measured in the magnetic field of 3 Tesla and zero magnetic field, respectively. Note that each scale of $1/T_1$ is different. A solid line indicates a T_1T=constant law for the reference.

almost the same as $T_c=92K$ in the zero magnetic field because the powder sample is preferentially oriented with the c-axis perpendicular to the magnetic field as reported already (Shimizu *et al.*, 1988) and the upper critical field with the c-axis perpendicular to the magnetic field H_{c2} has a quite large slope against the temperature with $-(dH_{c2}/dT) T_c > 43$ Tesla/T (Iye *et al.*, 1988). In T_1 measurement of ^{17}O, a significant distribution of T_1 has not been found, though the ^{17}O NMR spectrum has two main contribution from the CuO_2 and BaO layers.

Figure 5 shows the temperature dependence of $1/T_1$ for ^{17}O (solid circles) (Kitaoka *et al.*, 1988d) and ^{63}Cu in CuO_2 plane (the same data in Fig.2). Above T_c, $1/T_1$ of ^{17}O looks like a Korringa type (T_1T=constant) up to 150K as shown by a solid line, while it deviates upward from a linear temperature dependence above 150K. This behavior of T_1 of ^{17}O seems to be similar to that of ^{63}Cu in CuO chain. In contrast, $1/T_1$ of ^{63}Cu in CuO_2 plane shows no Korringa behavior in the whole temperature range, though a T_1T=constant behavior appears in the narrow temperature range between T_c and 70K for $La_{1.85}Sr_{0.15}CuO_4$ (see Fig.2).

Below T_c, there appears a further marked difference in the relaxation behavior between ^{17}O and ^{63}Cu. As seen in Fig.5, $1/T_1$ of ^{17}O (solid circles) exhibits a distinct enhancement just below T_c, $1/T_1$ of ^{63}Cu in CuO_2 plane showed a strong reduction without a hump below T_c. Thus, the relaxation behaviors of ^{17}O and ^{63}Cu have been found to be quite different to each other in spite of their close location.

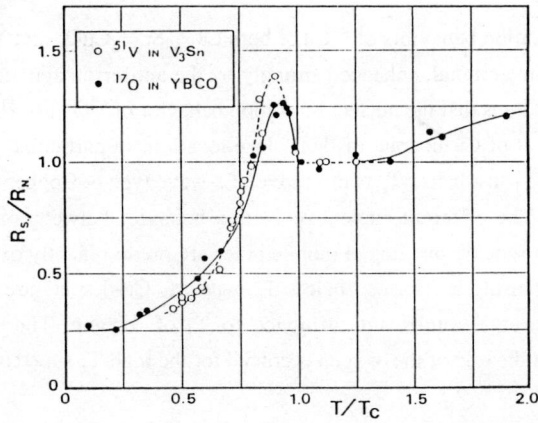

Fig. 6. Comparison of the enhancement of $1/T_1$ just below T_c between ^{17}O and ^{51}V in V_3Sn which is an example for a BCS case where T_1 was measured in the magnetic field. The relaxation rate R_s below T_c is normalized by the Korringa law above T_c.

Next, we focus on the enhanced behavior of $1/T_1$ of ^{17}O just below T_c and compare with the case of a BCS superconductor. As well known, in s-wave superconductor, the presence of "coherence factor" together with the increase of the density of state near the energy gap edge both resulting from the s-wave symmetry of Cooper pair enhance $1/T_1$ just below T_c around $T/T_c=0.91\sim0.93$ (Hebel *et al.*, 1959). In Fig. 6, we show the normalized temperature dependence of the relaxation rate for ^{17}O (solid circles) and ^{51}V of V_3Sn (open circles) which is an example for a BCS superconductor measured in the magnetic field. As seen in the figure, the enhanced behavior of $1/T_1$ just below T_c is similar to each other. Moreover, there is a common tendency that both $1/T_1$ seem to saturate at low temperature. This is considered to be due to an extrinsic relaxation process caused by the spin diffusion effect to vortex cores in the mixed state above H_{c1}. Thus, it is hard to observe an exponential decrease of $1/T_1$ even in BCS superconductor in case that the measurement has been made in the magnetic field. Instead, the presence of the enhancement of $1/T_1$ just below T_c provides a key signature that the superconductivity is of s-wave type. Taking into account that the states near Fermi level are composed of the oxygen p-band (Takahashi *et al.*, 1988), we have a picture that the doped p-hole are moving through an assembly of Cu-d spin moments, experiencing a local exchange interaction with Cu d-spins. With this scheme, we expect that the superconductivity is realized by the p-holes pairig with a dominant s-wave symmetry.

CONCLUSION

The nuclear relaxation behaviors of ^{63}Cu of both $La_{1.85}Sr_{0.15}CuO_4$ and $YBa_2Cu_3O_7$ have been found to be unconventional, enhanced strongly by the antiferromagnetic spin fluctuation. The most striking feature is that the nuclear relaxation behavior of ^{17}O has exhibited a quite different behavior from that of Cu in spite of their close location. In particular, the observation of the enhacement of $1/T_1$ just below T_c is indicative of s-wave type of Cooper pairing for the high-T_c superconductor. The different nuclear relaxation behavior between oxygen and Cu may be resolved within a scheme that doped mobile holes are predominantly oxygen p-like and hence p-holes Cooper pairing are formed below T_c, whereas Cu-d spins are antiferromagnetically fluctuating without a significant influence to T_1 of oxygen. The ^{17}O NMR study has demonstrated that the role of the oxygen is crucial for the high-T_c superconductivity.

REFERENCES

Butaud, P., Y. Kitaoka, P. Segransan, Y. Berthier, C. Berthier, and H. Katayama-Yoshida (1988), in preparation.

Cardona, M., R. Liu, C. Thomsen, W. Kvess, E. Schonherr, M. Bauer, L. Genwel and W.

Koning (1988). Effect of Isotopic Substitution of Oxygen on T_C and the Phonon Frequencies of High T_c Superconductors. *Solid State Commun.*, 63, 789-793.

Fujimori, A., E. Takayama-Muromachi and Y. Uchida (1987). Electronic Structure of Superconducting Cu Oxides. *Solid State Commun.*, 63, 857-860.

Hebel, L.C., and C.P. Slichter (1959). Nuclear Spin Relaxation in Normal and Superconducting Aluminium. *Phys. Rev. B*, 113, 1504 - 1519.

Horvatic, M., P. Segransan, C. Berthier, Y. Berthier, P. Butaud, J.Y. Henry, M. Couach and J.P. Chanimade (1988). NMR Evidence for Localized Spins on Cu (2) Sites from Cu NMR in $YBa_2Cu_3O_7$ and $YBa_2Cu_3O_{6.75}$ Single Crystals. *Phys. Rev. Lett.* (submitted).

Imai, T., T. Shimizu, T. Tsuda, H. Yasuoka, T. Takabatake, Y. Nakazawa and M. Ishikawa (1988). Nuclear Spin-Lattice Relaxation of $^{63,65}Cu$ at the Cu(2) Sites of the High T_C Superconductor $YBa_2Cu_3O_{7-y}$. *J.Phys. Soc. Jpn.*, 57, 1771-1776.

Ishida, K., Y. Kitaoka, and K. Asayama (1988a). Nuclear Quadrupole Resonance of Cu in Superconducting $(La_{0.925}Sr_{0.075})_2CuO_4$. *J.Phys. Soc. Jpn.*, (in press).

Ishida, K., Y. Kitaoka, K. Asayama, H. Katayama-Yoshida, Y. Okabe and T. Takahashi (1988b). NMR and NQR Studies of ^{17}O and ^{63}Cu in CuO_2 Plane of High T_C $YBa_2Cu_3O_{6.65}$ with $T_c=61K$. *J.Phys. Soc. Jpn.*, 57, 2897-2900.

Iye, Y., T. Tamegai, T. Sakakibara. T. Goto, N. Miura, H. Takeya and H. Takei (1988). The Anisotropic Superconductivity of $RBa_2Cu_3O_{7-y}$ (R: Y, Gd and Ho) Single Crystal. *Physica C*, 153-155, 26-31.

Kitaoka, Y., S. Hiramatsu, T. Kondo and K. Asayama (1988a). Nuclear Relaxation and Knight Shift Studies of Copper in $YBa_2Cu_3O_{7-y}$. *J.Phys. Soc. Jpn.*, 57, 30-33.

Kitaoka, Y., S. Hiramatsu, Y. Kohori, K. Ishida, T. Kondo, H. Shibai, K. Asayama, H. Takagi, S. Uchida, H. Iwabuchi and S. Tanaka (1988b). Nuclear Relaxation and Knight Shift Studies of ^{63}Cu in 90 K and 60 K Class $YBa_2Cu_3O_{7-y}$. *Physica C.*, 153-155, 83-86.

Kitaoka, Y., K. Ishida, T. Kobayashi, K. Amaya and K. Asayama (1988c). Magnetic Phase Diagram in $(La_{1-x}Ba_x)_2CuO_4$. *Physica C*, 153-155, 733-734.

Kitaoka, Y., K. Ishida, K. Asayama, H. Katayama-Yoshida, Y. Okabe and T. Takahashi (1988d). Possible evidence for p-hole pairing with s-wave like symmetry in high T_C superconductor $YBa_2Cu_3O_7$ with $T_C=92K$. -enriched ^{17}O NMR-. *Nature* (submitted).

Koyama, T., and M. Tachiki (1988). Theory of Nuclear Relaxation in Superconducting High T_C Oxides. *Phys. Rev. B* (in press).

Mali, M., D. Brinkmann, L. Pauli, J. Roos, H. Zimmerman and J. Hullinger (1987). Cu and Y NQR and NMR in the Superconductor $YBa_2Cu_3O_{7-y}$. *Phys. Lett.*, 124, 112-115.

Proceeding of the International Conference on High Temperature Superconductors and Materials and Mechanism of Superconductivity, Interlaken, Switzerland, Feb.28-Mar.4, (1988). *Physica C*, 153-155.

Shimizu. T., H. Yasuoka, T. Imai, T. Tsuda, T. Takabatake, Y. Nakazawa and M. Ishikawa (1988). Site Assignment for Cu NQR Lines in $YBa_2Cu_3O_{7-y}$ Superconductor. *J. Phys. Soc. Jpn.*, 57, 2494-2505.

Takahashi, T., H. Matsuyama, H. Katayama-Yoshida, Y. Okabe, S. Hosoya, K. Seki, H. Fujimoto, M. Sato, and H. Inokuchi (1988). Evidence from angle-resolved resonance photoemission for oxygen 2p nature of the Fermi-liquid states in $Bi_2CaSr_2Cu_2O_8$. *Nature*, 334, 691-692.

Takahashi, T., F. Maeda, H. Arai, H. Katayama-Yoshida, Y. Okabe, T. Suzuki, S. Hosoya, A. Fujimori, T. Shidara, T. Koide, T. Miyahara, M. Onoda, S. Shamoto and M. Sato (1987). Shynchrotron-radiation photoemission study of the high T_c superconductor $YBa_2Cu_3O_{7-y}$. *Phys. Rev. B*, 36, 5686-5689.

Warren, W.W. Jr., R.E. Walstedt, G.F. Brennert, G.P. Espinosa and J.P. Remeika (1987). Evidence for Two Pairing Energies from Nuclear Spin-Lattice Relaxation in Superconducting Ba_2YCu_3O. *Phys, Rev. Lett.*, 59, 1860-1863.

NEUTRON SCATTERING STUDIES OF MAGNETIC EXCITATIONS IN $La_{2-x}Sr_xCuO_4$

M. A. Kastner[1], A. Aharony[2], R. J. Birgeneau[1,3], Y. Endoh[4], K. Fukuda[5],
D. R. Gabbe[6], Y. Hidaka[7], H. P. Jenssen[6], T. Murakami[7], M. Oda[7],
P. J. Picone[6,8], M. Sato[5], S. Shamoto[5], G. Shirane[3], M. Suzuki[7],
T. R. Thurston[1], and K. Yamada[3,4]

[1]*Department of Physics, Massachusetts Institute of Technology,
Cambridge, MA 02139 USA*
[2]*School of Physics and Astronomy, Beverly and Raymond Faculty of Exact Sciences,
Tel Aviv University, Tel Aviv 69978 Israel*
[3]*Department of Physics, Brookhaven National Laboratory,
Upton, NY 11973 USA*
[4]*Department of Physics, Tohoku University, Sendai 980 Japan*
[5]*Institute for Molecular Science, Myodaiji, Okazaki 444 Japan*
[6]*Center for Materials Science, Massachusetts Institute of Technology,
Cambridge, MA 02139 USA*
[7]*NTT Electrical Communications Laboratory, Tokai, Ibaraki 319-11 Japan*
[8]*Defense Science Technology Organization, Adelaide, Australia*

ABSTRACT

The results of recent neutron scattering studies on $La_{2-x}Sr_xCuO_4$ are reviewed. For $x = 0$ the material is an antiferromagnetic insulator, and the spins in the CuO_2 sheets are well-described by the 2D $S = 1/2$ square lattice Heisenberg model. With increasing x the Néel state is destroyed and is replaced at low temperatures, at least for samples which are nonmetallic or have low superconducting T_c's, by a spin glass. The instantaneous correlations as well as those of the glass are incommensurate with the crystal lattice. The spin-spin correlation length in the doped CuO_2 sheets equals the average separation between the holes which ultimately carry the supercurrent.

INTRODUCTION

The discovery that antiferromagnetic spin correlations persist in superconducting $La_{2-x}Sr_xCuO_4$ shows clearly that magnetism plays an essential role in the superconductivity in the copper oxides. The static and instantaneous spin correlations in pure, nonmetallic La_2CuO_4 have been thoroughly studied and are the subject of a number of papers (Endoh *et al.*, 1988; Birgeneau *et* al., 1988b). We focus here on the behavior of these correlations when holes are added by the substitution of Sr for La.

PHASE DIAGRAM

The crystal and magnetic structure of La_2CuO_4 is shown in Fig. 1. At high temperatures, the structure is body-centered tetragonal, space group I4/mmm, with one La_2CuO_4 formula unit per primitive cell. At a temperature T_0, which is ~ 530 K in stoichiometric material, the crystal exhibits a second order transition to the orthorhombic phase, space group Cmca, which has two formula units per unit cell (Grande et al., 1977; Birgeneau et al., 1987). The structural transition involves primarily a staggered rotation of the CuO_6 octahedra about either a $[1\,1\,0]\equiv_T$ or $[1\,\bar{1}\,0]_T$ axis; here the subscript T implies that one is using the tetragonal unit cell. The identical structural transition occurs in the Sr-doped material $La_{2-x}Sr_xCuO_4$ but with T_0 decreasing continuously to 0 K as x increases to ~ 0.2 (Fleming et al., 1987; van Dover et al., 1987; Birgeneau et al., 1988b; Thurston et al., unpublished; Jorgensen et al., 1987). This is illustrated in the phase diagram, Fig. 2, for $La_{2-x}Sr_xCuO_4$. All of the interesting magnetic behavior occurs in the orthorhombic Cmca phase. Accordingly, from this point on we shall use exclusively the Cmca orthorhombic unit cell. Representative lattice constants at 5 K are a=5.339 Å, b=13.100 Å and c=5.422 Å.

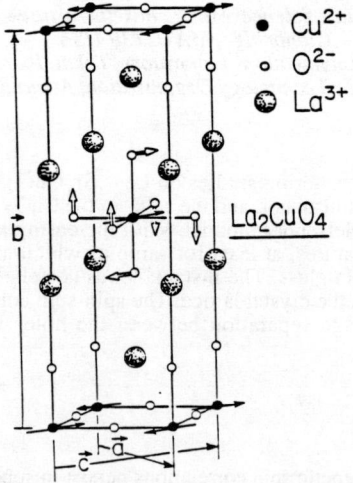

Fig. 1. Crystal and magnetic structure of La_2CuO_4; the arrows associated with the center oxygens indicate the direction of rotation in the orthorhombic phase.

Vaknin et al. (1987) showed that the spin $\vec{S}$ is along [0 0 1] while the antiferromagnetic modulation $\vec{\tau}$ is along [1 0 0]. Addition of holes to La_2CuO_4 has a drastic effect on the magnetic properties. We are, of course, especially interested in the magnetism in the superconducting regime. It is clear, however, that in order to understand the basic physics of the high-T_c materials it will be necessary to understand the evolution of the properties as the number of holes is varied. As indicated in Fig. 2, $La_{2-x}Sr_xCuO_4$ undergoes an insulator-metal transition for $x \simeq 0.05$. In the regime $x \leq 0.05$ for $T \leq 100$ K one typically observes variable range hopping conductivity, $\ln \sigma \sim (T_o/T)^{1/4}$, although in different samples the exponent may vary from 1/4 to 1/2 (Bednorz and Müller, 1986; Birgeneau et al., 1987; Jorgensen et al, 1987). In this regime the carriers are localized by disorder. For $x > 0.05$ the holes are delocalized and the material is then a metal at high temperatures and a superconductor at low temperatures. The superconducting state itself is very sensitive to the perfection of the material and, indeed, for many doped single crystals often only a small fraction (~ 10%) will be superconducting.

Shortly after the Bednorz-Müller discovery (Bednorz and Müller, 1986) samples of $La_{2-x}M_xCuO_4$ with $M = Ca^{2+}$, Ba^{2+} or Sr^{2+} were studied with a wide range of probes including bulk susceptibility (Fujita et al., 1987; Greene et al., 1987; Ishi et al., 1987; Johnston et al., 1987), NMR, NQR (Kumagai et al., 1987; Kitaoka et al., 1988), muon precession (Budnick et al., 1988; Uemura et al., in press) and neutron scattering (Shirane et al., 1987; Endoh et al., 1988; Birgeneau et al., 1988b). All probes indicated a rapid decrease of T_N with increasing hole concentration, whether introduced by increasing x or by oxygen excess, and the destruction of the Néel state for $x \geq 0.02$; the phase boundary is shown in Fig. 2. However, for samples with $0.02 < x < 0.05$ both NQR and μSR indicated some type of phase transition at low temperatures, $T \leq 10$ K. Indeed, the muon experiments indicated a freezing of the spins into a Néel-like state with a local moment of ~ 0.5 μ_B, close to that in pure La_2CuO_4 in the Néel state (Budnick et al., 1988; Harshman et al., 1988; Uemura et al., in press). We now know from neutron scattering experiments that this state has the quasi-static short range order corresponding to a spin glass; the spin-glass domain overlaps the superconducting region of the phase diagram as indicated in Fig. 2.

A phenomenological model for this unusual behavior has been proposed by Aharony et al. (1988) (Birgeneau et al., 1988a). A classical percolation model (Birgeneau et al., 1980; Cowley et al., 1980) cannot explain the destruction of the Néel state for such low concentrations as x~0.02; one must, therefore, consider explicitly the effects of the holes introduced by the doping.

Transport is carried by the electronic holes, which reside on the oxygen ions (Tranquada et al.,

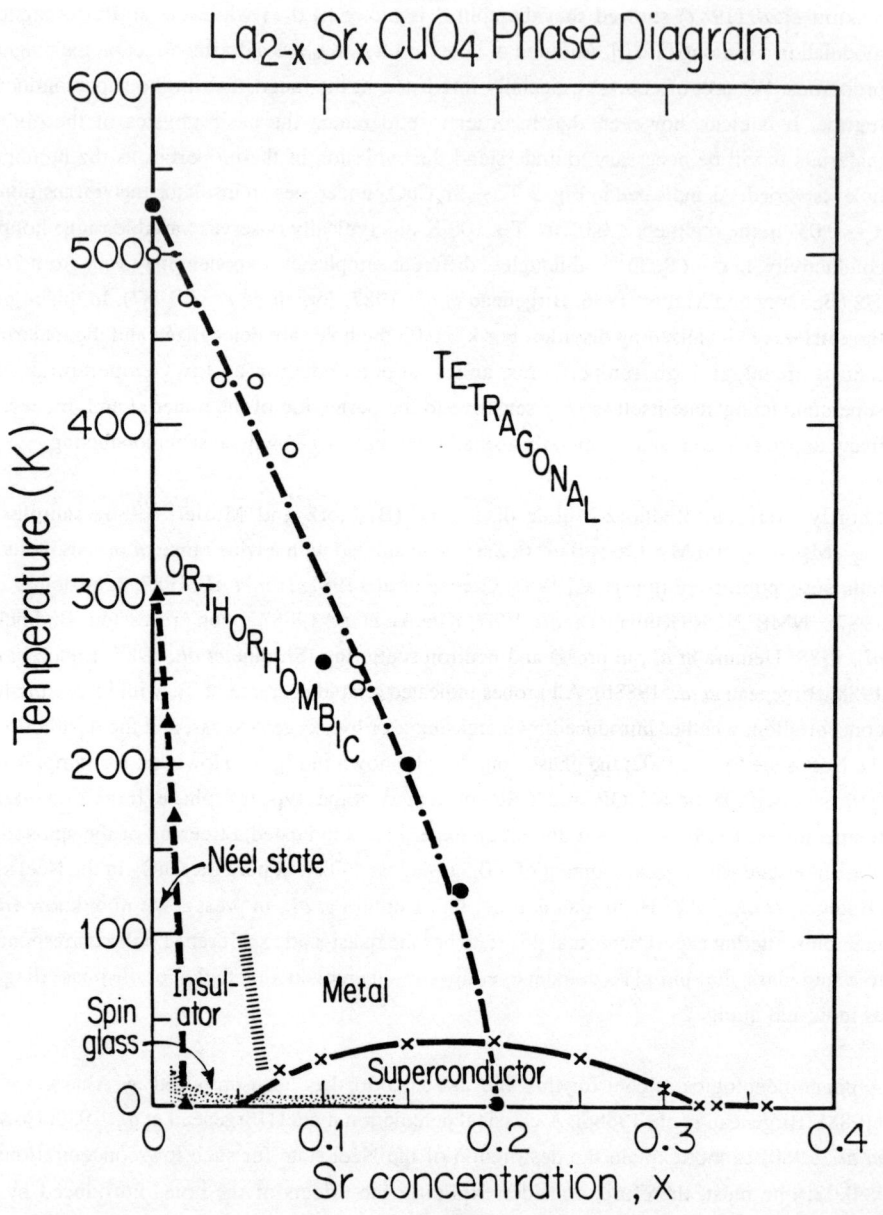

Fig. 2. Phase diagram of $La_{2-x}Sr_xCuO_4$. (See Fleming et al., 1987; van Dover et al., 1987; Fujita et al., 1987; Kumagai et al., 1987; Kitaoka et al., 1988; Budnick et al., 1988; Harshman et al., 1988; Uemura et al., in press; Green et al., 1987; Ishi et al., 1987; Johnston et al., 1987.)

1987, 1988; Nücker et al., 1988; Shen et al., 1987). For concentrations x ≤ 0.05, the holes are localized. The argument of Aharony et al. (1988) is as follows. Consider first an instantaneous configuration, with a single hole on one O^- ion. The spin of the hole, $\vec{\sigma}$, will have strong exchange interactions with the two neighboring Cu spins $\vec{S_1}$ and $\vec{S_2}$. Writing

$$H_\sigma = -J_\sigma \, \vec{\sigma} \cdot (\vec{S_1} + \vec{S_2}), \tag{1}$$

it is intuitively clear that, regardless of the sign of J_σ, the ground state of H_σ prefers $\vec{S_1} \parallel \vec{S_2}$. Quantum mechanically, the exact ground state of H_σ indeed has $S_{12} = 1$ (where $\vec{S}_{12} = \vec{S_1} + \vec{S_2}$, that is, $<\vec{S_1} \cdot \vec{S_2}> = 1/4$). It is thus reasonable to replace H_σ by a ferromagnetic (F) interaction, $\tilde{H}_\sigma = -K \, \vec{S_1} \cdot \vec{S_2}$, where $K = O(|J_\sigma|) \gg |J_{nn}|$. Here $J_{nn} \sim 1300$ K ~ 0.12 eV is the AF exchange interaction between neighboring Cu spins in the CuO_2 plane, and $K \gg |J_{nn}|$ because the Cu-Cu distance is twice that of Cu-O. The replacement of H_σ by $\tilde{H}_\sigma$ is exact for classical spins at low temperatures.

Since a strong F bond in the CuO_2 plane destroys the local AF order, it also influences the coupling to the neighboring planes. The Cu spins thus feel competing AF and F interactions. In the extremely localized case, the concentration of the F bonds would be x. As x increases, the localization length l_0 of each hole increases, and this will increase the effective concentration of F bonds.

Competing AF and F interactions are known to yield a sharp decrease in the Néel temperature T_N, a spin glass (SG) phase (Binder and Young, 1988) and a re-entrance from the AF to the SG phase (Aeppli et al., 1982) upon cooling, because of frozen random local moments. This yields the magnetic parts of Fig. 2. In the isostructural $K_2Cu_{1-x}Mn_xF_4$, the Cu ferromagnetism is lost at x $\simeq$ 0.2 corresponding to a concentration 0.36 of the very weak Cu-Mn and Mn-Mn AF bonds (Kimishita et al., 1986). As recently shown by Vannimenus et al. (unpublished) a large ratio K/|J| brings the threshold concentration down. The fact that $(l_0/a) \geq 3$ also renormalizes the threshold. Furthermore, quantum fluctuations also seem to lower the threshold, as indicated by preliminary Monte Carlo simulations (Morgenstern, private communication). All of these can plausibly explain why in doped La_2CuO_4 the Néel state vanishes at the low concentration x $\simeq$ 0.02 of holes.

INSTANTANEUS AND QUASIELASTIC SPIN CORRELATIONS

Shirane et al. (1987) showed that the spins in pure La_2CuO_4 are correlated instantaneously over very large distances, ≥ 400 Å at T_N. At first it appeared that this required a new phase.

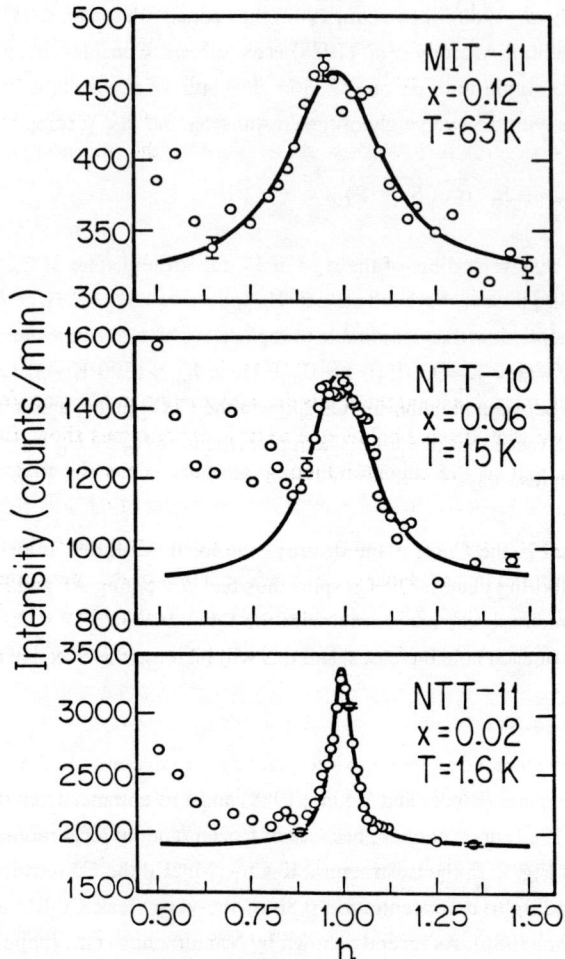

Fig. 3. Two-axis scans across the rod along (h,ζ,0) in NTT-11 (x = 0.02) at T = 1.6 K; NTT-10 (x = 0.06), T = 15 K; and MIT-11 (x=0.12), T = 63 K. In these scans $Q_\perp = \zeta \vec{b}^*$ varied from $0.1\ \vec{b}^*$ to $1.1\ \vec{b}^*$ as h varied from 0.5 to 1.5. The incoming neutron energy was 14.7 meV and the collimation was 40'-40'-40'-open. The solid lines are the results of fits to a single Lorentzian. The scattering intensity is normalized to the volumes of the crystals; the relative volumes were determined by measuring intensities of the transverse-acoustic phonons. Figure from Birgeneau et al. (1988b).

However, the calculations of Charkravarty, Halperin, and Nelson (1988) show that the large correlation length is simply the consequence of a large exchange energy in the paramagnetic phase of a 2D Heisenberg antiferromagnet. Indeed, if it were not for the quantum renormalization effects, which are large for $S = 1/2$, the correlation length would be macroscopic at room temperature. We discuss next how the addition of Sr affects the correlation length.

An extensive set of neutron experiments has been performed at Brookhaven on large single crystals of $La_{2-x}Sr_xCuO_4$ with x varying between 0.02 and 0.14 (Birgeneau *et al.*, 1988b; Birgeneau *et al.*, submitted). However, only some of the crystals with $x > 0.05$ are actually superconducting at low temperatures, and even in these the Meissner fraction varies from a few % to 80%. The explicit microscopic origin of this drastic variation in Meissner fraction is not yet fully understood. As we shall discuss below, the neutron experiments show, first, that the full Cu^{2+} moment is retained in the superconducting samples. Thus the magnetism coexists synergistically with the superconductivity. Second, the character of the spin fluctuations varies smoothly across the insulator-metal boundary. This is difficult to understand using current theoretical models. It should be noted that this smooth variation of the spin correlations through $x \simeq 0.05$ was already evident in the bulk susceptibility as x varied through 0.05 although the importance of this result was not generally recognized (Fujita *et al.*, 1987; Greene *et al.*, 1987; Ishi *et al.*, 1987; Johnston *et al.*, 1987). Indeed, many workers misinterpreted the 2D Cu^{2+} high temperature susceptibility as the Pauli susceptibility of the itinerant holes.

To measure the instantaneous spin correlations one must integrate $S(Q,\omega)$ over energy. This was done with 2-axis scans, which do not resolve energy, using a geometry in which $\vec{k}_f$ was ∥ $-\vec{b}^*$. In this way the energy integration is accomplished by collecting all neutrons with a fixed $\vec{q}_{2D}$, independent of total momentum. It was checked that for all samples the two axis scans at 300 K with $\vec{q}_{2D}$ fixed at 0 and $\vec{Q}_\perp$ varied exhibit a sharp peak at the position when $\vec{k}_f$ is along $-\vec{b}^*$. This shows that the fluctuations are predominantly 2D and dynamical at 300 K. This was confirmed by direct inelastic measurements for some of the samples.

From data like those in Fig. 3 one finds that the integrated intensity for all samples is independent of Sr concentration and is identical to that in pure La_2CuO_4 to an overall accuracy of ~ 20%. This implies that the full Cu^{2+} moment is retained in the doped samples. Of course, the neutrons are sensitive to all spins in the system, those of Cu^{2+} and those of O^-. The above observation must be qualified by the fact that the spin fluctuations extend up to about 0.27 eV and the high energy fluctuations are not rigorously included in the integration; however, the part of $S(\vec{Q})$ near $\vec{q}_{2D} = (100)$ and (001) is dominated by the low energy fluctuations which are properly included in the integration.

As is evident in Fig. 3, the magnetic scattering is centered about h = 1, the antiferromagnetic wavevector in pure La_2CuO_4, but it becomes progressively broader as the Sr^{2+} concentration, x, is increased, corresponding to shorter correlation lengths. Further, these profiles are, within the statistics, independent of temperature below 300 K. There is also some indication of a two-peaked character in the response, especially in the scan for x = 0.12.

Most recently, an extensive set of experiments (Birgeneau et al., submitted) has been carried out on two crystals, both with x $\simeq$ 0.11, T_c = 10 K and Meissner fractions of ~ 80%. These crystals, which are both large, 1 - 2 cm^3 in volume, and of high perfection, represent a significant accomplishment by Y. Hidaka and coworkers at NTT. It was discovered in these experiments that there was a striking thermal evolution in the distribution in energy of the scattering so that it was essential to separate the quasielastic ($|\Delta E|$ < 0.5 meV) and integrated inelastic ($|\Delta E|$ > 0.5 meV) contributions to $S(\vec{Q})$. Accordingly, the spectrometer was set up in the triple axis, energy resolving, mode and all scans were carried out twice, first detecting neutrons scattered off the pyrolytic graphite analyzer so that $|\Delta E|$ < 0.5 meV and second detecting neutrons passing straight through the analyzer. The effective reflectivity of the analyzer was measured to be 78% so that by subtracting 29% of the first scan from the second scan, one obtains precisely in the latter scan the intensity integrated over all energies with $|\Delta E|$ > 0.5 meV since the absorption by the analyzer is negligible. These latter data turn out to be particularly clean, with little contamination scattering.

Representative results are shown in Fig. 4. This shows the integrated inelastic ($|\Delta E|$ > 0.5 meV) and fitted total cross sections for the (h, h − 0.45, 0) scans across the 2D ridge at a sequence of temperatures. Two features are immediately evident. First, the scattering is broad and flat-topped with some indication of a two peaked structure. This incommensurate two-peaked structure was suggested in previous experiments (Birgeneau et al., 1988b). It was also discovered independently by Yoshizawa et al. (in press). Second, the total cross section, as measured in this particular cut through reciprocal space, varies only weakly with temperature from 350 K to 12 K. However, the spin fluctuations change from being predominantly inelastic at 350 K to predominantly quasielastic at 12 K. This quasielastic scattering comes from a glassy state with correlations between CuO_2 layers as well as within the layers. These correlations have a quite complicated geometry which is described in Birgeneau et al. (submitted). The spin glass behavior is induced by disorder and may be less prominent for materials with higher superconducting T_c's.

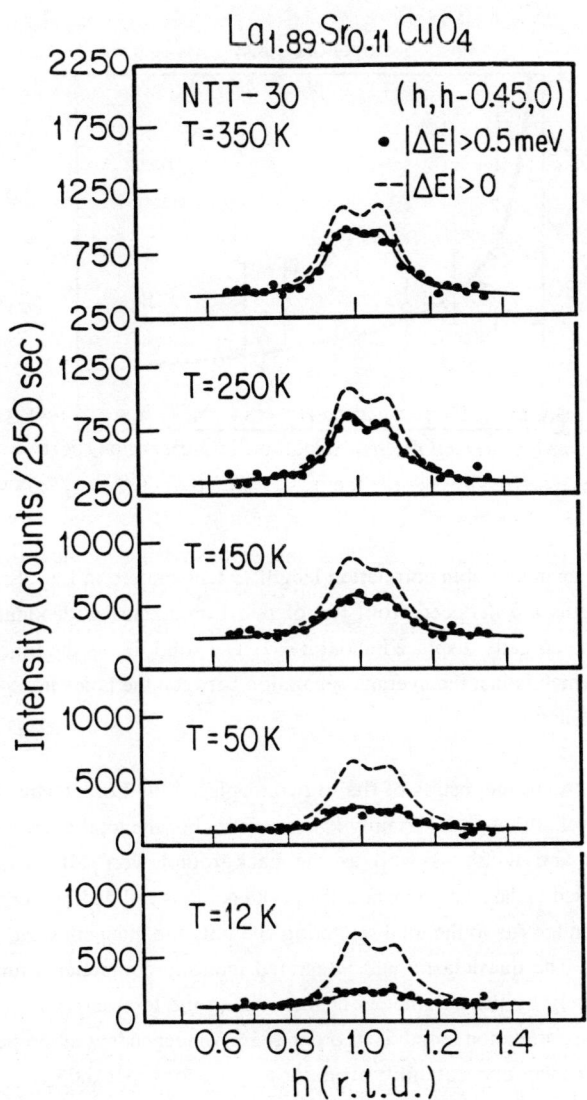

Fig. 4. Integrated inelastic ($|\Delta E| > 0.5$ meV) scattering for scans across the magnetic ridge along $(h, h - 0.45, 0)$; $E_i = 14.7$ meV and the collimator configuration is 40'-40'-40'-80'. The solid lines are the results of fits to two displaced 2D Lorentzians as discussed in the text. The dashed lines are the results of the best fits to the total scattering, elastic plus inelastic. Figure from Birgeneau *et al.* (submitted).

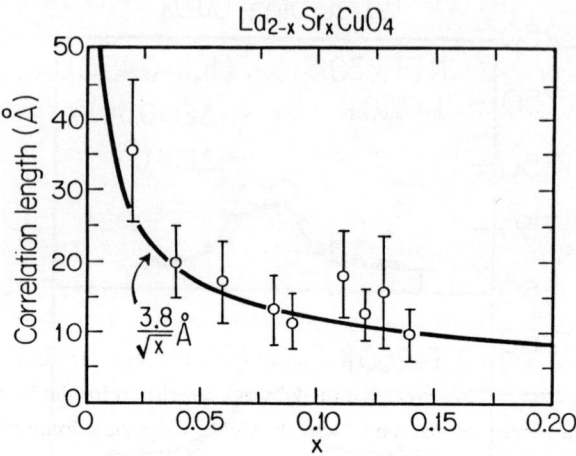

Fig. 5. Instantaneous spin correlation length vs temperature in $La_{2-x}Sr_xCuO_4$. The lengths are deduced from fits of two Lorentzians with identical widths symmetrically displaced about h = 1. The solid line is the function $3.8/\sqrt{x}$ Å which is just the average separation between the holes introduced by the Sr^{2+} doping.

The solid lines in Fig. 4 are the results of fits to two displaced 2D Lorentzians. Clearly this simple model works well although it certainly is not unique. For the total scattering, the peak positions, intensities and width as well as the background were all varied. For the quasielastic and integrated inelastic components the peak positions and widths were fixed at the values determined from the fits to the total scattering and only the intensities and background were allowed to vary. The quasielastic and integrated inelastic components turn out to be well-described separately by the parameters characterizing the total cross section. The 2D instantaneous spin-spin correlation length is of order 18±6 Å independent of temperature from 350 K to 5 K. The 2D incommensurability from these fits is of order 0.05 Å^{-1} to 0.08 Å^{-1}, although larger values are obtained from quasielastic scans $\perp$ the rod so the exact value of the incommensurability should be treated cautiously. At 350 K the total cross section, which corresponds to an integral from ~ kT to + 14.7 meV, is predominantly (~ 75%) inelastic while at 12 K the $|\Delta E| < 0.5$ meV component accounts for ~ 75% of the observed scattering.

We will discuss the dynamical behavior in more detail below. We consider first the instantaneous correlations in the samples studied to-date. In Birgeneau et al. (1988b) the two-axis scans were analyzed using a single Lorentzian profile since the data did not justify a more elaborate lineshape. However, the more recent experiments discussed above show clearly that the 2D correlations are incommensurate. Accordingly, all of the data presented in Birgeneau et al. (1988b) have now been re-analyzed using a two Lorentzian line-shape (Thurston, unpublished). In all cases the goodness-of-fit parameter χ^2 is improved, albeit at the expense of two additional adjustable parameters. The correlation lengths so-deduced are shown in Fig. 5. The solid line, $3.8/\sqrt{x}$ Å, is the average separation between the O$^-$ holes in the CuO$_2$ planes. Remarkably, this agrees with the measured correlation lengths quite well. It is evident therefore that the holes have an extraordinarily disruptive effect on the Cu^{2+} - Cu^{2+} antiferromagnetic state. This may be understood at least in part with the frustration model of Aharony et al. (1988) (Birgeneau et al., 1988a). However, that model was constructed for the localized regime; it is not clear within the context of that model why the Cu^{2+} spins should follow the O$^-$ hole spin so faithfully in the metallic state. Thus a more elaborate theory is most certainly required.

One final qualitative observation should be made on the instantaneous spin correlations. It is evident that profiles for the $x = 0.12$ sample (labelled MIT-11) shown at the top of Fig. 3 and those for the $x = 0.11$ sample shown in Fig. 4 (labelled NTT-30) are closely similar, if not identical. Indeed the fitted widths and incommensurabilities agree to within the errors. However, MIT-11 exhibits metallic resistivity down to ~ 100 K with an upturn at the lowest temperatures whereas, as discussed above, NTT-30 exhibits a superconducting transition at 10K. Thus the magnetic state is quite robust, depending primarily on the Sr^{2+} concentration, whereas the superconductivity itself is apparently quite delicate.

We consider finally the momentum and temperature dependence of the spin excitations themselves. So far there are only limited inelastic measurements in the doped samples. We show in Fig. 6, excitation *creation* scattering processes in NTT-10, $x = 0.06$ at $T = 120$ K for $E = 3, 6$ and 9 meV. The width in momentum space is identical to that obtained from the two axis scans (Fig. 3) while the peak intensity depends only weakly on energy. Studies as a function of temperature show, quite remarkably, that the profile is independent of temperature between 300K and 5K. The excitation *annihilation* scattering is simply related to that shown in Fig. 6 by the detailed balance factor, $e^{-E/kT}$, as expected (Thurston et al., submitted). It was also explicitly verified that the inelastic response function was independent of $\vec{Q}_\perp$, that is, the excitations are 2D in character (Thurston et al., submitted).

Figure 7 shows a series of scans in the superconducting sample NTT-35, $La_{1.89}Sr_{0.11}CuO_4$ for

E = ± 6 meV. The lineshape for excitation creation closely mirrors the integrated response (Fig. 4). Thus these represent excitations out of the slowly fluctuating ground state. Again, the excitation creation intensity is independent of temperature between 300 K and 5 K; this spans the region $kT \gg \hbar\omega$ to $kT \ll \hbar\omega$. The excitation annihilation cross section is related to the creation process by the detailed balance factor $e^{-E/kT}$ as is required by time reversal symmetry (Marshall and Lovesey, 1971). It has been explicitly verified (Thurston *et al.*, submitted) that there is no dispersion in the $\vec{b}$ direction at 6 meV, that is, the excitations are confined to the CuO_2 planes.

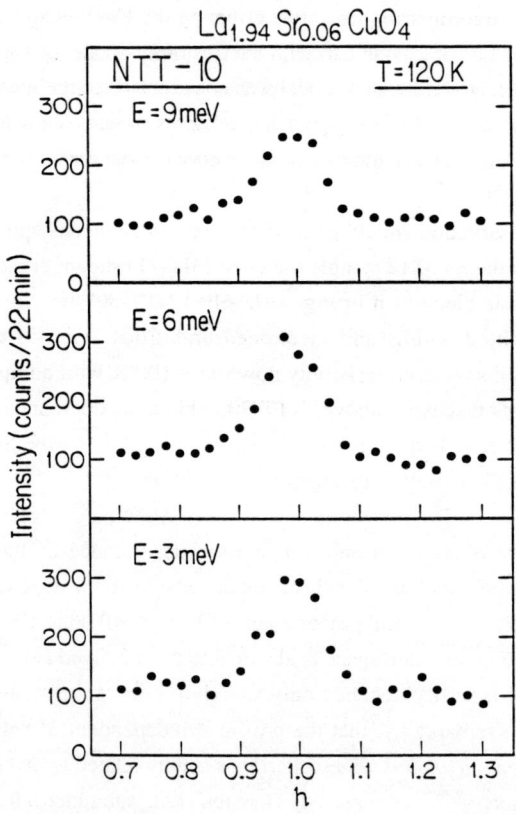

Fig. 6. Constant-energy scans across the 2D rod at T = 120 K in NTT-10 (x = 0.06). The outgoing neutron energy was 14.7 meV and the collimation was 40'-40'-40'-40'. The momentum transfer ⊥ the CuO_2 planes was held fixed at 0.6 O(b,→)*. Figure from Birgeneau *et al.* (1988b).

If the spin excitations were *bosons* as in pure La_2CuO_4 below T_N (Endoh et al., 1988) then the intensity for E = +6 meV would have changed by a factor of 5 between 300 K and 5 K. On the other hand the temperature dependence of the excitation intensity is only consistent with *Fermi* statistics if the chemical potential is much larger than 25 meV. Thus the excitation statistics remain to be understood.

Limited measurements have also been performed (Thurston et al., submitted) for energies varying between 4 meV and 18 meV. The lower limit of 4 meV is set by background considerations while above 18 meV phonon scattering processes dominate. The data at this point are incomplete but they do show that the excitation intensity depends only weakly on energy. Because of the broad distribution in energy the scattering at any given energy is quite weak. The count rate in Fig. 7 at 6 meV is 6 counts per minute; further, NTT-35 is a high quality single crystal 2 cm^3 in volume. It is not surprising, therefore, that similar inelastic neutron scattering experiments have not yet been successfully executed in other high temperature superconductors.

CONCLUSIONS

La_2CuO_4 itself represents a close realization of the 2D, S = 1/2 square lattice Heisenberg antiferromagnet. However, the magnetic correlations in the CuO_2 sheets are remarkably sensitive to the presence of holes on the oxygen ions. The underlying mechanism for this appears to be the strong coupling of the O^- hole spin to the Cu^{2+} spins on either side. This yields a net Cu^{2+} - Cu^{2+} ferromagnetic coupling which frustrates the CuO_2 antiferromagnetic order. Since the Cu^{2+} - Cu^{2+} exchange integral is ~ 0.12 eV in magnitude, the Cu^{2+} - O^- exchange integral may well exceed 0.5 eV. The instantaneous spin-spin correlation length in the doped CuO_2 sheets equals the average separation between the holes. The spin fluctuations at room temperature in heavily doped samples are primarily dynamic. However, even at 350 K there is significant weight in the low energy, $|E| < 0.5$ meV, component, and these low energy fluctuations are correlated three dimensionally. As the temperature is lowered the weight in the low energy component grows progressively and µSR experiments indicate a freezing of all spins in the whole sample below ~ 4 K. Thus the superconductivity occurs in the presence of a slowly fluctuating 3D incommensurate spin fluid. This glassy behavior probably results from disorder and may be absent in materials with higher T_c.

It is clear that the novel magnetism and the phenomenon of high-T_c superconductivity are inextricably bound up together. At the minimum, the magnetism provides a large energy scale necessary for the high T_c's. Theories are still at an early stage of development (Anderson, 1987; Anderson et al., 1987; Emery, 1987; Lee and Read, 1987; Hirsch, 1987; Kivelson et al., 1987;

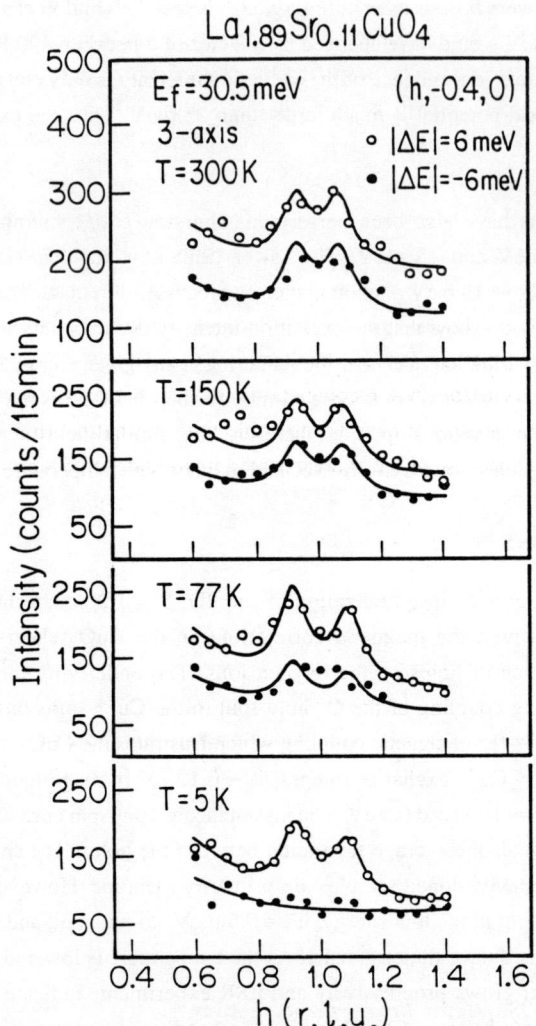

Fig.7. Constant energy scans across the 2D rod along (h,−0.4,0) in NTT-35, $La_{1.89}Sr_{0.11}CuO_4$. The outgoing neutron energy was 30.5 meV and the collimation was 40′-40′-40′-80′. The lines for the +6 meV data are the results of fits to two symmetrically displaced Lorentzians. The lines for -6 meV are calculated from the +6 meV fits assuming detailed balance. The open circles represent excitation creation and the closed circles excitation annihilation scattering processes.

Gros et al., 1987;Thouless, 1987; Schrieffer et al., 1988). Indeed, at the most basic empirical level, one wonders if it is a coincidence that the magnetic and superconducting coherence lengths are so similar in magnitude. Manifestly, any realistic model must explain both the magnetism and the superconductivity.

It is clear that many more neutron experiments remain to be done on these systems. First, it is imperative that high resolution measurements be carried out on the spin excitations at energies comparable to the superconducting gap energy. Second, it is obviously important that experiments as detailed as those in $La_{2-x}Sr_xCu_2O_4$ be carried out on the other CuO_2 superconductors. Such experiments place great demands on crystal growth techniques. We recall that in a high quality crystal of superconducting $La_{1.89}Sr_{0.11}CuO_4$ of volume 2 cm^3 the signal rate for 6 meV excitations was 6 counts per minute. If the crystal were, say 0.5 cm^3 in volume, the experiment would not have been doable with current neutron sources. It seems likely, nevertheless, that because of the unprecedented importance of this problem the requisite advances in crystal growth will indeed be achieved. We anticipate, therefore, extensive neutron studies of these systems for the indefinite future.

ACKNOWLEDGEMENTS

The work at MIT was supported by the National Science Foundation Grants Nos. DMR 85-01856, DMR 87-19217, and DMR 84-15336. Research at Brookhaven is supported by the Division of Materials Science, US Department of Energy under contract DE-AC02-76CH00016. This work was supported by the US - Japan Cooperative Neutron Scattering Program and a Grant-In-Aid from the Japanese Ministry of Education, Science and Culture.

REFERENCES

Aeppli, G., et al., (1982). *Phys. Rev. B*, 25, 4882.
Aharony, A., et al., (1988). *Phys. Rev. Lett.*, 60, 1330.
Anderson, P. W. (1987). *Science*, 235, 1196.
Anderson, P. W., et al., (1987). *Phys. Rev. Lett.*, 58, 2790.
Bednorz, J. G. and Müller, K. A. (1986). *Z. Phys. B*, 64, 189.
Binder, K. and Young, A. P. (1988). *Rev. Mod. Phys.*, 58, 801.
Birgeneau, R. J., Endoh, Y., Hidaka, Y., et al., (submitted). *Phys. Rev. Lett.*
Birgeneau, R. J., et al., (1980). *Phys. Rev. B*, 21, 317.
Birgeneau, R. J., et al., (1987). *Phys. Rev. Lett.*, 59, 1329.

Birgeneau, R. J., Kastner, M. A., and Aharony, A. (1988a). *Z. Phys. B*, 71, 57.

Birgeneau, R. J., et al., (1988b). *Phys. Rev. B*, 38, October 1.

Budnick, J. L., et al., (1988). *Europhys. Lett.*, 5, 647.

Chakravarty, S., Halperin, B. I., Nelson, D. R. (1988). *Phys. Rev. Lett.*, 60, 1057 and to be published.

Cowley, R. A., et al., (1980). *Phys. Rev. B*, 21, 4038.

Emery, V. J., (1987). *Phys. Rev. Lett.*, 58, 2794.

Endoh, Y., et al., (1988). *Phys. Rev. B*, 37, 7443.

Fleming, R. M., et al., (1987). *Phys. Rev. B*, 35, 7191.

Fujita, T., et al., (1987). *Jpn. J. Appl. Phys.*, 26, L402.

Grande, V. B., Müller-Buschbaum, Hk., and Schweizer, M. (1977). *Z. Anorg. Allg. Chem.*, 428, 120.

Greene, R. L., et al., (1987). *Solid State Commun.*, 63, 379.

Gros, C., Joynt, R., Rice, T. M. (1987). *Z. Phys. B*, 68, 425.

Harshman, D. W., et al., (1988). *Phys. Rev. B*, 38, 852.

Hirsch, J. E., (1987). *Phys. Rev. Lett.*, 59, 228.

Ishi, H., et al., (1987). *Physica*, 148B, 419.

Johnston, D. C., et al., (1987). *Phys. Rev. B*, 36, 4007.

Kastner, M. A., et al., (1988). *Phys. Rev. B*, 37, 111.

Kimishita, Y., et al., (1986). *J. Phys. Soc. Japan*, 55, 3574.

Kitaoka, Y., et al., (1988). *Physica C*, 153-155, 733.

Kivelson, S. A., Rokhsar, D. S., and Sethna, J. P. (1987). *Phys. Rev. B*, 35, 8865.

Kumagai, K., et al., (1987). *Physica B*, 148, 480.

Lee, P. A. and Read, M. (1987). *Phys. Rev. Lett.*, 58, 2691.

Marshall, W. and Lovesey, S. W. (1971). *Theory of Neutron Scattering*. Oxford.

Morgenstern, I., private communication.

Nücker, N., et al., (1988). *Phys. Rev. B*, 37, 5158.

Schrieffer, J. R., Wen, X.-G., and Zhang, S.-C. (1988). *Phys. Rev. Lett.*, 60, 944.

Shen, Z.-X., et al., (1987). *Phys. Rev. B*, 36, 8414.

Shirane, G., et al., (1987). *Phys. Rev. Lett.*, 59, 1613.

Thouless, D. J., (1987). *Phys. Rev. B*, 36, 7187.

Thurston, T. R., (unpublished work).

Thurston, T. R., Axe, J. D., Birgeneau, R. J., et al. (unpublished work).

Thurston, T. R., et al., (submitted). *Phys. Rev. B*.

Torrance, J. B., et al., (1988). *Phys. Rev. Lett.*, 61, 1127.

Tranquada, J. M., et al., (1987). *Phys. Rev. B*, 36, 5263.

Tranquada, J. M., et al., (1987). *Phys. Rev. B*, 35, 7187.

Uemura, Y. J., *et al.*, *J. de Physique* (in press) and references therein.
Vaknin, D., *et al.*, (1987). *Phys. Rev. Lett.*, 58, 2802.
van Dover, R. B., Cava, R. J., Batlogg, B., and Rietman, E. A. (1987). *Phys. Rev. B*, 35, 5337.
Vannimenus, J., *et al.*, (unpublished).
Yoshizawa, H., *et al.*, (in press) *J. Phys. Soc. Japan*.

$Bi_2Sr_2Y_{1-x}Ca_xCu_2O_{8+\delta}$: THE MAGNETIC PARTNER OF $Bi_2Sr_2CaCu_2O_8$
A NMR AND μSR STUDY.

R. De Renzi, G. Guidi, P. Carretta, *G. Calestani and °S. J. F. Cox

Dipartimento di Fisica, Viale delle Scienze, I-43100 Parma
*Istituto di Strutturistica Chimica, Viale delle Scienze, I-43100 Parma
°Rutherford Appleton Laboratory, Chilton OX11 oQX-UK

ABSTRACT

We present NMR and μSR evidence for antiferromagnetic ordering in $Bi_2Sr_2Y_{1-x}Ca_xCu_2O_{8+\delta}$ for x<0.05. Magnetization curves are obtained from ZF μSR for three different compositions.

A striking peculiarity of the high T_c superconducting copper perovskites is the existence of a magnetically ordered partner for most of them. Many authors have speculated on the role of frustrated antiferromagnetic interaction (e.g. Aharony *et al.*, 1988; de Jongh, 1988) in the formation of the Cooper pairs, as an alternative to the more conventional phonon coupling scheme. This might or might not be the key to the explanation of high critical temperatures; however the understanding of the mechanism by which small stoichiometric variations may induce such drastic changes in the conductive and magnetic properties of this class of materials is an open challenge. Neutron scattering, μSR and NMR on compounds of the family of $La_{2-x}Sr_xCuO_4$ (Nishilara *et al.*, 1987; Birgenau *et al.*) and of $YBa_2Cu_3O_{7-y}$ (Brewer *et al.*,1988; Rossat-Mignod *et al.*, 1988: Yamada *et al.*) have already explored the magnetic side of their phase diagrams, whose complete manifestation is the formation of an antiferromagnetic order.

The ordered magnetic state can be reached by two apparently distinct mechanisms: either by substituting one of the original divalent components with a trivalent one (like in the $La_{2-x}Sr_xCuO_4$ case), or by reducing the oxygen content (like in the $YBa_2Cu_3O_{7-y}$ and in the La_2CuO_{4-y} cases). Both mechanisms are likely to induce and enhancement of the electron density at the copper sites. Alternatively they can be viewed as a mean of decreasing the electron-hole concentration in the copper-oxygen planes, thereby releaving the magnetic frustration introduced by each hole in its surrounding.

The lack of a magnetic counterpart for the new class of materials based on Bi and Tl constituted until now a puzzling exception to the case discussed above. We dedicated our efforts to the identification of a magnetic counterpart for $Bi_2Sr_2CaCu_2O_8$. Contrary to the case of $YBa_2Cu_3O_{7-y}$ this compound is known to have a well defined oxygen stoichiometry, so that oxygen deficiencies cannot be induced in a single phase material. Considerations on the atomic radius of different trivalent substituents convinced us that Y was a candidate likely to induce magnetic ordering.

A batch of $Bi_2Sr_2Y_{1-x}Ca_xCu_2O_{8+\delta}$ samples (x=0, 0.02, 0.05, 0.33 and 0.67) was prepared by solid state reaction of stoichiometric quantities of Bi_2O_3, $SrCO_3$, $CaCO_3$, Y_2O_3 and CuO. Mixed powders of the components were calcined for 8 hours at 800°C and then fired a few more times at increasing temperatures for a total of 40 hours, each time after grinding and remixing. This annealing procedure took place at 880°C for the x=0, 0.02 and 0.05 materials while it was performed at 850°C for the rest of the batch.

The powder X-ray diffraction pattern of the $Bi_2Sr_2YCu_2O_{8+\delta}$ sample, shown at the top of Fig.1, indicates that the compound is isostructural with $Bi_2Sr_2CaCuO_8$. The measured lattice paremeters are a=5.468(3) Å, b=5.426(3) Å and c=30.181(6) Å; the reduction of the c-axis is in qualitative agreement with the decrease of the atomic radius of Y with respect to Ca. No appreciable contamination by other phases is detected, within the sensitivity of the diffractometer. A first indication of magnetic order in $Bi_2Sr_2Y_{1-x}Ca_xCu_2O_{8+\delta}$, for small values of x, comes from copper NMR. In the perovskite compounds the two abundant isotopes, ^{63}Cu and ^{65}Cu, give rise to a broad spectrum, centered around their Larmor frequencies in the applied magnetic field, which is due to the powder average of their quadrupolar interaction with the electric field gradient at the nucleus. This spectrum, well characterized both for $YBa_2Cu_3O_{7-y}$ (Mali *et al.*, 1987; Guidi *et al.*, 1988) and for $Bi_2Sr_2CaCu_2O_8$ (Carretta *et al.*, 1988), still survives in $Bi_2Sr_2Y_{.33}Ca_{.67}Cu_2O_{8+\delta}$ and in $Bi_2Sr_2Y_{.67}Ca_{.33}Cu_2O_{8+\delta}$, but is totally absent in the pure $Bi_2Sr_2YCu_2O_{8+\delta}$ specimen. The complete disappearance of the resonance in the pure $Bi_2Sr_2YCu_2O_{8+\delta}$ case is similar to the total loss of signal from the antiferromagnetically ordered Cu(2) in $YBa_2Cu_3O_6$ (Guidi *et al.*, 1988) and it can be attributed to a large hyperfine interaction which shifts the signals to a higher frequency range. However such shifted signals have been searched for in zero applied field in the range 60-100 MHz, with no success until now.

μSR provides direct evidence of the magnetic ordering, which is only hinted by the negative

NMR results. We studied only the three specimen with the lower values of x by this technique. The experiments were performed at the Rutherford Appleton Laboratory, on the ISIS muon beamline, using the DIZITAL spectrometer. The muon source provides 26 MeV/c muons, 100% spin polarised (Eaton et al., 1988).

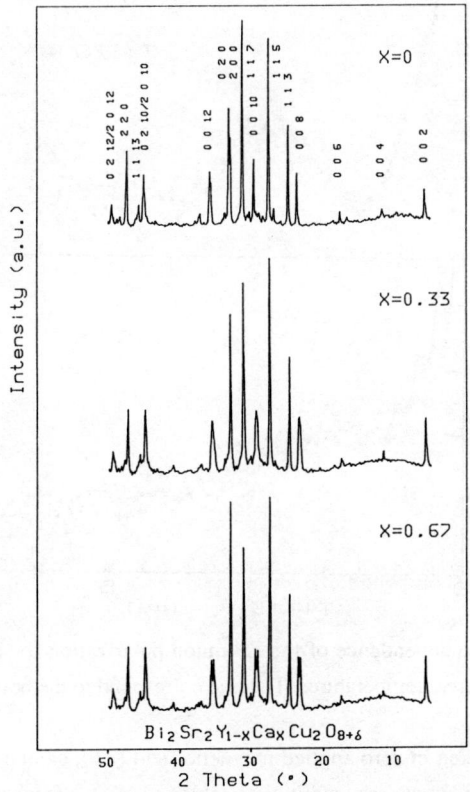

Fig. 1. X-ray powder diffraction patterns of the three samples used for NMR.

In a µSR experiment the spin of the muons precesses inside the sample under the influence of the local field at the side of implantation, which is generally an interstitial lattice site. This precession is revealed, by means of single particle counting techniques (Schenck, 1985), as a wave superimposed on the time histogram of the muon decay.

Figure 2 shows the time dependence of the muon spin polarization for the x=0 sample at two representative temperature. This quantity is extracted from the µSR histograms by taking the ratio $P_\mu(t) = (N_f(t) - \alpha N_b(t))/(N_f(t) + \alpha N_b(t))$, where $N_{f,b}$ are the counting rates in detectors which are respectively forward and backward with respect to the direction of the initial muon polarization (α is just a normalization constant).

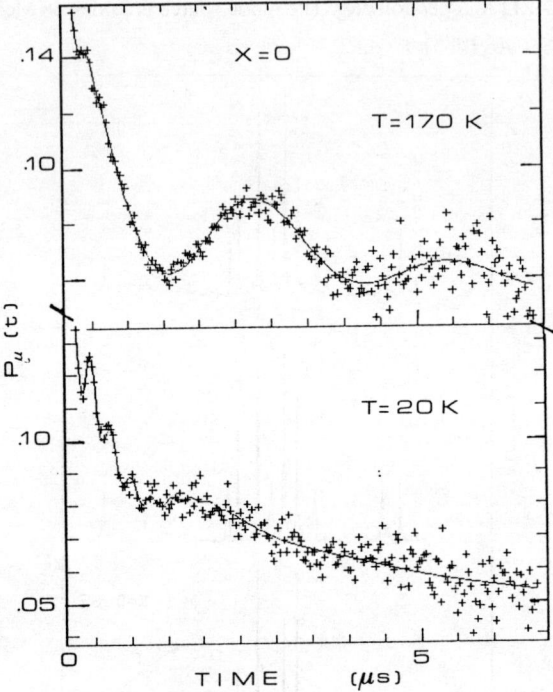

Fig.2. Time dependence of the ZF muon polarization for $Bi_2Sr_2YCu_2O_{8+\delta}$ at two different temperatures. The line correspond to the best fit to the data.

The measurements, taken in zero applied magnetic field (ZF), can be analyzed in terms of at least three components, which are readily detectable in Fig. 2: a fast precession, at a frequency of a few MHz, which decays rapidly, a lower frequency precession of about 0.4 MHz and a constant term with a slow exponential damping. Amplitudes, frequencies and relaxation parameters are temperature dependent, so that the higher frequency is more evident at 20K, while the lower frequency is easier to see at 170K. The first two signals can only be due to a non vanishing spontaneous magnetization, probably arising from the ordered copper magnetic moments, which produce net local fields at the muon sites, mainly of dipolar nature. The fact that two distinct frequencies are observed indicates that at least two different site of implantation are available for the muon; the amplitude of the signals reflects their respective stopping probability. The higher frequency, which corresponds to a local field at the muon site of about 300 Gauss, accounts for 5% of all muons around 170K and 10% below 120K. The lower frequency, corresponding to 30 Gauss, accounts for 36.5% of the muons above 150K, but only for 20% below 140K. It must be noted that the powder average in a ZF measurement does not influence the observed frequencies, which are proportional to the magnitude of the internal fields, but it affects their amplitudes, which is reduced at most to 2/3 of the ideal single crystal value in a P_μ plot like that of Fig. 2.

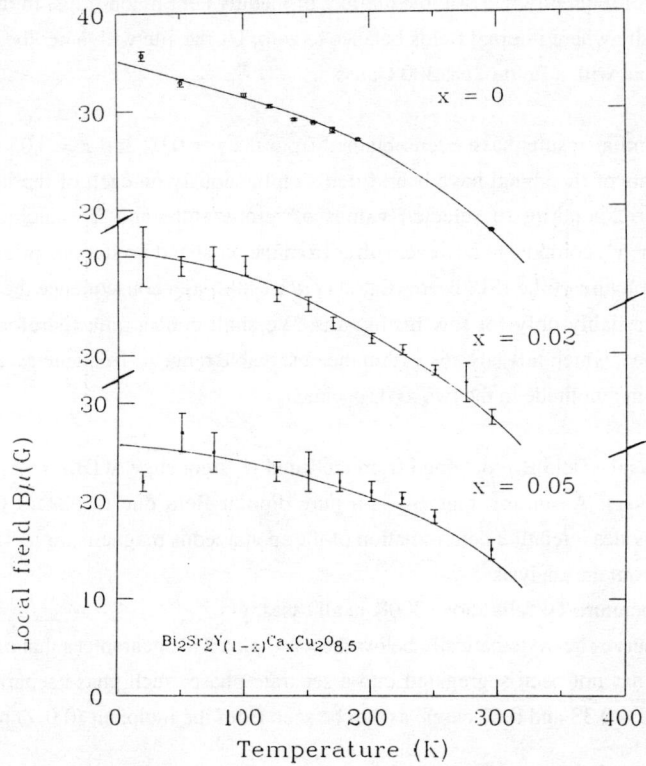

Fig. 3.   Local field B_μ at the U_1 muon site. The lines are just guides to the eye.

The third signal observed in Fig. 2, which is at zero frequency, corresponds to 38% of all implanted muons at all temperatures and has possibly two distinct origins. Due to the powder nature of the samples a fraction of the implanted muons experience a local field which is parallel to their initial spin direction. This fraction gives rise to a non oscillating term which amounts at most to one third of the amplitude of all precessing signals. Most of the observed zero frequency signal has such an origin, but a small fraction remains, which probably comes from a separate muon stopping site where the local field is negligible. A zero dipolar field could be the balance of two equal and opposite contributions from two magnetic sublattices. This fact is strongly suggestive of antiferromagnetic ordering, as is also expected by analogy with La_2CuO_4 and $YBa_2Cu_3O_6$.

The existence of two or more distinct stopping sites for the muons in these complicated perovskite crystals is not at all surprising, especially in view of the superstructures observed (Matsui et al., 1988), although not completely characterized, which lower the exact symmetry of the lattice and consequently increase the number of inequivalent muon traps. In the following, we label B the site where internal fields balance to zero, U_1 the site with a net local field of 30 Gauss and U_2 that with a field of ca. 300 Gauss.

Qualitatively similar results have been obtained from the x = 0.02 and x = 0.05 samples. All three components of the signal have been fitted simultaneously on each of the individual P_μ histograms, corresponding to selected values of temperature and Y stoichiometry. The parameters of the U_2 component however suffer from the passband limitations originated by the peculiar time structure of the ISIS beam (Eaton et al., 1988); as a consequence their values can be determined reliably only for few histograms. We shall concentrate therefore on the U_1 contribution alone, which falls always within the observable range of frequencies, although it is greatly reduced in amplitude in the two x≠0 specimen.

Figure 3 displays the field B_μ, obtained from the fitted U_1 frequency, ν ($B_\mu = \nu/\tau\mu$, with $\tau\mu$ = 13.55 KHz/Gauss). Assuming that B_μ is a pure dipolar field due to distant Cu magnetic moments, it provides a relative determination of the spontaneous magnetisation. The following points emerge from the analysis:
- The Neél temperature T_N falls above 300K in all cases.
- The two x≠0 curves lie systematically below the x = 0 plot. This guarantees that on average the Ca component has not been segregated into a separate phase; such phase separation instead occured in the x = 0.33 and 0.67 cases, as can be seen from the incipient $(0,0,\ell)$ peak splitting in Fig. 1.
- $B_\mu(T=0)$ decreases with increasing Y concentration. The value of the Cu moments seems to be directly affected by the substitution. This is understandable if the magnetic moment is delocalized.

In conclusion our results underline the role of magnetic interactions in high T_c superconducting cuprous perovskites.

REFERENCES

Aharony, A., R.J. Birgenau, A. Coniglio, M.A. Kastener and H.E. Stanley (1988). *Phys.*

Rev. Lett., 60, 1330.
Birgenau, R.J. et al. private communication.
Brewer, J.H. et al. (1988). Phys. Rev. Lett., 60, 1073.
Carretta, P., C. Bucci, R. De Renzi, G. Guidi, C. Vignali, G. Calestani and F. Licci, to be published.
de Jongh, L. J. (1988). *Solid State Commun.*, 65, 963.
Eaton, G.H., A. Carne, S. F. J. Cox, J.D. Davies, R. De Renzi, O. Hartmann, A. Kratzer, C. Ristori, C.A. Scott, G.C. Stirling and T. Sundqvist (1988). *Nucl. Instrum. Methods*, A269, 483.
Guidi, G., C. Bucci, P. Carretta, R. De Renzi, R. Tedeschi, C. Vignali and F. Licci (1988). *Solid State Commun.*, 68, 759.
Mali, M., D. Brinkmann, L. Pauli, J. Ross, H. Zimmermann and J. Hulliger (1987), *Phys. Lett. A*, 124, 112.
Matsui, Y., H. Maeda, Y. Tanaka, S. Horiuchi (1988), *Jpn. J. Appl. Phys.*, 27, L372.
Nishihara, H., H. Yasuoka, T. Shimuzu, T. Tsuda, T. Imai, S. Sasaki, S. Kanbe, K. Kishio, K. Kitazawa and K. Fueki (1987) *J. Phys. Soc. Japan*, 56, 4559.
Rossat-Mignod, J., P. Burlet, M.J.G. M. Jurgens, J.Y. Henry and C. Vettier (1988). *Physica C*, 152, 19.
Schenck, A. (1985). *Muon spin rotation spectroscopy.*, A. Hilger, Bristol.
Yamada, Y., et al., submitted to *J. Phys. Soc. Japan*.

LOW FIELD MICROWAVE ABSORPTION IN CERAMIC $YBa_2Cu_3O_{7-\delta}$

F. Bordi, S. Onori, A. Rosati and E. Tabet

Physics Laboratory, Istituto Superiore di Sanità
Viale Regina Elena 299, 00161 Roma, Italy

ABSTRACT

The surface resistivity R(H) and its magnetic field derivative dR/dH of a superconducting $YBa_2Cu_3O_{7-\delta}$ ceramic pellet were measured with an ESR spectrometer. The behaviour of dR/dH as a function of the modulation amplitude would suggest a stepped microwave absorption of the pellet in the presence of a magnetic field. This conclusion is discussed in the context of the models for granular superconductors.

The electromagnetic interactions at microwave frequency with the electronic structure of the superconducting specimen surface layer reflect in a rather accurate way the peculiar features of the superconducting state. This applies to the most general features of the superconductive transition as, for instance, the coherent structure of the collective electron state as well as to more specific aspects as those related to the pathways of magnetic field penetration inside a given specimen. Electromagnetic absorption studies have been therefore largely applied in superconductor physics through the extensive utilization of different experimental techniques. In particular Electron Paramagnetic Resonance (EPR) technique has proved to be an extremely sensitive tool for the detection of magnetic field-dependent microwave absorption. By using a phase-sensitive technique to measure the complex reflection coefficient of the cavity the field derivative of the sample resistance and reactance can be determined (Friedberg and Strandberg, 1969; Sridhar and Kennedy, 1988). This type of measurements was widely performed on type I and type II superconductors allowing fundamental parameters such as critical temperature, critical magnetic fields, penetration depth, coherence lenght, G-L parameter etc. to be evaluated (Di Crescenzo *et al.*, 1973). Detailed studies on mixed and intermediate states, surface superconductivity, flux flow and irreversible effects (critical state, flux creep, flux jumps) were also successfully performed with microwave technique (Kim and Stephen, 1969). From a more practical point of view it should also be stressed that EPR technique is a non destructive one and that resistivity measurements can be performed without making any contacts or having any percolative path as in conventional four probe measurements. As for the special case of the high T_c ceramic superconductors there are also some other reasons which make EPR technique particularly suitable to study their properties. In particular the following should be mentioned: 1) at liquid nitrogen temperatures thermal phonon effects are dominant so that quantities like specific heat, thermal conductivity, sound attenuation can give less information on the superconducting state as compared to conventional superconductors. On the contrary e.m. radiation essentially interacts with charge carriers undergoing the superconductive transition; 2) the high T_c oxides are caracterized by the presence of Cu^{2+} paramagnetic ions which were thought to play an important role in the mechanism of superconductivity, even if recent experiments on single cristals (Albino O de Aguiar *et al.*, 1988) strongly support the idea that the g=2 copper ion resonant signal (Shaltiel *et al.*, 1987) should come from impurity phases; 3) a strong field dependent microwave absorption takes place at low magnetic field (Blazey *et al.*, 1987) and is generally attributed to the presence of microstructures such as clusters of small homogeneous superconducting regions weakly coupled by Josephson junctions in which magnetic flux can penetrate. The maximum of the modulated absorption signal for zero field-cooled samples was found to be in the 0.5-100 Oe range depending on sample. No definitive conclusions can be drawn about the temperature dependence of the position of the maximum because contradictory results are present in the literature (Blazey *et al.*, 1987; Fonkis *et al.*, 1988)while its amplitude strongly increases as temperature decreases (Khachaturyan *et al.*, 1987). Field-cooled samples showed magnetic memory effects produced by flux trapping: the low field peak amplitude decreases and shifts to higher fields (Khachaturyan *et al.*, 1987). Despite the much experimental work done, some of the most relevant features of the low field signal are not still completely understood and the underlying physical picture is not yet clarified; among these features one can quote: i) the strong dependence of the shape and amplitude of the modulated absorption signal on the amplitude of the modulation field, Hm, which is not understandable in the framework of the used phase sensitive detection technique; ii) the noise, which is present for low modulation amplitudes (Hm < 250 mOe in our case) and is well above the instrumental level.

In this paper the surface resistivity, R(H), of a ceramic $YBa_2Cu_3O_{7-\delta}$ pellet and its field derivative were simultaneously measured at various modulation field amplitudes and frequencies. A conventional EPR spectrometer Varian E112, working in the X band (around 9.4 GHz), was used in this study. The sample, a small bar with dimension 1x3x8 mm3 obtained from an $YBa_2Cu_3O_{7-\delta}$ pellet, was placed inside a quartz tube in the centre of a rectangular TE102 cavity; details of sample preparation and characterization are reported elsewere (Paternò *et al.*, 1988). The field derivative of the reflected power from the cavity was obtained by the usual technique of magnetic field modulation and lock-in detection; at the same time the reflected power was also measured directly from the d.c. output of the signal preamplifier. Temperature was held constant at 78K, by a nitrogen gas-flow cryostat.

In Fig. 1 we compare the signal obtained from the lock-in detector at two different modulation amplitudes Hm. At intermediate amplitudes the shape of the curves changes gradually between the two extremes shown. The noise, well above the instrumental level, rapidly increases as Hm decreases. On the contrary the measured integral absorption curves do not show any dependence on Hm. In Fig. 2 typical absorption curves are shown. The signal profiles both of the derivative and the absolute absorption are independent of the modulation frequency in the range examined (270 - 105 Hz). As it is evident from comparison of Fig. 1 a) and b) and Fig. 2, the lock-in output gives the correct field derivative of the reflected power only if Hm is large enough, whereas by lowering Hm the consistency between the absolute absorption curves and their field derivatives obtained from lock-in is lost. A marked asymmetry connected with the sign of the field gradient is evident in the absolute curves. An analogous behaviour has been observed in d.c. resistivity measurements (Sun Shifang et al., 1988) and has been explained in terms of pinning and depinning of magnetic fluxons at the grain boundaries and of the corresponding onset of an inhomogeneous field distribution in the sample.

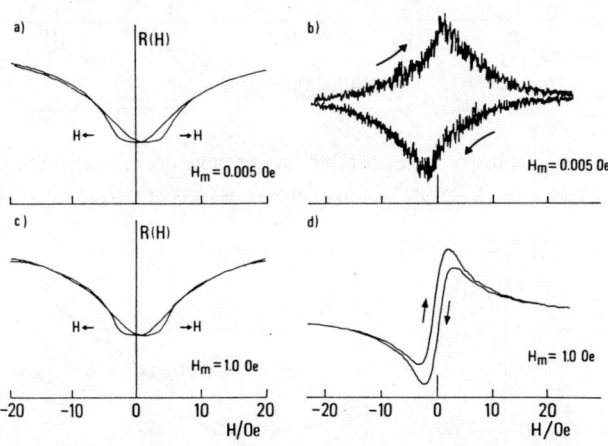

Fig.1 dR/dH in arbitrary units for increasing and decreasing magnetic field recorded a) at Hm=0.005 Oe and b) at Hm=1 Oe

This asymmetry is reflected in the asymmetry of the derivative curves recorded at high Hm values, in fact the peak in the increasing field direction from zero is systematically higher than the corresponding one for decreasing field. A plot of peak amplitude v.s. Hm for the lock-in output reveals an interesting feature (Fig.3). According to the lock-in detector theory, as far as Hm is smaller than the linewidth, the amplitude of the derivative signal should exhibit a linear dependence on Hm. In our case the linewidth is about 5 Oe and the dependence of the amplitude on decreasing Hm is indeed linear for Hm < 1.2 Oe, but at modulation lower than about 250 mOe linearity is lost. Some authors, looking at the integral absorption curves calculated from the lock-in output invoked an "indeterminate state" of the sample as H approaches to zero at low Hm (Glarum et al., 1988).

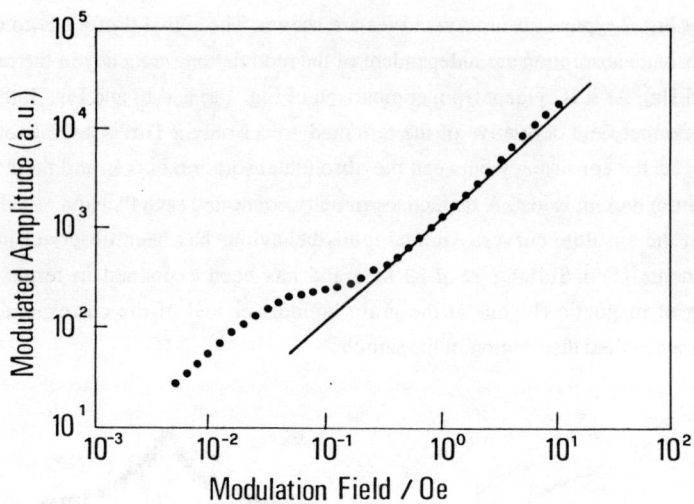

Fig.2 R(H) in arbitrary units for increasing and decreasing magnetic field. R(H) is insensitive to amplitude and frequency variation of the modulation field.

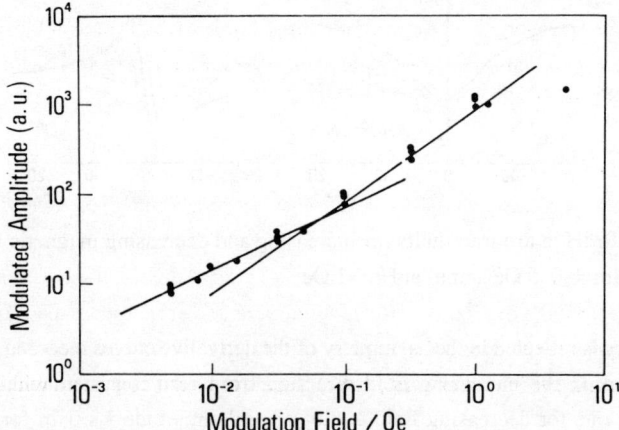

Fig.3 Modulated amplitude v.s. modulation field amplitude for a) normal field sweep, and b) staircase field sweep.

Looking at the measured integral absorption curves it is clear that there is no "indetermination" in the state of the sample. The main point is on the contrary that the derivative response exhibits two regimes as a function of Hm.

We suggest that this behaviour may reflect the presence of a step structure in the absorption curves arising from a macroscopic flux quantization. This hypothesis can, in our opinion, explain in a coherent manner the experimental results. On one hand the spikes due to the leading and trailing edges of the steps may be responsible for the inherent noise observed on the derivative curves. On the other hand, as a simple computer simulation easily shows, the derivative of a "stepped" absorption curve is strongly deformed when the derivation step is smaller than the mesas of the absorption curve. The noise, stonger in the proximity of the peaks, can mask the smaller peak of an asymmetric derivative curve, that can even appear "single peaked". However the presence of a small peak in the nearly symmetric position with respect to the resolved one, by a careful inspection of poor filtered derivative curves, can not be excluded. Moreover in these conditions the peak-to-peak amplitude is no longer linear as a function of Hm. This suggests that the width of the mesas DH may be extimated from the value H* in which the derivative amplitude v.s. Hm deviate from linearity. In our case from Fig.3 curve a) one obtains for ΔH a value of about 250 mOe. In order to be more confident of the evaluation of mesas whidth from H* we have swept the magnetic field with a 1000 step staircase sweep, each step being 75 mOe wide. Fig.3 curve b) shows that in this case the derivative amplitude v.s. Hm moves away from linearity exactly around 75 mOe.

It has been recently pointed out (Clem, 1988) that most of the e.m. properties of the bulk high T_c superconductors can be understood by treating the material as an array of weakly interacting superconducting grains. In our case the typical grain dimension is around 10 μm . The system is then described as an ensemble of superconducting loops, each of them enclosing a specific area: we shall denote with S its projected value in the plane normal to the applyed magnetic field. Ebner and Stroud (1985) have shown that such an ensemble, in the presence of a magnetic field, exhibits well known features of a disordered system with frustration. As long as the distribution of S among the loops is sharply peaked around a preferred value S0 the energy will present a periodic structure, whith the field separation between two adjacent minima given by F0/S with F0=hc/2e. A random distribution of the loop directions partially washes out the energy periodicity, whereas a random S distribution would completely destroy it. From our measurements one gets S ~ 80 μm2, which would suggest the onset of an effective intergranular structure, as also supported by critical current measurements (Kwak *et al.*, 1988). A different picture, based on intragranular Josephson loops (Blazey *et al.*, 1987; Müller *et al.*, 1987), seems to be not consistent with our data.

REFERENCES

Albino O. de Aguiar J., A.A. Menowsky, J. van den Berg and H.B. Brom, (1988). ESR in YBaCuO single crystals. *J. Phys.C: Solid State Phys.*, 21, L237-L241.
Blazey, K.W., K.A. Müller, J.G. Bednorz, W. Berlinger, G. Amoretti, E. Buluggiu, A. Vera and F.C. Matacotta, (1987). Low-field microwave absorption in the superconducting copper oxides. *Phys. Rev. B*, 36, 7241-7243.
Clem, J.R., (1988). Granular and superconducting-glass properties of the high-temperature superconductors. *Physica C*, 153-155, 50-55.
Di Crescenzo, E., P.L. Indovina, S. Onori and A. Rogani, (1973). Temperature and thickness

dependence of critical magnetic fields in lead superconducting films. *Phys. Rev. B*, 7, 3058-3065.

Ebner, C. and D. Stroud, (1985). Diamagnetic susceptibility of superconducting clusters: Spin-glass behaviour. *Phys. Rev. B*, 31, 165-171.

Fonkis, V., O. Dobbert, K.P. Dinse, M. Lehnig, T. Wolf, W. Goldacker, (1988). Fine structure and hysteresys in low-field microwave absorption of YBCO superconductors. *Physica C*, 156, 467-472.

Friedberg, C.B., and M.W.P. Strandberg, (1969). Microwave magnetoresistance measurements. *J. Appl. Phys.*, 40, 2475-2479.

Glarum, S.H., J.H. Marshall and L.F. Schneemeyer, (1988). Field dependent microwave absorption in high-T_c superconductors. *Phys. Rev. B*, 37, 7491-7495.

Khachaturyan, K., E.R. Weber, P. Tejedor, A.M. Stacy and A.M. Portis, (1987). Microwave observation of magnetic field penetration of HT_c superconducting oxides. *Phys. Rev. B*, 36, 8309-8314.

Kim, Y.B. and M.J. Stephen, (1969). Flux flow and irreversible effects. In: *Superconductivity* (R.D. Parks, Ed.), Vol. II, Chap. 19, pp. 1107-1166.

Kwak, J.F., E.L. Venturini, P.J. Nigrey,and D.S. Ginley, (1988). Evidence for homogeneous superconducting grains in high-T_c oxides. *Phys. Rev. B*, 37, 9749-9752.

Müller, K.A., M.Takashige and J.G.Bednorz, (1987). Flux trapping and superconductive glass state in LaCuOBa. *Phys. Rev. Lett.*, 58, 1143-1146.

Paternò, G., C. Alvani, S. Casadio, U. Gambardella and L. Maritato, (1988). Small field behaviour of critical current in YBaCuO sintered samples. *Appl. Phys. Lett.*, 53, 609-611.

Shaltiel, D., J. Genossar, A. Grayevsky, Z.H. Kalman, B. Fischer and N. Kaplan, (1987). ESR in new high temperature superconductors. *Solid State Comm.*, 63, 987-990.

Sridhar, S. and W.L. Kennedy, (1988). Novel technique to measure the microwave response of high T_c superconductors between 4.2 and 200 K. *Rev. Sci. Instrum.*, 59, 531-536.

Sun Shifang, Zhao Yong, Pan Guoqjang, Yu Daoqi, Zhang Han, Chen Zuyao, Qian Yitai, Kuan Weiyan and Zhang Qirui, (1988). The behaviour of negative magnetoresistance and hysteresys in YBaCuO. *Europhys. Lett.*, 6, 359-362.

MATERIALS

OZONE ANNEALING OF YBCO SUPERCONDUCTORS: TOWARD THE MAXIMUM OF DIAMAGNETIC T_c AND MINIMUM OF ΔT_c

F. Celani[*], L. Liberatori[*], R. Messi[+], S. Pace[o], A. Saggese[*] and N. Sparvieri[=]

[*]INFN-Laboratori Nazionali di Frascati, 00044 Frascati Italy
[+]Dip. di Fisica Univ. di Roma II, Tor Vergata, 00173 Roma Italy
[o]Dip. di Fisica Univ. di Salerno, 84100 Salerno Italy
[=]Selenia SpA, Direzione Ricerche, 00131 Roma Italy

ABSTRACT

A.C. low frequency susceptibility measurements, using the "inductance variation" method, have been performed on sintered YBCO pellets fabricated using a new process. This process is based on a modified citrate pyrolysis method and on the use of ozone in all the thermal treatments. The diamagnetic behavior of this material have been studied as a function of the temperature changing both the amplitude of the a.c. magnetic field (0.2, 2, 8 Gauss) and the test frequency (120, 1K, 10K Hertz). At 0.2 Gauss the width (10%-90%) of the diamagnetic transition of the ozone samples is as low as 2K with a diamagnetic T_c of 98K. This width value obviously enlarges with the increase of the applied field: at 2 Gauss, with almost the same T_c, ΔT_c increases up to 6K. The behaviour at 8 Gauss clearly shows the contributions to the diamagnetic signal due both to the intergrain Jj coupling and to the isolated grains. The ozone annealed samples are compared with both oxygen annealed samples and samples made following the conventional calcination - refiring procedure: the last two show larger transition widths and a lower stability against water.

KEYWORDS

Superconductivity; High-T_c; YBCO; Ozone; A.C. Susceptibility.

INTRODUCTION

After the discovery of superconducting oxides (Bednorz *et al.*, 1986), a race toward the materials characterization and the highest critical temperature T_c arose (Hor *et al.*, 1987).

Due to the preparation easiness of sintered samples, showing the correct transition temperature by resistivity measurements, the experimental works on sintered pellets conduced sometimes to unsettled results. Indeed, it is obvious that the existence of some superconducting path with the correct critical temperature, checked by resistivity transition, does not insure the correct stoichiometry of the most volume of the sample and does not allow the detection of spurious phases. It is well known that every sintered sample shows a low critical current density due to the presence of spurious material on the grain surfaces. For instance, these surface problems lead to bad tunnel measurements. In the same way, any "surface" measurement on sintered pellets should be done with a lot of care to be sure that the material under measurement is actually superconducting. In order to avoid these problems, many efforts are spent for fabricate and characterize thin films and single crystals. In any case troubles for making large crystals and very good thin films still lead to interest in good quality sintered samples for both basic research and technological applications. For these reasons we have developed a new fabrication process of $Y_1Ba_2Cu_3O_{7-\delta}$ based both on a modified citrate pyrolysis method and on the use of an ozone enriched oxygen atmosphere (Celani et al., 1988a,b,c,).The samples show both good mechanical and diamagnetic properties (Celani et al., 1988d,e) and the presence of a preferential texture of microcrystals (Celani et al., 1988b).

The optimization of the fabrication process needs the characterization of many samples, so that it is necessary a simple investigation method, able to determinate the pellets goodness. In spite of the simplicity the method requires accuracy, repeatability and has to be easy to use.

We believe that the a.c. susceptibility measurements, using the inductance variation method as function of frequency, temperature and amplitude of the oscillating magnetic field is a simple and powerful method for a first characterization of the fabricated samples.

In this paper the main steps of the fabrication process together with the measurements apparatus are described. After we report measurements of superconducting diamagnetic transitions and discuss the role of screening current and of intergrain Josephson junctions. Comments and conclusions close the paper.

FABRICATION PROCESS AND MEASUREMENT APPARATUS

Most of YBCO fabrication methods use the calcination processes, where, starting from powders (oxides, carbonates, etc.), the homogeneous mixing of Y, Ba, Cu in the correct stoichiometry is obtained only mechanically and by thermal diffusion. In different methods a more intimate mixing is obtained using solutions.
Our fabrication process modifies the usual pyrolytic method, well known in the past, and recently applied to ceramics superconductors (Flokstra et al., 1987). Since the YBCO features are strongly dependent on the oxygen stoichiometry, to reach a complete and homogeneous oxidation of the compound, we use in all thermal treatments an oxygen atmosphere enriched with ozone which shows a very strong oxidation power.

The YBCO preparation starts from copper and yttrium oxides and from barium carbonate. After powder grinding, the addition of nitric acid transforms these powders into three "nitrate solutions" with some excess of nitric acid. The solutions are mixed together and the addition of citric acid forms metallorganic compounds avoiding the barium precipitation. The subsequent addition of NH_4OH drives the pH to the value of about 6.8. After the warming-up of this liquid the pyrolytic reaction starts. This reaction is exothermic and generates flames.

The final product is a fine black powder with granulometry of about 50 - 100 nm. The powder is then annealed at 950 C into an alumina crucible in presence of 100 l/h flux of oxygen with roughly 1% ozone. After 12 hours the furnace is slowly cooled down to room temperature with a cooling rate of about 50 C/hour. The compound obtained just after the first thermal cycle is almost completely superconducting as proved by later described diamagnetic measurements. The powder is ground, filtered, pressed and annealed again to obtain sintered samples of desired shapes. Usually cylindrical samples with diameter of ~20 mm and thickness ranging from 2 up to 7 mm are obtained.

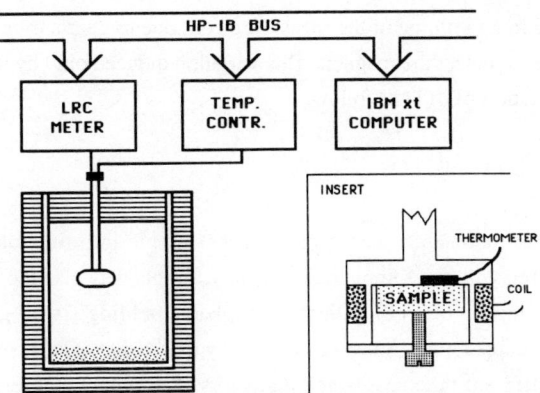

Fig. 1. Experimental Set-Up for measuring the a.c. diamagnetic characteristic vs. temperature. The insert shows the sample holder with the pick-up coil and the thermometer.

All these samples have been characterized by using the home made equipment of Fig. 1. The insert is made of a PVC sample holder with a notch, where a silicon diode thermometer (DT470-13 Lake Shore) in good thermal contact with the pellet is located.

The thermometer and the sample are almost thermally insulated from the external world. Coaxial to the sample, as shown in the insert of Fig.1, there is a coil made of about 200 turns of copper wire, whose inductance (~1mH) is measured as function of the temperature. This insert is dipped into a dewar initially cooled at 77K by a small amount of liquid nitrogen. The thermal leaks of the dewar determines a slow thermal drift of the insert (about 1K/min) toward room temperature. The silicon diode and the coil are connected respectively to a temperature controller (DRC-91C Lake Shore) and to a LCR meter (HP4262A Hewlett Packard). The LCR meter can work only at three frequencies and, in order to maximize the resolution, in the autorange mode it works at fixed amplitudes of the a.c sinusoidal field. These amplitudes have been measured by a Hall probe and for 120Hz, 1KHz, 10KHz, are respectively equal to 8G, 2G, 0.2G. To overcome these constraints sometime a lock-in amplifier and an a.c. current generator have been used. As shown in Fig. 1, the complete measuring apparatus is controlled by a personal computer through the HP-IB bus.

SAMPLES CHARACTERIZATION

Below the critical temperature the superconducting shielding current of the pellet reduces the magnetic field inside the coil and leads to an inductance reduction. Because the pellets will never fill completely the inner volume of the measuring coil, due to geometrical factors, some linked flux exists. In order to detect the magnetic flux variation induced only by the samples, we define the inductance variation ΔL_s (T) as follow:

$$\Delta L_s (T) = L_v (T) - L_s (T) \tag{1}$$

where L_s (T) is the inductance measured with the sample into the holder and L_v (T) is the inductance measured with the sample holder empty. In this way also the stray flux is canceled out. In order to compare ΔL_s (T) with the complete shielding signal generated by an ideal superconductor, the apparatus has been calibrated by measuring at 4.2K several cylindrical lead samples having different thicknesses and diameters. The obtained reference signal ΔL_r have been interpolated in order to obtain the ideal signal as a function of the dimensions of cylindrical samples. ΔL_s (T) is then compared with the signal corresponding to an identical dimension lead sample by the ratio $\Delta L\%$ (T) :

$$\Delta L\% (T) = \frac{\Delta L_s (T)}{\Delta L_r} 100 \tag{2}$$

The method has been tested by a high quality bulk niobium cylinder, obtaining the 100% expulsion percentage within the experimental accuracy.

The detailed temperature dependence of ΔL% (T) for YBCO pellets has been performed in the temperature range 77-300K; further measurements were made at 4.2K. All measurements have been repeated at different values of frequency with the corresponding different amplitudes of the a.c. applied magnetic field.

In order to prove the effectiveness of the ozone during the sample preparation, a series of pellets have been made following the same fabrication process respectively with or without the use of ozone during all thermal treatments. Fig. 2 reports a series of curves obtained for different measuring conditions.

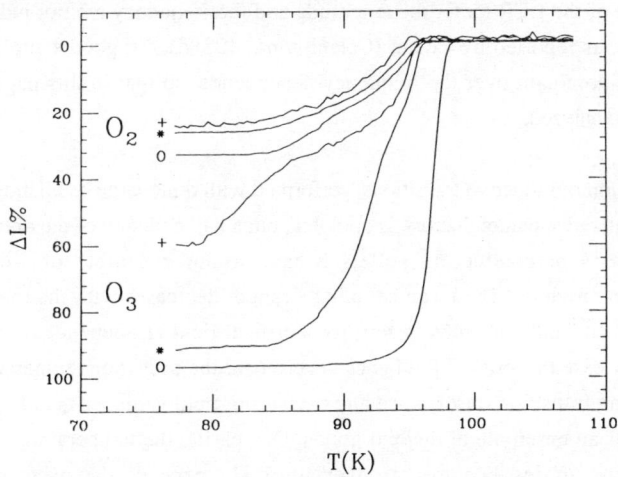

Fig. 2. Inductance variation ΔL% at different conditions for samples annealed in oxygen and ozone atmosphere. (+ freq. 120Hz, a.c. field 8 Gauss; * freq. 1KHz, a.c. field 2 Gauss; o freq. 10KHz, a.c. field 0.2 gauss).

Measured samples were stored for more than six months using only small closed plastic boxes and were cycled many times between room and liquid nitrogen temperature without any care against water and moisture. The difference between ozone and oxygen is evident. The ozone transition curves are sharper, and reach the complete shielding value few degrees below the transition temperature. The curves enlarge only for a.c. magnetic fields of the order of tenth of Gauss. Indeed at 0.2 Gauss the diamagnetic superconducting transition width is only ~2K, while it becomes ~7K and ~16K respectively for a.c. magnetic field of 2 and 8 Gauss. Otherwise, in this temperature range close to the transition temperature and in the presence of the same a.c. magnetic fields, the oxygen made samples show a smaller inductance variation and a broad transition for all values of the a.c. applied magnetic field.

A.c. susceptibility measurements are sensitive both to superconducting shielding currents and to eddy currents induced into spurious normal conducting regions which may be present inside the material. Since the eddy currents are frequency dependent, it has been suggested (McCallum *et al.*, 1982) that the dependence of the a.c. susceptibility on frequency is a good method to detect the presence of normal conducting regions. However someone in the literature reports different explanations of this frequency dependence.

Due to the use of the LCR meter the amplitude and the frequency are not independent. In our measurements, as reported by Celani (Celani *et al.*, 1988d), for good samples the amplitude dependence is dominant over the frequency dependence, so that in this paper only the main contribution is analyzed.

The interpretation of inductive transitions, performed with quite large oscillating magnetic fields on granular sintered superconductors, is complex, but a naive picture of the obtained data can be done. Sintered superconducting pellets behave as an ensemble of weakly connected superconducting regions. The mean coupling strength decreases with the magnetic field and, due to thermal fluctuations, goes to zero for a critical field H_{cj} dependent on the fabrication method. In any case this critical field goes to zero near the transition temperature. Close to the superconducting transition onset, since any measuring field is higher than H_{cj}, the sample can be considerd as an ensemble of disjoint grains. Decreasing the temperature the links between superconducting grains become stronger until H_{cj} exceeds the measuring field and a macroscopic Josephson current can flow through grains and generates the shielding effect of the whole sample. In other words the magnetic field interferes with the intergrain Josephson junctions destroying the weaker couplings, so that measurements with different magnetic fields give information on the intergrain couplings strength.

Other effects can generate a broad transition such as:
a) the existence of a distribution of the transition temperature due to a bad sample quality,
b) the presence of small values of effective lower critical field induced by the granular nature of the sample and by a large number of small impurity regions.

In Fig. 2 the measurements at low field on ozone samples show a behavior like an ideal bulk superconductor with a $\Delta L\%$ (T) of about 100% just 4K below T_c. In the 0.2 G curve the grains and junctions transitions are indistinguishable. The curves at higher fields show a double transition due to the breaking of grain coupling.

On the contrary, the oxygen sample have a broader diamagnetic transition, so that only at very low magnetic fields and at liquid helium temperature shows an almost complete shielding. Moreover, the onset of the transition temperature of ozone samples is about 2K higher than the corresponding oxygen one.

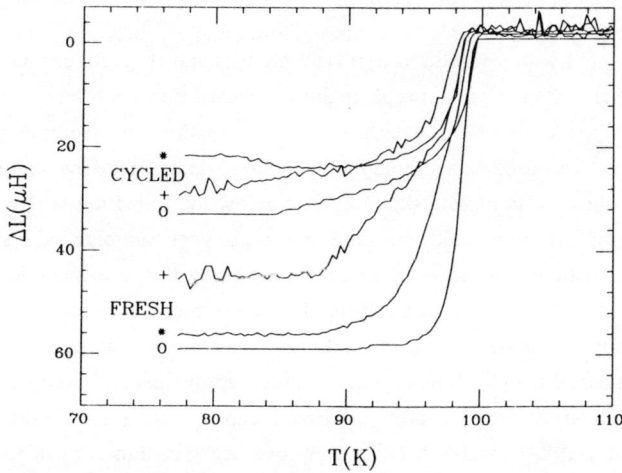

Fig. 3. Inductance variation of a sample obtained by the conventional calcination - refiring procedure. (+ freq. 120Hz, a.c. field 8 Gauss; * freq. 1KHz, a.c. field 2 Gauss; o freq. 10KHz, a.c. field 0.2 gauss).
Fresh: sample as prepared
Cycled: sample after several thermal cycles.

In order to test the sample stability the measurements were repeated several times with thermal cycling between room and liquid nitrogen temperature without cares against moisture and humidity. Beside samples made by pyrolysis, pellets made with different processes have been measured. The number of sample analyzed is not sufficient to judge any other fabrication method, however it is possible to demonstrate the weakness of low density samples made with the standard calcination - refiring procedure. Indeed Fig. 3 reports the measurements done as soon as the calcinated sample was fabricated, and after several thermal cycles. The irregular form of the measured pellets prevents the previously described volume normalization; for this reason Fig. 3 reports the inductance variation in µH.

Nevertheless the inductance variation shows a remarkable difference between the first measurements, when the material is fresh done, and the measurements after several thermal cycles. On the contrary, a large number of measurements, performed in the same environment on an old ozone sample does not show, up to now, noticeable degradation of the diamagnetic behaviour.

CONCLUSION

Samples fabricated with a modified pyrolytic process and respectively with the presence or absence of ozone in the oxygen atmosphere during thermal treatments have been analyzed by the temperature dependence of the a.c. susceptibility. In these measurements the inductance variation method was used as a simple method able to determinate the sample quality. Ozone samples have shown a complete shielding just below the transition temperature with "quite large" measuring magnetic field. On the contrary, oxygen samples have shown very broad diamagnetic transitions. Moreover, thermal cycling and water deteriorate low density samples fabricated by calcination processes. After many thermal cycles old ozone samples have diamagnetic transition better than calcinated samples after only few thermal cycles. In this way, both the goodness of the pyrolytic method and the improvement of the quality by using ozone enriched oxygen atmosphere has been confirmed. Our results have been recently corroborated (Berkeley *et al.*, 1988) by the dramatic improvements in thin film deposition techniques determined by the presence of an ozone atmosphere.

REFERENCES

Bednorz, J.C. and K.A. Muller, (1986). Possible high T_c superconductivity in the Ba-La-Cu-O system. *Z. Phys.*, B64, 189-193.

Berkeley, D.D., B.R. Johnson, N. Anand, K.M. Beauchamp, L.E. Conroy, A.M. Goldmann, J. Maps, D. Mauersberger, M.L. Mecartney, J. Morton, M. Tuominen and Y-J, Zhang, (1988). Ozone processing of MBE grown $YBa_2Cu_3O_{7-x}$ film. *Proc. of the Appl. Sup. Conf. 88* to be published on *IEEE Trans. on Magn.*

Celani, F., R. Messi, S. Pace and N. Sparvieri, (1988a). Metodo di preparazione di superconduttori ceramici ad alta temperatura critica caratterizzati da buone proprietá diamagnetiche. *Il Nuovo Saggiatore*, 4, 7-11.

Celani, F., L. Fruchter, C. Giovannella, R. Messi, S. Pace, A. Saggese and N. Sparvieri (1988b). Torque measurements of textured YBCO sintered pellets. *Proc. of the App. Sup. Conf. 88* to be published on *IEEE Trans. on Magn.*

Celani, F., W.I.F. David, C. Giovannella, R. Messi, V. Merlo, S. Pace, A. Saggese and N. Sparvieri, (1988c). Pyrolytic citrate synthesis and ozone annealing: two key steps toward the optimization of sintered YBCO (1988). To be published on *VUOTO*.

Celani, F., L. Liberatori, A. Saggese, S. Pace, N. Sparvieri, C. Giovannella and R. Messi (1988d). Bulk superconductivity on ozone annealed YBCO samples by a.c. screening currents measurements. *Proc. of II International Workshop on the properties of ceramic and their measurements* to be published on *Material Chemistry and Physics*.

Celani, F., R. Messi, N. Sparvieri, S. Pace, A. Saggese, C. Giovannella, L. Fruchter and C. Chappert (1988e). On the field cooled susceptibility of superconducting YBCO samples. to be published on *Journal de Physique-Colloque*.

Flokstra, J., G.J. Gerritsma, D.H.A. Blank, and E.G. Keim (1987). X-ray, ESR and XPS on YBCO prepared by citrate synthesis. *Proc of the European Workshop on: High Tc superconductors and potential Applications*, 1, 429-430.

Hor, P.H. R.L. Meng, Y.Q, Wang, L. Gao, Z.J. Huang, J. Bechtold, K. Forster, and C.W. Chu (1987). Superconductivity above 90K in the Square-Planar Compoud System $ABa_2Cu_3O_{6+x}$ with A=Y, La, Nd, Sm, Eu, Gd, Ho, Er, and Lu. *Phys. Rev. Lett.*, 58, 1891-1894.

McCallum, R.W. W.A. Karlsbach, T.S. Radhakrishnan, F. Pobell, R.N. Shelton and P. Klavins (1982). Evidence for impurity phase superconductivity in $EuMo_6S_8$ under pressure. *Solid State Comm.*, 42, 819-822.

PHOTOEMISSION

PHOTOEMISSION ON HIGH T_C SUPERCONDUCTORS

P. Steiner, S. Hufner, A. Jungmann, V. Kinsinger and I. Sander

Fachbereich Physik, Universität des Saarlandes
D-6600 Saarbrücken, Germany

ABSTRACT

XPS and UPS valence band spectra of CuO, Cu_2O and several superconducting cuprates are compared to theoretical band structure calculations. This shows that for CuO and the cuprates the interpretation of these spectra has to take into account electronic correlations in the final "two hole" state of photoemission. Core level spectra for the Bi-Sr-Ca-Cu-O system show very simple features while for Tl-Ba-Ca-Cu-O systems complicated behaviour is observed probably due to the multiphased nature of these samples. The influence of contaminations and oxygen loss on the shape on the core level spectra is discussed for the case of the $Y_1Ba_2Cu_3O_{7-x}$ system.

INTRODUCTION

Since the discovery of high temperature superconductivity in the cuprates spectroscopic methods have been extensively used (Wendin, 1987; Fuggle *et al.*, 1988; Ramaker, 1988; Haas, 1989; Proceedings Interlaken, 1988) to elucidate the physical properties of these outstanding materials, as is clearly demonstrated in the present proceedings. Here we review our XPS and UPS photoemission results on these ceramic cuprates in comparison to results on the simpler semiconducting oxides Cu_2O and CuO. Most of these results are presented in detail elsewhere. Therefore we summarize here the main conclusions and include some unpublished results.

In the first part we compare valence band data for Cu_2O and CuO and the cuprates to one electron band structure theory. In the second part we discuss typical features observed in XPS core level spectra of these samples.

Most of the samples were in the form of pellets prepared by the usual ceramic procedures. They are scraped in situ by a diamond file to remove surface contaminations. The only contamination detected by core level photoemission was carbon of the order of a few mol.%, the amount depending on the quality of the sample (e.g. porosity etc.) and the preparation procedures. The Cu_2O, CuO, La_2CuO_4 and $Y_1Ba_2Cu_3O_{7-x}$ samples were single phased, as detected from x-ray powder diffraction patterns, while the Bi-Sr-Ca-Cu-O and the Tl-Ba-Ca-Cu-O sample contained detectable amounts of other phases, due to the well known difficulties in the preparation of these compounds. The superconductivity of the samples have been measured resistively before and after the photoemission experiments with typical values of 90K for the $Y_1Ba_2Cu_3O_{7-x}$ systems, around 80K for Bi-Sr-Ca-Cu-O and between 100-125K for Tl-Ba-Ca-Cu-O, depending on the composition.

VALENCE BAND SPECTRA

Valence band photoemission has been in the past a very powerful tool for the investigation of the electronic properties of matter. A large amount of data on metals and alloys, especially angular resolved photoemission data on single crystals of metals have shown that there is very good agreement between these data and state of the art single particle band structure calculations. Exceptions are systems like e.g. containing 4f-elements or some 3d metals like Ni, where strong correlations between the electrons are responsible for final state effects in the photoelectron spectra which lead to deviations from the ground state band structure calculations, which treat these correlations only in an average way.

The situation for oxides is much less clear. But it has recently been shown in experiments on simgle crystals of "metallic" oxides of Ti_2O_3 and V_2O_3 (Smith and Henrich, 1988) that at least for the electrons a few eV around the Fermi energy a band model for their electronic properties seems to be appropriate. For the superconducting cuprates it has already been argued earlier (see also the contributions of e.g. A. Fujimori and G.A Sawatzky in these proceedings) that strong electron correlations between the Cu-3d electrons make a direct comparison of the photoemission valence band spectra of these cuprates to single particle band structure calculations at least a problem, if not impossible.

In Fig. 1 we show XPS and UPS photoelectron valence band data for the simple oxides Cu_2O and CuO (Steiner *et al.*, 1988b), which both are semiconductors with a band gap of about 2 to 1.5 eV. From the photoelectron cross sections (Scofield, 1976; Goldberg *et al.*, 1981) we expect that for the XPS regime the spectra are dominated by the Cu-3d derived density of states and that for He-I the O-2p derived features are larger.
For He-II the cross sections are nearly equal for the $Cu-3d^{10}$ and $O-2p^6$ shells, so tha He-II spectra should reflect quite nicely the total density of states, if we neglect the energy dependence of the matrix elements.

	XPS	He-II	He-I
hν [eV]	1254	40.8	21.2
$\frac{\sigma(\text{Cu-3d})}{\sigma(\text{O-2p})}$	20	1.1	0.5

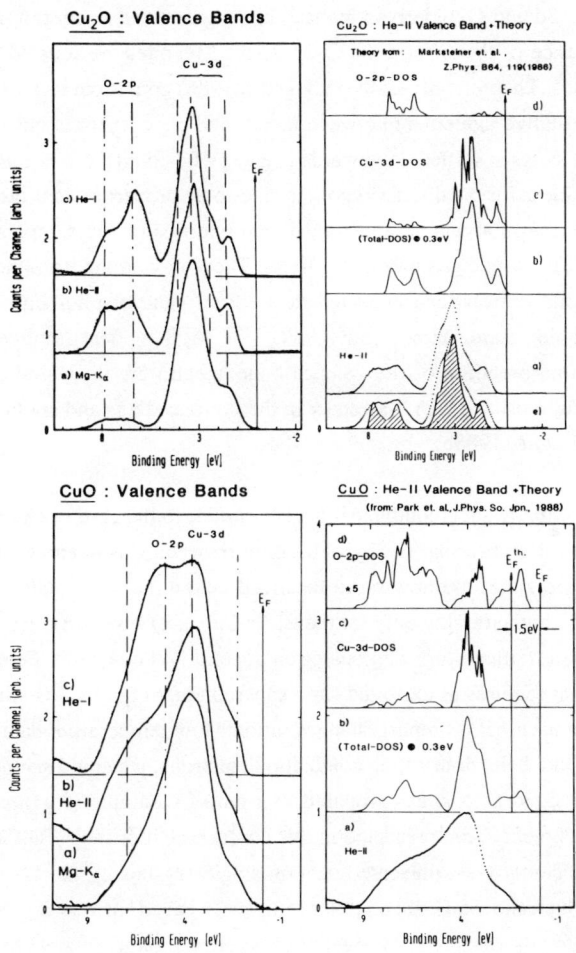

Fig. 1. XPS (hν = 1254 eV), He-I (hν = 21.2 eV) and He-II (hν = 40.8 eV) valence band spectra of Cu_2O and CuO and comparison to single particle band structure theory (Marksteiner et al., 1986; Park et al., 1988).

Comparison of the spectra for the simple oxide Cu_2O, with a closed $Cu-3d^{10}$ and $O2p^6$ shell, to single particle band-structure calculations (Markesteiner et al., 1986) shows that band structure theory works very well for this case (for Cu_2O the Fermi energy of the theory has been located into the middle of the band gap for the comparison with the He-II data), where in the final state of the photoemission process we have only "one hole" either in the Cu-3d or O-2p shell and electron correlation effects play only a minor role. The situation is more complicated for CuO, with a ground state configuration which is probably a mixture of the "one hole" states Cu $(3d^9)$ and Cu $(3d^{10})L^{-1}$, where L^{-1} represents a hole in the oxygen ligand shell. Due to the hybridazation the Cu-3d and O2p derived features are smeared out in energy. But as deduced from the hv dependence of the spectra the O-2p derived features are located mainly 1-2 eV below the Cu-3d states. Theoretically CuO (Park et al., 1988) comes out as a metal, which it is of course not. A qualitative agreement between theory and experiment is only obtained if the theoretical density of states is shifted downwards in energy by about 1.5 eV. But this procedure makes it of course useless for the discussion of the electronic properties near the Fermi energy, which determine the thermodynamic and transport properties of the samples, one is really interested in when discussing e.g. superconductivity. These deviations from theory have been attributed to the strong correlations expected in the final state of photoemission, which is essentially a "two hole" state, namely $3d^8$, $3d^9L^{-1}$ or $3d^{10}L^{-2}$. The implications of these correlations on the interpretation of the valence band spectra are discussed in detail in the contributions of A. Fujimori and G.A. Sawatzky in these proceedings and are therefore omitted here (see also Steiner et al., 1988b).

The situation for the cuprate superconductors is very similar to the case of CuO (see Steiner et al., 1988b), with the difference that the hybridazation between Cu-3d and O-2p states is still stronger and the Cu-3d states are now more localized below the O-2p states. Even so these samples are metals in photoemission only for the Bi-Sr-Ca-Cu-O system a clear Fermi edge has been observed in room temperature photoelectron spectra. Whether this Fermi edge shows temperature dependent features as expected for a superconductimg metal is still a question of debate. While Chang et al., 1988 report changes in their low temperature data, which can be explained by a BCS like behaviour of the conduction electrons, we could see no anomalies in the He-I spectra measured at 20K as compared to a pure Cu sample (*), in agreement with findings of F.U. Hillebrecht et al., presented in this conference. For the $Y_1Ba_2Cu_3O_{7-x}$ derived systems there have recently been presented data on single crystals by the Los Alamos group (Arko et al., 1988) which show a Fermi edge for samples "cleaved" at very low temperatures of about 20K, which already disappears irreversibly when the samples are heated to slightly elevated temperatures of only 60K. This is explained as arising from the well known sensitivity of these samples to oxygen loss, which might be even more prunuced show up in surface sensitive experiments like photoemission. So the question whether these oxide superconductors behave like normal metals at least for electrons close to the Fermi energy has to be clarified by further experiments on well characterized samples.

We conclude this section by presenting in Fig.2 the XPS valence band data for different superconducting cuprates and compare them to theory.

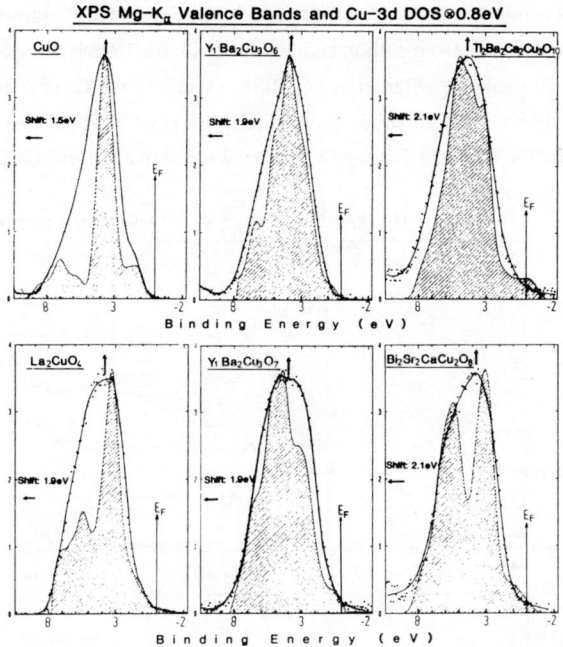

Fig. 2. XPS valence band spectra of CuO and several superconducting cuprates. The shaded area is the theoretical Cu-3d density of states shifted to lower energies as indicated in the figure.

Similar as for CuO a qualitative agreement between the XPS data and the Cu-3d derived theoretical density of states is only obtained when the theoretical results are shifted to considerable lower energies by about 2eV, which again demonstrates the strong electronic correlations in these systems, at least for electrons several eV below the Fermi energy.

CORE LEVEL SPECTRA

XPS core level spectroscopy has intensively been used to clarify the chemical bonding in the high T_c cuprates (Wendin, 1987; Fuggle et al., 1988; Ramaker, 1988; Haas, 1989; see also other contributions from this conference). Here we limit ourselves to only a few remarks.

Many core level spectra taken from these sample are often found to show rather complicated features and it is by far unclear whether these complications represent problems of sample inhomogeneity, of contaminations, or due to the extreme surface sensivity of photoelectron spectroscopy. To our knowledge still now only Bi-Sr-Ca-Cu-O samples, which could be cleaned easily completely from carbon contaminations have shown "simple" shapes for the core level spectra of all elements (Steiner *et al.*, 1988) including the O-1s spectrum. Even for CuO contaminations by water or absorbed oxygen (Steiner *et al.*, 1988b) can be a problem, which shows up in a satellite of the O-1s spectra at about 2 eV above the main line at 529.4 eV.

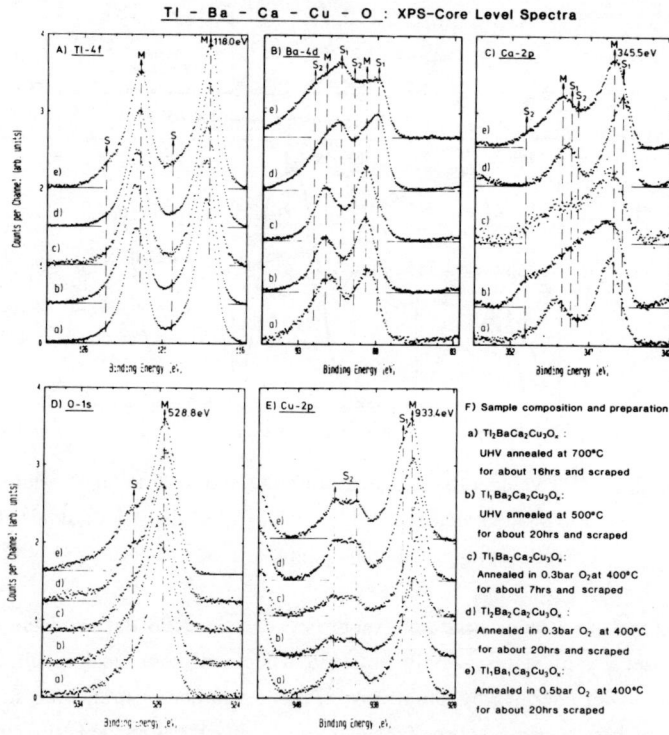

Fig. 3. Core level spectra of several Tl-Ba-Ca-Cu-O superconductors with a T_c between 100K and 125K.

Core level spectra for a series of different Tl-Ba-Ca-Cu-O samples are presented in Fig. 3. As can be seen, most of the spectra show complicated line shapes, depending not very systematically on sample composition or sample treatment. Since these samples are known to be multiphased it is impossible to decide whether most of the complicated line shapes are due to intrinsic properties of the superconducting phase or arise from impurity phases or even surface and grain boundary contaminations. At least the Cu-2p spectra always show the satellites typical for a Cu-$3d^9$ ground state.

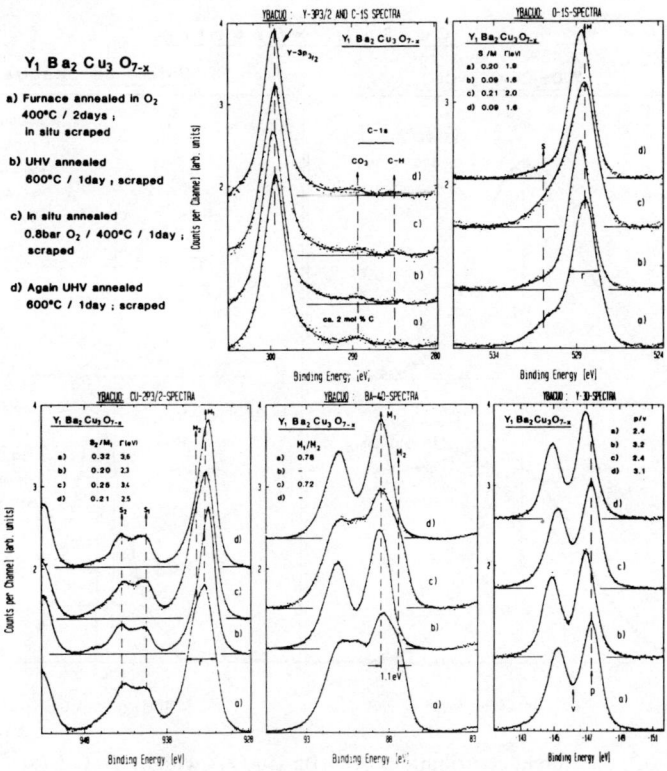

Fig. 4.   Core level spectra of $Y_1Ba_2Cu_3O_{7-x}$ for different in situ treatments of the sample.

Fig. 4 shows the core level spectra of a well oxygenated $Y_1BaCu_3O_{7-x}$ sample ($x \approx 0$), for different in situ treatments. The bulk density of the sample was about 90%. For all sample treatments a sample contamination from carbonates and/or hydro-carbons of the order of a mol.% was detected. The well oxygenated sample a) with $x \approx 0$ shows a shoulder M_2 in the Cu-$2p_{3/2}$ line which is attributed to the $3d^9L^{-1}$ ground state of a formal trivalent Cu (III) (often called Cu^{3+}) in the Cu-O chains. After oxygen loss of the sample ($x \approx 1$) this feature has completely disappeared (Fig. 4b,d) and the Cu-2p line shows a contribution of a $3d^{10}$ configuration (monovalent Cu (I) or Cu^{1+}), as expected for the case that the oxygen in the Cu-O chains are removed. This behaviour and a decomposition of the Cu $2p_{3/2}$ line into the different contributions with CuO as a reference is shown in more detail in Fig. 5. Here $NaCuO_2$ (Steiner et al., 1987) serves as a reference of a Cu (III) oxide ($3d^9L^{-1}$ ground state and not $3d^8$) and CuO_2 for a Cu (I) oxide ($3d^{10}$).

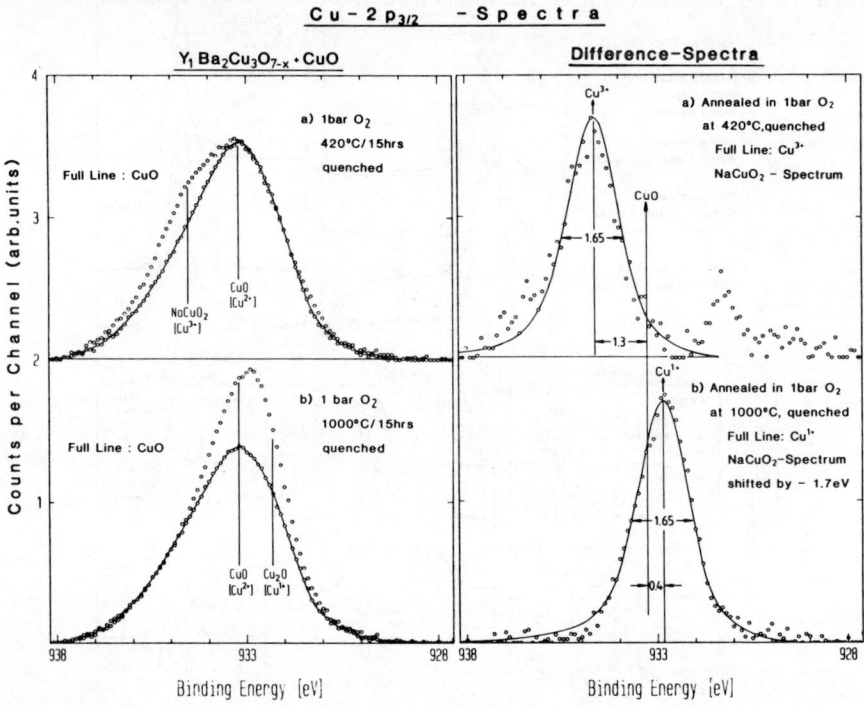

Fig. 5. Different contributions of $Y_1Ba_2Cu_3O_{7-x}$ with a) $x \approx 0$, b) $x \approx 1$.

As has already been demonstrated earlier (Steiner et al., 1988a) the Ba-4d spectra show a sharp feature at low binding energies (M2 in Fig. 4) which disappears with decreasing oxygen content. This behaviour is reversible, when the sample is oxygenated (Fig.4c) and deoxygenated (Fig. 4d) in situ and may therefore stand as an indication of the oxygen content in the sample volume analysed by photoemission.

The Y-3d spectra show a narrowing and a small shift of about 0.5 eV to higher binding energies, when going from $x \approx 0$ to $x \approx 1$ which may reflect inhomogeneities in the sample near

$x \approx 0$. The shoulder S in the O-1s spectra in Fig. 4 has nearly disappeared for a deoxygenated sample ($x \approx 1$) obtained by heating the sample to elevated temperatures around 600°C in the UHV system. It reappears after oxygenation. But since this affords high oxygen pressures at rather low temperatures (300-400°C) it is more likely that this satellite S repesents absorbed H_2O or other oxygen containing species due to the porosity of the sample (the purity of the oxygen used was 99.995%) than intrinsic properties, in agreement with our experience for CuO. But it can also be seen from Fig. 4 that there is a loss in intensity at the low binding energy side of the O-1s line, when going from $x \approx 0$ to $x \approx 1$, which shows that the chain oxygens, which are lost during deoxygenation, have binding energies in the low binding energy

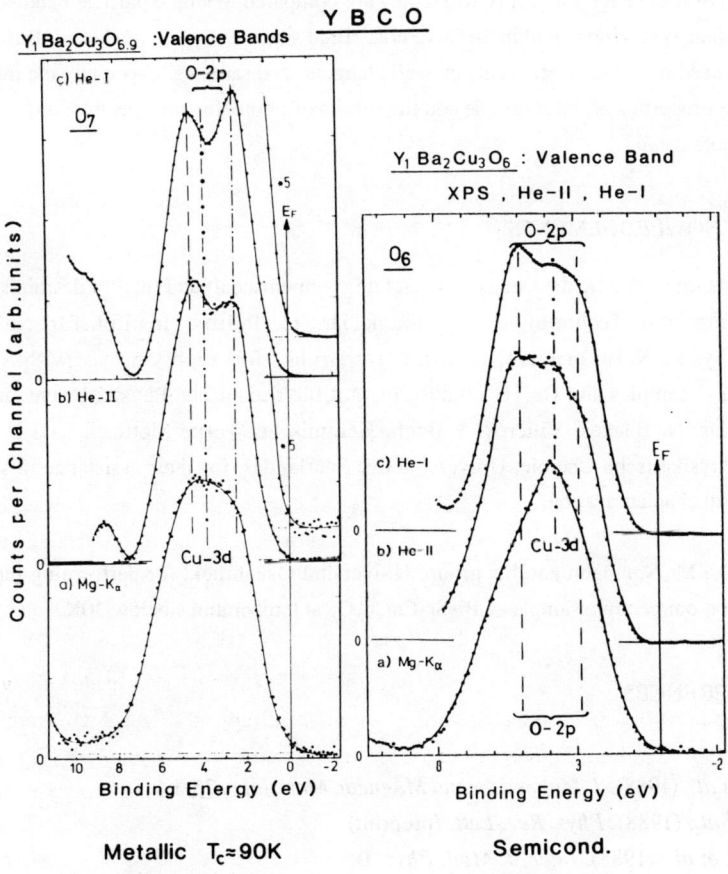

Fig. 6. Valence band spectra of $Y_1Ba_2Cu_3O_7$ and $Y_1Ba_2Cu_3O_6$

region of the line. This is in agreement with results from high-energy-electron-loss spectroscopy reported by J. Fink et al., in this conference. This oxygen loss can also be followed in the valence band spectra, shown in Fig. 6. From the hv dependence of the spectra and from the shape of the He-I spectra we can conclude that the oxygen atoms, which are lost during deoxygenation have a density of states concentrated more towards the direction of the Fermi energy.

CONCLUSION

Photoemission spectra on Cu oxides and cuprate superconductors have shown that in the interpretation of these data, especially when they are compared to single particle band structure theory, electronic correlations within the "two hole" final states cannot be neglected. But there is also a strong need for more experiments on well characterized samples. Especially the influence of the surface properties of these samples on the results of photoelectron spectroscopy has to be clarified in more detail.

ACKNOWLEDGEMENTS

This work was supported by the Deutsche Forschungsgemeinschaft and the Bundesministerium für Forschung und Technologie. We thank Dr. C. Politis, Institut für Nukleare Festkörperphysik, Kermforschungszentrum Karlsruhe, for supplying us with several Tl-Ba-Ca-Cu-O samples and Dr. H. Schmitt, Institut für Technische Physik, Universität des Saarlandes, Dr. N. Backes, Villeroy & Boch, Keramische Werke Mettlach, and S. Junk, Institut für Physikalische Chemie, Universität des Saarlandes, for their assistance in sample preparation and characterization.

*We thank Dr. M. Neumann and his group, Universitat Osnabrück, for performing the UPS experiments on our ceramic sample of $Bi_2Sr_2CaCu_2O_8$ at temperatures below 30K.

REFERENCES

Arko, A. J. et al., (1988). *J. Magnetism and Magnetic Materials* , 75, L1.
Chang, Y. et al., (1988). *Phys. Rev. Lett.* (preprint).
Fuggle, J. C. et al., (1988). *Inter. J. Mod. Phys.* B.
Goldberg, S.M. et al., (1981) *J. Electron Spectr.*, 21, 285.
Haas, K.C. (1989). *Solid State Physics*, 42.

Markesteiner, P.*et al.*, (1986). *J Phys. B*, 64, 119.

Park, K. T. *et al.*, (1988) Technical Report of ISSP, Ser. A, Nr. 1960, May 1988, and *J.Phys. Soc. Japan.* (preprint).

Proceedings of the Interlaken Conference 1988 on "High Temperature Superconductivity", and the other contributions of the present proceedings.

Ramaker, D.E.(1988). *Adv. Chem. Phys.*

Scofield, J.H. (1976). *J. Electron Spectr.*, 8, 129.

Smith, K. E. and V.E. Henrich, (1988). *Phys. Rev. B*, 38, 5965.

Steiner, P. *et al.*, (1987). *Z. Physik B*, 67, 497.

Steiner, P. *et al.*, (1988). *Physica C*, 156, 213.

Steiner, P. *et al.*, (1988a). *Z. Physik B*, 69, 449.

Steiner, P. *et al.*, (1988b). *Z. Phys. B*, (in press).

Wendin, G. (1987). *J. de Physique Coll.*, 48 (C-9), 483.

PHOTOEMISSION SPECTROSCOPY OF $Bi_2CaSr_2Cu_2O_{8+\delta}$ IN THE NORMAL AND SUPERCONDUCTING STATE

F.U. Hillebrecht, J. Fraxedas, L. Ley, J. Trodahl and J. Zaanen

*Max-Planck-Institut für Festkörperforschung,
Heisenbergstrasse 1, D-7000 Stuttgart 80,
Federal Republic of Germany*

ABSTRACT

XPS and UPS valence band spectra of $Bi_2CaSr_2Cu_2O_{8+\delta}$ are reported and compared to spectra predicted within the local density approximation. The Cu d - O p band dominates the spectra at all photon energies, although it is found 1 - 2 eV lower than predicted. We argue that this results from electronic correlations which enhance the anisotropy of the ground state charge distribution. The only change observed upon entering the superconducting state is the expected sharpening of the Fermi edge. We find no evidence for the formation of a gap of 15 meV of larger.

KEYWORDS

High-T_c superconductors; photoemission.

Photoemission is a standard technique for the study of surfaces of solids which among many other techniques has been applied to the new high T_c superconductors. The aim of these and other experiments is to assemble all experimental evidence on these materials in order to understand the physical mechanism leading to superconductivity at temperatures higher than in conventional superconductors. In this context photoemission can provide a picture of the electronic structure, or more precisely of the one-particle Green's function corresponding to removing an electron from the system. In simple cases this corresponds closely to the density of electronic states as obtained in a ground state electronic structure calculation. The by now well-documented disagreement between ground state band structure calculations and photoemission spectra for the high T_c systems tells us that these materials are highly correlated systems. We present here data for $Bi_2CaSr_2Cu_2O_{8+\delta}$ with a transition temperature of 85K, which again show the above mentioned disagreement with band structure theory, and argue on the basis of previous theoretical work that correlation effects are symmetry-dependent. A complete account of the experiments is given elsewhere (Hillebrecht *et al.*, 1988).

In Fig. 1 photoemission spectra taken with different photon energies are compared to the spectra predicted by local density approximation (LDA) band structure calculations (Marksteiner et al., 1988). The differing photoemission sections of various constituents have been included in the calculation, and this is responsible for the photon-energy dependence of the predicted spectra.

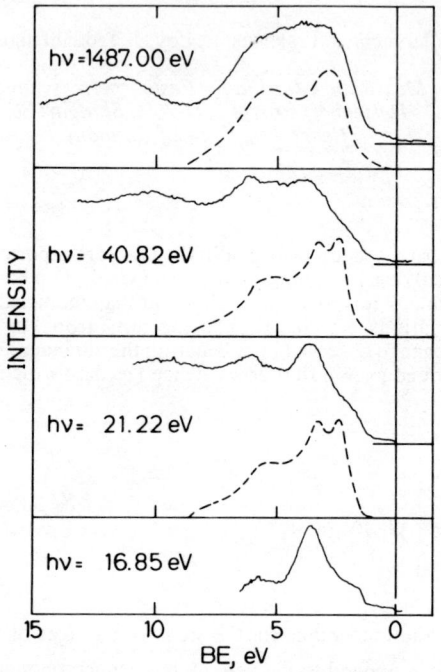

Fig. 1 Photoemission spectra at near normal emission from clean $Bi_2CaSr_2Cu_2O_{8+\delta}$ faces at four photon energies. All binding energies are referred to the Fermi level. The dashed curves are spectra predicted by a local density calculation, and have been shifted 1.7 eV towards higher binding energy to bring the 1487 eV prediction into line with the measured spectrum.

The photoemission spectrum is for all energies dominated by the Cu d - O p hybrid states. The measured width of these bands is in agreement with calculation, but the position is not: the theoretical bands are 1 - 2 eV closer to the Fermi level (E_F) than measured. Such differences are not uncommon; they are usually ascribed to the effects of exchange and correlation which are not adequately taken into account by a local density calculation.

We now present a qualitative argument concerning the influence of correlation on the electronic structure of BiCaSrCuO. It is known from calculations for Ni (Oles and Stollhoff, 1984) that an anisotropic charge distribution is made even more anisotropic by electronic correlation. The argument relies on noting that there is a contribution to the correlation energy from the increased phase space for charge fluctuations if the system is more anisotropic. The charge distribution in the Cu-O planes of BiCaSrCuO is in fact characterized by a considerable amount of anisotropy. This anisotropy is enhanced by shifting the bands of other than $d_{x^2-y^2}$ symmetry away from the Fermi level, leading to the observed position of the main Cu d - O p bands. The states of $d_{x^2-y^2}$ character are not so strongly affected except possibly for a moderate narrowing. In this picture the states remaining at E_F are then two bands of mostly $d_{x^2-y^2}$ character. This provides a suggestion as to why a two-band Hubbard model like the one of Emery (Emery, 1987) appears to capture several aspects of the electronic structure of high T_c materials fairly well.

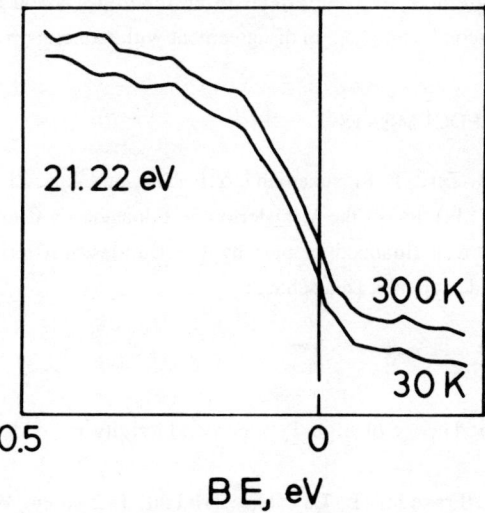

Fig. 2. Fermi edge of $Bi_2CaSr_2Cu_2O_{8+\delta}$ at temperatures of 300 and 30K. The only change is a sharpening of the edge at low temperature as a result of changes in the Fermi function.

Although the origin of superconductivity in these materials is uncertain, one can expect a correlation between the transition temperature and the energy gap in the superconducting state. According to the BCS theory the gap is 3.5 $k_B T_c$, i.e. about 23 meV for this material, and tunneling measurements suggest a gap in the range of 15 - 23 meV (Vieira et al., 1988). This is so large that it raises the hope of observing the gap directly in high resolution photoemission, and indeed one observation has already been reported (Imer et al., 1988). To eject a "superconducting" electron, i.e. an electron from a paired state, one first has to break the pair by raising both electrons to the Fermi level of the normal state. For two electrons, this requires an energy Δ. Therefore, in a photoemission spectrum the high kinetic energy threshold normally at E_F should be shifted to lower kinetic energy by the full gap energy. In Fig. 2 we show photoemission data taken with a total resolution of 70 meV above and well below the superconducting transition of $Bi_2CaSr_2Cu_2O_{8+\delta}$. It is clear that there is no sign of a shift although a shift of 15 meV would be seen clearly. The gap may be smaller than the BCS value or it may be anisotropic, but if the gap is isotropic, has at least the BCS value, and extends to the surface, then we should have observed it in our spectra. Our failure to do so may be the result of a surface region that is not characteristic of the bulk.

In summary our valence band spectra are in qualitative agreement with LDA bandstructure calculations, although there are clear effects of correlation. Data at the Fermi edge show no evidence of the superconducting gap, in disagreement with tunneling measurements.

ACKNOWLEDGEMENTS

We are grateful to A. Zettl, P. Pinsukanjana, Y.F. Yan, and Z.X. Zhao for providing us with samples. One of us (J.F.) thanks the Ministerion de Educacion y Ciencia (Spain) for financial support. This work was financed in part by the Bundesministerium für Forschung und Technologie der Bundesrepublik Deutschland.

REFERENCES

Emery, V.J., (1987). Theory of high-T_c superconductivity in oxides, *Phys. Rev. Lett.*, 58, 2794.

Hillebrecht, F.U., J. Fraxedas, L. Ley, H.J. Trodahl, J. Zaanen, W. Braun, M. Mast, H. Petersen, M. Shaible, L.C. Bourne, P. Pinsukanjana, and A. Zettl, Experimental electronic structure of $Bi_2CaSr_2Cu_2O_{8+\delta}$, *Phys. Rev. B*, in print.

Imer, J.-M., F. Patthey, B. Dardel, W.-D. Schneider, Y. Baer, Y. Petroff, and A. Zettl, High-resolution photoemission study of the low-energy excitations reflecting the superconducting state of Bi-Sr-Ca-Cu-O single crystals, unpublished.

Marksteiner, P., S. Massidda, J. Yu, A.J. Freeman, and J. Redinger, (1988). Calculated photoemission and x-ray emission spectra of $Bi_2Sr_2CaCu_2O_8$, *Phys. Rev. B*, 38, 5098.

Oles, A.M. and G. Stollhoff, (1984). Correlations effects in ferromagnetism of transition metals, *Phys. Rev. B*, 29, 314.

Vieira, S., M.A. Ramos, M. Vallet-Regi, and G.U. Gonzalez-Calbet, (1988). Tunneling measurements of the energy gap $Bi_4Ca_3Sr_3Cu_4O_{16+\delta}$, *Phys. Rev. B*, 38, 9395.

ELECTRONIC STRUCTURE OF $YBa_2Cu_3O_{7-\delta}$ AND $Bi_2Sr_2CaCu_2O_{8+\delta}$ BY X-RAY PHOTOEMISSION AND AUGER SPECTROSCOPIES

A. Balzarotti, M. De Crescenzi, N. Motta, F. Patella, J. Perrière[+],
F. Rochet[+] and Murali Sastry[*]

*Dipartimento di Fisica, II Universita' di Roma
Via E. Carnevale, I-00173 Roma - Italy.*

ABSTRACT

X-ray photoemission and Auger spectra of the Cu 2p core level in $YBa_2Cu_3O_{7-\delta}$ and of the O 1s level in $YBa_2Cu_3O_{7-\delta}$ and $Bi_2Sr_2CaCu_2O_{8+\delta}$ are measured as a function of the oxygen concentration and of temperature. A molecular cluster description of the Cu states is introduced which includes the contribution of trivalent copper. Correlation energies for copper and oxygen valence holes are estimated. An oxygen excess is likely to modify the surface stoichiometry of both materials.

KEYWORDS

High-T_c superconductors; electronic properties; copper oxides; X-ray spectroscopy; Auger spectroscopy.

INTRODUCTION

Many discrepancies among experimental results on the high-temperature copper oxide superconductors $YBa_2Cu_3O_{7-\delta}$ (YBCO) and $Bi_2Sr_2CaCu_2O_{8+\delta}$ (BSCCO) are related to the dependence of the properties of this compound on stoichiometry. Both electronic properties and crystallographic structure are strongly affected by oxygen content, which, in turn, may easily vary during measurements due to the experimental conditions experienced by the sample. Pressed powder samples are particularly affected by surface degradation e/o contamination. The surface sensitivity of the experimental techniques used for studying the electronic properties, as XPS, UV-photoemission and Auger spectroscopy, is a further difficulty in discriminating among intrinsic and extrinsic effects, a difficulty which will be overcome by the availability of reliable single crystals.

Nevertheless, a few general conclusions can be assessed:

i) the failure of single particle density of states calculations in describing the measured electronic structure, a fingerprint of the presence of highly correlated electrons (Fujimori et al., 1987; Thiry et al., 1988; van der Marel et al., 1988; Gourieux et al., 1988; Kurtz et al., 1987; Steiner et al., 1988; Balzarotti et al., 1987; Sarma et al., 1987), and

ii) the existence of highly mobile holes favoured by the mixed-valence of copper (Gourieux et al., 1988; Steiner et al., 1988; Eskes and Sawatzky; Martin et al., 1988; Balzarotti et al., 1988, 1988a; de Groot et al., 1987)

Open questions still remain on:

i) the nature of the observed O 1s core level peaks and their relative intensities,

ii) the presence at particular sites of copper or oxygen atoms in different oxidation states.

The present investigation supports the idea of dynamic valence fluctuations, both at Cu and O sites, induced by the formation of nearly degenerate charged complexes $(Cu^{3+}O^{2-})^+$ and $(Cu^{2+}O^-)^+$ (Martin et al., 1988) in both YBCO and BSCCO. The effective Cu^{3+} content which can be estimated spectroscopically is 0.32 for YBCO (δ=0.15-0.2), 30% higher than that determined from neutron diffraction data, and 0.24 for BSCCO films, indicating in both cases a large covalence of the Cu-O bond. Formal Cu^{3+}, however, is not present in bulk BSCCO. Intensity, spectral shape and energy position of the structures largely depend on oxygen content, while modifications with temperature (T=80K) in the superconducting phase are minor except for the Cu $L_{2,3}VV$ Auger spectrum.

EXPERIMENTAL

The samples were prepared with the standard ceramic technique from Y_2O_3, $BaCO_3$ and CuO in stoichiometric ratios for YBCO and Bi_2O_3, $CaCO_3$, $SrCO_3$ and CuO for BSCCO. Resistivity as a function of the temperature was measured and yielded a transition temperature of 92±2 K and 90±2 K for YBCO and BSCCO, respectively. The lattice structure of YBCO was determined by neutron powder diffraction measurements at 300K. The results indicated an orthorhombic structure with oxygen stoichiometry in the range 6.8-6.9 ($\delta \simeq$ 0.15-0.20) (The neutron diffraction characterization has been performed at the ISIS neutron facility of the Rutherford Appleton Laboratories, England. The data analysis was based on the Cambridge Crystallography Soubroutine Lybrary (CCSL); Balzarotti et al., 1988).

$Bi_2Sr_2CaCu_2O_{8+\delta}$ superconducting films were grown on MgO oriented single crystals using bulk sintered powders as target for ablation with YAG laser (λ= 1.06 nm) followed by a KDP frequency doubling crystal. The films were annealed in 1 atm of pure oxygen in the temperature range 860-880 °C. Their composition, checked by Rutherford backscattering (RBS) analysis, was found to be close to the 2:2:1:2.

XPS and Auger measurements were performed in ultrahigh vacuum at a pressure in the 10^{-10} Torr range using a twin anode (Al and Mg) x-ray source. Electrons were monitored by a VG-CLAM 100 hemispherical analyzer and the overall energy resolution was 1.1 eV.
Before each set of measurements the samples were scraped in situ to expose fresh superconducting grains of correct stoichiometry. Spectra taken after different scrapes were quite reproducible but not identical since the exposed surface has a distribution of grains different from scrape to scrape.

RESULTS

Cu

In Fig. 1 are shown the 2p core levels of Cu in YBCO compared to the ones measured in CuO. The two spin-orbit components are followed, at ~ 9 eV higher binding energy, by the associated

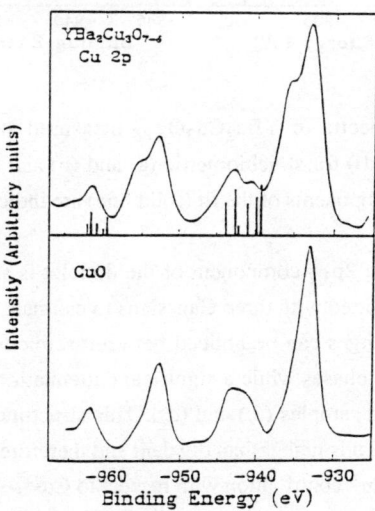

Fig.1. Cu 2p XPS spectra of $YBa_2Cu_3O_{6.85}$ and CuO after background subtraction. The multiplet components of the satellites are also shown. Notice the peak on the high binding energy side of the spin-orbit doublet in $YBa_2Cu_3O_{6.85}$.

satellites which show the multiplet structure characteristic of two pd-electrons. The marked difference between the two spectra is the large asymmetry on the high binding energy side of the main peak which in the sample of Fig. 1 shows up as a separate structure. This peak is very sensitive to the oxygen stoichiometry and tends to be washed out quickly for $\delta > 0.3$.

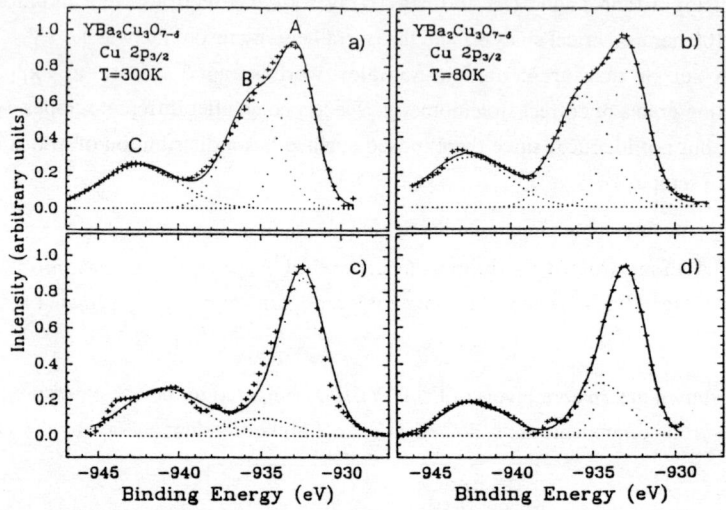

Fig.2. Cu $2p_{3/2}$ XPS spectra of $YBa_2Cu_3O_{6.85}$ measured at T=300K (a) and (c) and T=80K (b) and (d)) for stoichiometric (a) and (b) and oxygen deprived surfaces. The Gaussian components of the fit (solid line) are indicated by dots.

In the set of data of Fig. 2, the $2p_{3/2}$ component of the doublet is shown at T= 300K and T= 80K. The lineshape has been fitted with three Gaussians to estimate the relative weights of the structures. No significant changes can be noticed between semiconducting ((a) and (c)) and superconducting ((b) and (d)) phases while a significant attenuation of peak B is observed in both cases in oxygen deprived samples ((c) and (d)). This structure, as discussed later, arises because the oxidation state of Cu is higher than divalent and therefore the area ratio R= B/(A+C) represents an estimate of the Cu^{3+} contribution with respect to Cu^{2+}.

The stoichiometric ratio $Cu^{3+}/Cu^{2+}= (1-2\delta) / (2+2\delta)$ is plotted in Fig. 3 as a function of δ (broken line). The experimental R points are systematically higher than expected and agreement is found for an over-concentration of Cu^{3+} of the order of 30% as shown by the curve $Cu^{3+}/Cu^{2+}= (1-2\delta+d) / (2+2\delta-d)$ with d = 0.3 (full line).

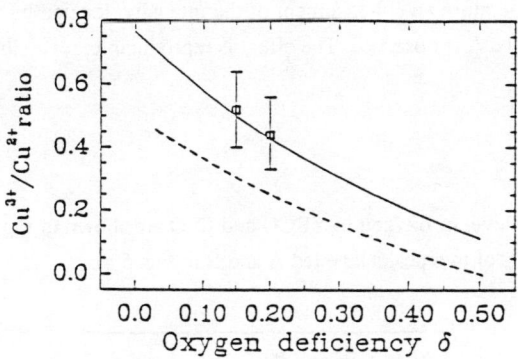

Fig.3. Average Cu^{3+}/Cu^{2+} ratio as a function of the oxygen vacancy concentration ∂ Dashed line represents the stoichiometric ratio; solid line is calculated including an extra charge fluctuation d=0.3.The experimental points are evaluated by the XPS 2p spectra of Cu.

The L_3VV Auger spectra of YBCO at T=300K and T=80K are shown in Fig. 4, after background subtraction. The main structure peaks at 918.0 eV with smaller features on both sides. On the low kinetic energy side is also expected the contribution of the $L_2L_3M_{4,5}$ Coster-Kronig preceded transition (van der Laan *et al.*, 1981) which is easily observed in metallic copper but barely seen in all Cu-based superconducting materials (Sarma and Rao, 1988; Heinrich *et al.*, 1987). The energy position of the main peak coincides, within experimental uncertainty, with that of CuO, indicating that copper is predominantly in the formal 2+ valence state.

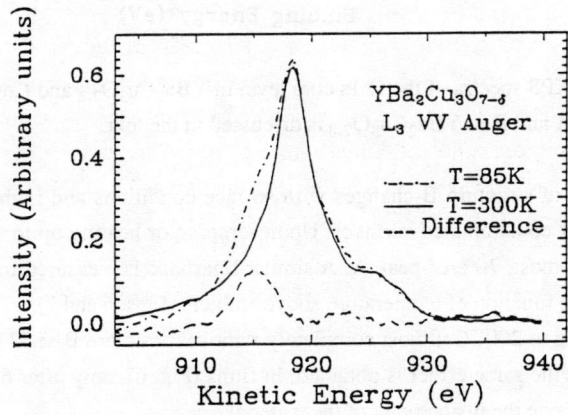

Fig.4. Cu $L_3M_{4,5}M_{4,5}$ Auger spectrum of $YBa_2Cu_3O_{7-\delta}$ measured at 300K, at 85K and their difference spectrum. Background has been subtracted.

On lowering the temperature an enhancement of the intensity, located by the difference curve (broken line) at ~915.0 eV, is observed. The effect is reproducible, reversible and is not present in CuO.

O

Spectra of the core 1s level of oxygen in YBCO and CuO are shown in Fig. 5 a) and b). The 1s level in YBCO consists of two peaks labelled A and B in Fig. 5 a).

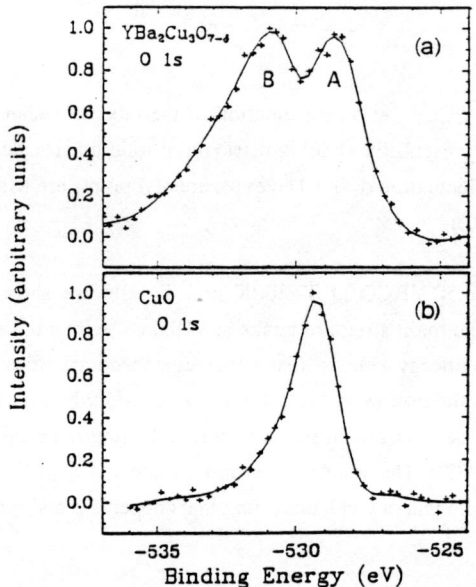

Fig.5. XPS spectra of the O 1s core level in $YBa_2Cu_3O_{7-\delta}$ and CuO. The origin of peaks A and B in $YBa_2Cu_3O_{7-\delta}$ is discussed in the text.

The intensity of structure B changes with surface conditions and is the dominant feature in air-exposed or contaminated surfaces. Upon scraping or heating up to 100 °C, its intensity is reduced to, at most, 70% of peak A. A similar lineshape is measured in Bi-Sr-Ca-Cu-O films and pellet, as a function of temperature above ambient (Figs. 6 and 7).
A mild heating to 200 °C almost completely removes structure B at 531.5 eV in bulk samples (Fig. 7) while the same effect is obtained, in films (Fig. 6), only after 6h heating at 880 °C, a temperature above the fusion point of the material.

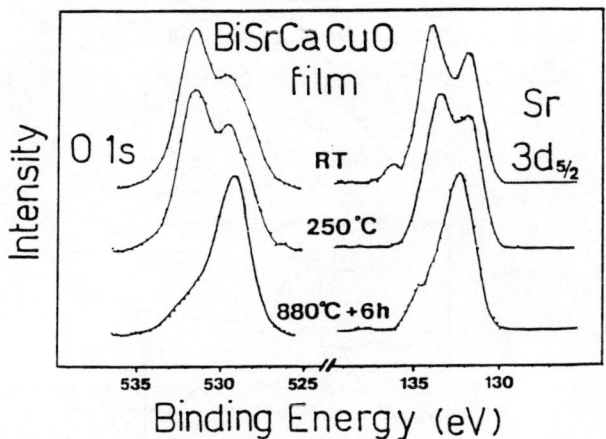

Fig.6. XPS spectra of the O 1s and Sr $3d_{5/2}$ core levels in $Bi_2Sr_2CaCu_2O_{8+\delta}$ films, grown by laser ablation, at different temperatures above ambient. The bottom spectrum is taken after 6h heating at a temperature above the fusion point of the material.

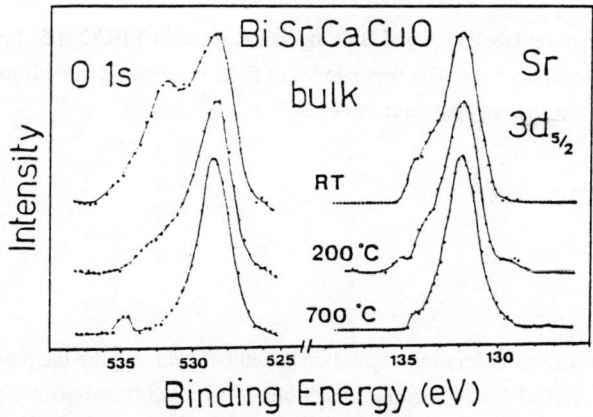

Fig.7. XPS spectra of the O 1s and Sr $3d_{5/2}$ core levels in bulk samples of $Bi_2Sr_2CaCu_2O_{8+\delta}$ at different temperatures above ambient. The bottom spectrum is taken after 6h heating at a temperature above the fusion point of the material.

In Figs. 6 and 7 is also shown the concomitant behaviour of the $3d_{5/2}$ core level of Sr.

In Fig. 8 the $KL_{2,3}L_{2,3}$ Auger lines of oxygen in YBCO and CuO are compared.

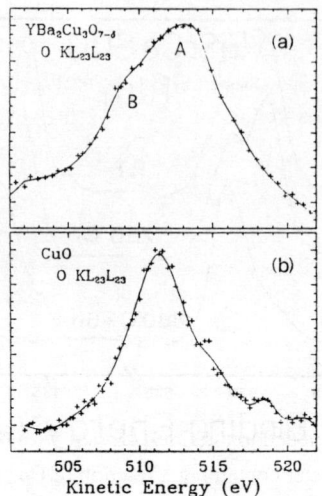

Fig.8. O $KL_{2,3}L_{2,3}$ Auger spectrum of $YBa_2Cu_3O_{7-\delta}$ and CuO after background subtraction.

A double structure, labelled A and B, is again measured in YBCO at 514 eV and 507 eV kinetic energy (KE), respectively. The dependence of peak B on surface condition is analogous to that of peak B of the 1s core spectrum shown in Fig. 5 a).

DISCUSSION

<u>Cu.</u>

Charge neutrality considerations, based on an ionic model, require the presence of mixed-valent elements in YBCO for an oxygen stoichiometry δ ranging between 0.5 and 1. The average valence of Cu, which exists in different oxidation states, would be in this case, $(7-2\delta)/3$ (2.23 for $\delta = 0.15$), implying a static mixture of $1-2\delta$ Cu^{3+} and $2(1+\delta)$ Cu^{2+} ions per formula unit. Alternatively, assuming divalent Cu ions and a fixed formal valence for the other cations, $1-\delta$ sites should be occupied by monovalent oxygens.

The calculated electron affinity of Cu^{3+} in these systems is very close to the ionization potential of O^{2-} (Martin et al., 1988), therefore, in the cluster formed by a central Cu atom surrounded by its nearest neighbours oxygens, the presence of either Cu^{3+} or O^- species will determine the formation of the nearly degenerate charged complexes $(Cu^{3+}O^{2-})^+$ (a hole on the Cu^{2+} ion) and $(Cu^{2+}O^-)^+$ (a hole on the O^{2-} ion). The near degeneracy is responsible for the high mobility of the hole which can easily transfer between Cu and O. The preferential site of formation of these complexes in the lattice would be the Cu atoms on the linear chains between two BaO planes, due to the smaller Cu-O distance along the c axis (~ 1.80 Å) with respect to the Cu-O distances in the ab-plane (~ 1.96 Å) (David et al., 1987).

The oxidation state of copper, higher than divalent, determines a larger covalence of the Cu-O bond (Fuggle et al., 1988). Ultraviolet photoemission measurements show increasing hybridization of the O and Cu valence states on going from cuprous oxide to cupric oxide to Cu-based superconducting materials (Steiner, 1989).

In a molecular orbital description applied to the CuO_4 cluster, within a configuration interaction scheme, the ground state of the cluster containing a fraction of Cu^{3+} ions, includes, besides the $3d^9$ and $3d^{10}\underline{L}$ states of Cu^{2+}, also the $3d^8$, $3d^9\underline{L}$ and $3d^{10}\underline{L}^2$ states of Cu^{3+}, where $\underline{L}$ and $\underline{L}^2$ denote one and two holes on the ligand oxygens, respectively (Fujimori et al., 1987; Gourieux et al., 1988; Steiner et al., 1988; Balzarotti et al., 1988). The $3d^8$ component can be excluded from the basis set since the high correlation energy of the Cu d-electrons moves this state to ~ 10 eV excitation energy with respect to the Fermi level (Gourieux et al., 1988). The divalent-like charge transfer state $3d^9\underline{L}$ gives the major contribution, as expected for a larger copper to oxygen hybridization. In x-ray photoemission and absorption measurements on $NaCuO_2$ (Gourieux et al., 1988; Steiner et al., 1988), generally assumed as a prototype compound for Cu^{3+}, the $3d^8$ configuration cannot be clearly observed, confirming the tendency to covalence with increasing oxidation state. Because of the double ionization on the ligand, the weight of the $3d^{10}\underline{L}^2$ component is expected to be small (Gourieux et al., 1988), and, therefore, it has been disregarded in the present calculation. The resulting groundstate of the cluster is then described as $\alpha\,|3d^9\rangle + \beta\,|3d^9\underline{L}k\rangle + \gamma\,|3d^{10}\underline{L}\rangle$ where k denotes the conduction electron needed to have the same number of electrons in the cluster states.

The final states probed in a core-level photoemission experiment are obtained removing a core electron (c). The $|3d^9\rangle$ and $|3d^{10}\underline{L}\rangle$ initial states which are separated by the charge transfer energy $\Delta' = \langle 3d^{10}\underline{L}|H|3d^{10}\underline{L}\rangle - \langle 3d^9|H|3d^9\rangle$ (~ 1.2 - 2.0 eV), in the final state will be separated by the Coulomb interaction Q between the core-hole and the d-electron. These states constitute the main $\underline{c}3d^{10}\underline{L}$ peak and the $\underline{c}3d^9$ satellite at ~ 9 eV higher binding energy in the 2p-core photoemission spectrum of divalent Cu (peaks A and C of Fig.2 a)). The excited $\underline{c}3d^9\underline{L}k$, located close to the satellite, is not observed. A peak, instead, is detected at ~ 2.5 eV higher binding energy from the main $\underline{c}3d^{10}\underline{L}$ peak for YBCO (structure B of Fig. 2 a), ie. at the same binding energy as expected for the decaying of the unscreened $\underline{c}3d^9\underline{L}$ into the well-screened shake-down $\underline{c}3d^{10}\underline{L}^{2-}$ state. Its energy coincides also with that of the $2p_{3/2}$ component in $NaCuO_2$ (Steiner et al., 1988).

The eigenenergies of the final states relative to the $\underline{c}3d^9$ level are obtained on diagonalizing the hamiltonian matrix:

$$\begin{vmatrix} -E & T & 0 \\ T & -E-Q+\Delta & T' \\ 0 & T' & -E-Q+U_{hh} \end{vmatrix} \quad (1)$$

where $T=<3d^9|H|3d^{10}\underline{L}>$ and $T'=<3d^{10}\underline{L}|H|3d^{10}\underline{L}^2>$ are mixing matrix elements in the final state, U_{hh} is the Coulomb repulsion of two delocalized O holes and Δ is the charge transfer integral. The best fit of the Cu $2p_{3/2}$ spectra of YBCO and BSCCO is obtained for the set of parameters listed in Table I.

Table I. Measured and calculated (values in brackets) relative binding energy positions ΔE and satellite-to-main peak intensity ratios I_A/I_C (see Fig.2 a) for the Cu $2p_{3/2}$ core level and electronic structure parameters (defined in the text).

SAMPLE	ΔE_{AB}(eV)	ΔE_{AC}(eV)	I_C/I_A	T	T'	Δ	U_{hh}(eV)	Q	n_d
$YBa_2Cu_3O_{6.85}$	2.8±0.6 (2.83)	10.7±1.0 (10.2)	0.35±0.2 (0.29)	2.8	0	0.0	2	8.5	9.5
$YBa_2Cu_3O_{6.80}$	2.4±0.5 (2.32)	9.7±0.6 (9.70)	0.32±0.2 (0.28)	2.7	0	0.0	1.5	8.1	9.5
$Bi_2Sr_2CaCu_2O_{8+\delta}$ film	0	8.7±0.3 (8.60)	0.33±0.05 (0.34)	2.1	0	0.0	0	7.5	9.5
$Bi_2Sr_2CaCu_2O_{8+\delta}$ bulk	0	8.7±0.3 (8.73)	0.48±0.05 (0.49)	1.5	0	0.0	0	8.2	9.5

We remark on the lack of hybridization between the Cu^{3+} and Cu^{2+} cluster states (T'=0) and the small (U_{hh}= 1.5-2.0 eV) Hubbard repulsion between O holes. The values of the transfer integral T, the satellite-to-main peak ratio I_C/I_A and the number of d-electrons in the groundstate n_d, are close to those estimated from valence band photoemission spectra and are consistent with a strong electronic correlation at the copper site and a large covalence of the Cu-O bond.

Contribution of the Cu^{3+} states can also be inferred by the analysis of the Cu L_3VV Auger spectrum of YBCO shown in Fig. 4.

The final state of the core electron photoemitted from the solid in an XPS experiment, coincide with the initial state of an Auger process involving the same initial core hole. The final CVV Auger states can be reached by filling the core hole with a valence electron and ejecting another electron from the valence band. The divalent $\underline{c}3d^9$ and $\underline{c}3d^{10}\underline{L}$ initial states of Cu^{2+} will, then, give rise to the $3d^7$, $3d^8\underline{L}$ and $3d^9\underline{L}^2$ Auger final states observed also in CuO (Balzarotti et al., 1987; Fleish and Mains, 1982) and Cu dihalides (van der Laan et al., 1981). The $3d^8\underline{L}$ configuration contains multiplet terms arising from two equivalent d-holes (^1S, ^3P, ^1D, ^1G, ^3F), the most intense being the ^1G term at 918.0 eV KE. The $3d^9\underline{L}^2$ contribution occurs on the high KE side of the main $3d^8\underline{L}$ term, at an energy distance U_{dd}-Δ, where U_{dd} is the correlation energy between d-electrons.

In order to estimate U_{dd}, we have used the expression for the two holes spectral function (Cini, 1977; Sawatzky, 1977),

$$D(E) = \frac{D_0(E)}{(1 - U_{dd} F(E))^2 + (\pi U_{dd} D_0(E))^2} , \qquad (2)$$

where $D_0(E)$ is the self-convolution of the one-particle density of states and F(E) is the Hilbert transform of $D_0(E)$. Using the valence band density of state calculations of Temmmermann et al. (1987), we found a good agreement with the measured Auger spectrum for $U_{dd} \sim 7$ eV, of the same order of that found in other Cu oxides (Thuler et al., 1982). This value brings the contribution of the $3d^9\underline{L}^2$ state at ~ 925 eV KE where an enhancement of the Auger intensity is observed. The $3d^7$ component is expected on the low KE side of the main $3d^8\underline{L}$ peak, at an energy distance $2 U_{dd}$-Q (≥ 5 eV). The structure is broadened by the multiplet splittings of the configuration and crystal field effects and cannot be clearly identified in the spectra. The peak at 915 eV is, then, more likely due to the $3d^8\underline{L}^2$ final state which can be reached from the $\underline{c}3d^{10}\underline{L}^2$

initial state of Cu^{3+}. The temperature dependence of this structure, reported also in other studies (Balzarotti *et al.*, 1987; Sarma and Rao, 1988), is, at present, not completely understood. A plausible explanation could be either changes in oxygen stoichiometry with temperature which would modify the hybridization of the Cu-O initial states or the effect of residual chemisorbed species.

O.

A debated question in the study of the electronic properties of superconducting oxides is the interpretation of the observed O 1s photoemission line (Fig. 5 a)). Whether the intrinsic structure should consist of a single or a double peak is still a controversial point. The non equivalence of the four oxygen sites is expected to split the core line in different components. Such components, at energies < 531.0 eV (Nücker *et al.*) are not resolved in the XPS spectrum and are responsible for the linewidth (~ 1.7 eV) of peak A (528.5 eV). Although an intrinsic contribution at 531.5 eV cannot be excluded, undoubtedly, the sensitivity of peak B (531.5 eV) to surface conditions suggests the presence of a large extrinsic component at the same energy. By comparison with CuO, where mainly a single line is detected at ~ 529.0 eV (Fig. 5 b)), peak A, in YBCO, is likely to be that associate with the Cu-O bond.

Hydroxide formation, O chemisorbed on defect sites or BaO segregated at the surface are a number of possibilities which could be invoked for the presence of a second component (B). Notice that an over-stoichiometry of O in the surface region, which is not characterized by current techniques of structural analysis, could also account for the excess of Cu^{3+} detected by X-ray spectroscopy.

Similar conclusions are also suggested by the study of films and bulk samples of $Bi_2Sr_2CaCu_2O_{8+\delta}$ (Rochet *et al.*) where, contrary to YBCO, a stoichiometric excess of oxygen could be present in the structure. The O 1s spectrum (Figs. 6 and 7) is again formed by two peaks at the same binding energies as in YBCO, suggesting that the chemical environment of the O atoms is similar in the two cases. The low binding energy component at ~ 529.2 eV is the highest in the case of bulk ceramic samples while the high binding energy component (531.5 eV) dominates in the case of films up to temperatures as high as 750 °C. We explain this discrepancy by the possibility of retention of the excess oxygen in the film surface because of its textured orientation. In such a case, formal Cu^{3+} is evidenced by the examination of the Cu L_3VV spectra (Rochet *et al.*). The high binding energy O 1s component depends again on the surface stoichiometry of the sample, which is altered by heating above the fusion temperature (880 °C) and its strong weakening at 880 °C is also correlated to the disappearance of formal

Cu^{3+} (Rochet *et al.*). We emphasize the difference between strongly bound oxygen of the film surfaces and losely bound oxygen of the ceramic grains which can leave the surface by moderate heating at 200 °C. In contrast to the bulk case, changes of the Sr $3d_{5/2}$ core level in the film are correlated to those of the O 1s core level. This might suggest that fluctuating oxygens are situated close to the Sr planes.

In the O $KL_{2,3}L_{2,3}$ Auger spectrum of YBCO (Fig. 8 a)) we can distinguish two components A and B at 514.0 eV and 507 eV KE respectively, while a single structure at 512 eV is measured in CuO (Fig. 8 b)). The intensity of peak B varies with oxygen content like peak B in the O 1s spectrum (Fig. 5) and therefore can be interpreted as its Auger counterpart.

The most intense component of the $KL_{2,3}L_{2,3}$ line at 514.0 eV is the $^1D_2$ term of the configuration $2s^22p^4$ (two equivalent holes in the L shell) of the O^{2-} ion. The other multiplets of the KLL line (KL_1L_1, $KL_1L_{2,3}$) lie outside the spectral range examined.

In a three step model of the Auger process, the kinetic energy E_k of the CVV electron is given by

$$E_k(CVV) = E_b(C) - 2E_b(V) - U, \qquad (3)$$

where $E_b(C)$ and $E_b(V)$ are the binding energies of the core hole and of the centroid of the valence band, respectively, and U is the correlation energy of two valence holes. We can evaluate by this formula, the Coulomb interaction U_{pp} of two p-holes at the oxygen sites giving rise to structures A and B in the spectra. By the measured binding energy of the O 1s level and from the energy position of the oxygen valence band centroid (~ 4 eV) we obtain the values U_{pp}= 4-5 eV and U_{pp}= 14 eV for the 514.0 eV and 507.0 eV Auger structures, respectively. The first value is characteristic of O bound to transition metals (Legaré *et al.*, 1985) and is the effective U for the oxygen 2p-holes in the Cu-O cluster. The second value, characteristic of non-transition metal oxides, suggests that the component of the Auger profile at 507.0 eV KE could originate in part from surface oxides as BaO or Y_2O_3 and in part from excess oxygen or oxygen chemisorbed on defect sites in the surface region, with a larger intra-atomic screening.

CONCLUSIONS

Core level spectra of $YBa_2Cu_3O_{7-\delta}$ and $Bi_2Sr_2CaCu_2O_{8+\delta}$ show additional features characteristic of a formal valence state of copper larger than two. Local charge neutrality analysis suggest the possibility of a dynamic charge fluctuation in the Cu-O clusters of the linear chain due to the presence of Cu^{3+} and O^- species. Based on this evidence, we have developed a molecular description of the Cu states using a basis set formed by the $|3d^9>$ and $|3d^{10}\underline{L}>$ states of divalent copper and the $|3d^9\underline{L}k>$ divalent-like state of the Cu^{3+} configuration. A number of parameters, which define the molecular state, are computed and found in good agreement with the experiments and with previous estimates. Effects of surface degradation have been found to be particularly severe for the oxygen spectra.

This work has been supported by the Centro Interuniversitario di Struttura della Materia (CISM) and by the Gruppo Nazionale di Struttura della Materia (GNSM) of the Consiglio Nazionale delle Ricerche (CNR).

+ Groupe de Physique des Solides de l'Ecole Normale Superieure, Universitè Paris 7, Tour 23, 2 Place Jussieu, 75251 Paris, France.
* ICTP Postdoctoral fellow at Rome II University

REFERENCES

Balzarotti, A., M. De Crescenzi, C. Giovannella, R. Messi, N. Motta, F. Patella and A. Sgarlata, (1987). *Phys. Rev. B*, 36, 8285.

Balzarotti, A., M. De Crescenzi, N. Motta, F. Patella and A. Sgarlata, (1988). *Phys.Rev. B*, 38, 6461.

Balzarotti, A., M. De Crescenzi, N. Motta, F. Patella and A. Sgarlata, (1988a) *Proc. of the Int. Adriatico Res. Conf.*, Trieste July 1988, ed. S.Lundqvist, E.Tosatti, M.P.Tosi, and Yu Lu (World Sci. Singapore), to be published and *Int. J. Mod. Phys.B* (to be published).

Cini, M., (1977). *Solid State Commun.*, 24, 681.

David, W.I.F., W.T.A. Harrison, J.M.F. Gunn, O. Moze, A.K. Soper, P. Day, J.D. Jorgensen, D.G. Hinks, M.A. Beno, L. Soderholm, D.W. Capone II, J.K. Schuller, C.U. Segre, K. Zhang, and J.D. Grace, (1987). *Nature*, 327, 310.

de Groot, R.A., H. Gutfreund, and M. Weger, (1987). *Solid State Commun.*, 63, 451.

Eskes, H., and G.A. Sawatzky, to be published.

Fleish, T.H., and G.J. Mains, (1982). *Appl. Surf. Science*, 10, 51.

Fuggle, J.C., J. Fink and N. Nucker, (1988). *Proc. of the Int. Adriatico Res. Conf.*, Trieste July 1988, ed.S. Lundqvist, E. Tosatti, M.P. Tosi, and Yu Lu (World Sci. Singapore), to be published and *Int. J. Mod. Phys. B*, to be published.

Fujimori, A., E. Takayama-Muromachi, Y. Uchida, and B. Okai, (1987). *Phys. Rev. B*, 35, 8814.

Gourieux, T., G. Krill, M. Maurer, M.F. Ravet, A. Menny, H. Tolentino, and A. Fontaine, (1988). *Phys. Rev. B*, 37, 7516.

Heinrich, B., K. Myrtle, N. Alberding, W.B. Xing, O. Rajora, K.B. Urquhart, and S. Gygax, (1987). *Proc. 34th AVS-AIP Conf.*, to be published.

Humbert, P., and J.P. Deville, (1985). *J. Electron Spectrosc. Relat. Phenom.*, 36, 131.

Kurtz, R.L., R. Stockbauer, D. Mueller, A. Shih, L. Toth, M. Osofsky, and S.A. Wolf, (1987). *Phys. Rev. B*, 35, 8818.

Legaré, P., G. Maire, B. Carriere, J.P. Deville, (1977). *Surf. Sci.*, 68, 348.

Martin, R.L., A.R. Bishop, Z. Tesanovich, (1988). *Proc. of the Symposium on High Temperature Superconductors*, Los Angeles, in press.

Nücker, N., H. Romberg, X.X. Xi, J. Fink, B. Gegenheimer, Z.X. Zhao, to be published.

Rochet, F., A. Balzarotti, M. De Crescenzi, N. Motta, Murali Sastry, F. Patella, J. Perrière, F. Kerhervé, G. Hauchecorne, to be published.

Sarma, D.D., K. Sreedhar, P. Ganguly, and C.N.R. Rao, (1987). *Phys. Rev. B*, 36, 2371.

Sarma, D.D., and C.N. Rao, (1988). *Solid State Commun.*, 65, 47.

Sawatzky, G.A., (1977). *Phys. Rev. Lett*, 39, 504.

Steiner, P., S. Hufner, V. Kinsinger, I. Sander, B. Siegwart, H. Schmitt, R. Schulz, S. Junk, G. Schwitzgebel, A. Gold, C. Politis, H.P. Muller, R. Hoppe, S. Kemmler-Sack, and C. Kunz, (1988). *Z. Physik B*, 69, 449.

Steiner, P., (1989). this volume.

Temmerman, W., Z. Szotek, P.J. Durham, J.M. Stocks, P.A. Sterne, (1987). *J. Phys. F: Metal. Phys.*, 17, L319.

Thiry, P., G. Rossi, Y. Petroff, A. Revcolevschi, and J. Jegoudez, (1988). *Europhys. Lett.*, 5, 55.

Thuler, M.R., R.L. Benbow, and Z. Hurych, (1982). *Phys. Rev. B*, 26, 669.

van der Laan, G., C. Westra, C. Haas, and G.A. Sawatzky, (1981). *Phys. Rev. B*, 23, 4369.

van der Marel, D., J. van Elp, G.A. Sawatzky and D. Heitmann, (1988). *Phys. Rev. B*, 37, 5136.

INTERFACE FORMATION BETWEEN METALS AND HIGH-T_C SUPERCONDUCTORS

C. Laubschat, E. Weschke, M. Domke, O. Strebel, J.E. Ortega, and G. Kaindl

Freie Universität, Fachbereich Physik
Arnimallee 14, 1000 Berlin 33, Germany

ABSTRACT

The results of Photoemission studies using synchrotron radiation on the chemical and electronic structure of interface between ceramic superconductors ($YBa_2Cu_3O_{7-\delta}$, $Bi_5Sr_3Ca_2Cu_4O_y$) and metals (Ag, Rb, Mg) are presented. The interfaces with Ag are non-reactive, with no noticeable changes in the stoichiometry of the superconductor at the interface. On the other hand, Mg and Rb reduce the ceramic superconductors inducing a metal-semiconductor transition in the interfacial region.

KEYWORDS

Interfaces; Photoemission; Passivation; $Ag/YBa_2Cu_3O_{7-\delta}$; $Mg/YBa_2Cu_3O_{7-\delta}$; $Rb/YBa_2Cu_3O_{7-\delta}$; $Ag/Bi_5Sr_3Ca_2Cu_4O_y$; $Mg/Bi_5Sr_3Ca_2Cu_4O_y$

INTRODUCTION

The discovery of ceramic high-T_c superconductors has prompted intense research into their fundamental properties and the underlying mechanisms of superconductivity. Microelectronic applications of these materials, however, will critically depend on their integration with existing processing techniques. In this respect, a detailed understanding of interface of these materials with metals, semiconductors and insulators on a microscopic basis is of considerable interest. In particular, interfaces with metals are important in connection with chemical passivation of the surface and electrical contact formation.

Similar as in the related case of metal/semiconductor interfaces, electron spectroscopies are well suited for such studies due to their tunable surface sensitivity. Recently, a series of photoemission (PE) and inverse-PE studies of the formation of interfaces between metals and the high-T_c superconductors $La_{2-x}Sr_xCu_2O_4$, $YBa_2Cu_3O_{7-\delta}$, and $Bi_2Ca_{1-x}Sr_{2+x}Cu_2O_8$ was performed (see table 1). As is evident from this table, most of the interfaces studied were found to be reactive; the observed reaction processes are essentially characterized by the removal of oxygen from the substrate due to oxidation of the overlayer.

Table 1. Overview of metal/high-T_c superconductor interfaces studied so far by electron-spectroscopic methods. R and N denote reactive and non reactive interfaces, respectively. Refs.: [a] (Wagener et al., 1988); [b] (Laubshat et al., 1988a); [c] (Laubschat et al., 1988b); [d] (Meyer III et al., 1987a); [e] (Meyer III et al., 1987b); [f] (Hill et al., 1988); [g] (Hill et al., 1987); [h] (Gao et al., 1987); [i] (Gao et al., 1988a); [k] (Meyer III et al., 1988); [*l*] (Gao et al., 1988b). The entries without reference are from the present work.

	$La_{2-x}Sr_xCuO_4$	$YBa_2Cu_3O_7$	Bi-Ca-Sr-Cu-O
Cu	R [f]	R [a]	
Ag		N [a,b,c]	N
Au	N [d]	N [a]	
Pd		R [a]	
Fe	R [g]	R [h]	
Ti	R [e]		
La		R [*l*]	
Rb		R [b,c]	R
Mg		R	R
Al		R	
Ge	R [i]		
Bi		R [k]	N [k]

In this article we report on the formation of interfaces between the high-T_c superconductor $YBa_2Cu_3O_{7-\delta}$ and the metals Ag, Mg, and Rb. Furthermore we present the results of experiments with interfaces between $Bi_5Sr_3Ca_2Cu_4O_y$ and Ag and Mg. In both cases, the interfaces with Mg and Rb are characterized by strong chemical reactions: The divalent Cu(II) species is reduced as indicated by the quenching of the 12.4-eV satellite; additional valence-band features signal an oxidation of Mg and Rb, respectively. For $Bi_5Sr_3Ca_2Cu_4O_y$, we additionally observe a reduction of trivalent Bi as is evident from the occurence of a further component in the Bi-5d core-level PE spectra. As a consequence of these chemical reactions, a metal-semiconductor transition takes place in both cases in the ceramic material adjacent to the interface caused by oxygen removal from the substrate.

The interfaces with Ag were found to be non-reactive up to coverages of $\simeq$ 40 Å for $YBa_2Cu_3O_{7-\delta}$ and $\simeq$ 33 Å for $Bi_5Sr_3Ca_2Cu_4O_y$, respectively. A passivation experiment performed with $Bi_5Sr_3Ca_2Cu_4O_y$ revealed that a Ag overlayer of $\simeq$ 25-Å thickness on a ceramic pellet is not suitable for preventing the surface of the supeconductor from reacting with subsequently deposited Mg.

EXPERIMENTAL

For our experiments, polycrystalline ceramic materials were used. Samples of $YBa_2Cu_3O_{7-\delta}$ were prepared by the usual ceramic route and characterized by X-ray diffraction as single phase material (orthorhombic) in each case. They displayed positive Meissner effect and relatively sharp (< 1K) resistivity transitions as measured with a standard four-probe technique. T_c values of $\simeq$ 92K were found for all samples. The $Bi_5Sr_3Ca_2Cu_4O_y$ materials were prepared by heating stoichiometric present the results of experiments with interfaces between $Bi_5Sr_3Ca_2Cu_4O_y$ and Ag and Mg. In both cases, the interfaces with Mg and Rb are characterized by strong chemical reactions: The divalent Cu^{2+} species is reduced as indicated by the quenching of the 12.4-eV satellite; additional valence-band features signal an oxidation of Mg mixtures of CuO, Bi_2O_3, $CaCO_3$, and strontium acetate in air at 830°C for 2 days. After grinding and additional sintering at 830°C, the samples were quenched to room temperature. X-ray diffraction as well as electron-microprobe analysis showed the resulting materials to be 85% single phase, with zero electrical resistance below 84K.

Photoemission experiments were performed at the Berliner Elektronenspeicherring für Synchrotronstrahlung (BESSY) using the TGM3 monochromator ($YBa_2Cu_3O_{7-\delta}$) in the energy range from 70 to 700 eV. The samples were mounted on a stainless steel plate by spot-welded strips of Ta foil. The samples were cleaned in situ by repeated scraping with a diamond file. Interfaces were prepared in the case of Ag and Mg by evaporating the pure metals (4N) from a resistively heated tungsten coil, with coverages measured by a quartz-crystal microbalance. In case of Rb a commercial Rb_2CrO_4 dispenser (SAES) was used; in this case, coverages were roughly estimated fron intensity ratios of core-level PE lines from absorbate and substrate atoms, respectively. The pressure during the measurements was in the low 10^{-10}-mbar range rising briefly to $1 \cdot 10^{-9}$ mbar during metal evaporations. The $Ag/YBa_2Cu_3O_{7-\delta}$ measurements were performed at room temperature, whereas during all other experiments the substrate was cooled with liquid N_2 to a temperature of $T \simeq 150K$.

INTERFACES WITH $YBa_2Cu_3O_{7-\delta}$

Fig. 1 shows valence-band PE spectra of the $Ag/Yba_2Cu_3O_{7-\delta}$ interface for increasing Ag coverages. The spectra were taken at a photon energy of 139 eV, where the Ag-5s PE cross section is negligibly small and Ag-4d emission is strongly reduced by a Cooper minimum. Therefore, the spectra reflect mainly emission from the substrate. They are dominated by strong emission from the Cu-3d/O-2p derived valence band with its centroid $\simeq$ 4 eV below E_F. At this photon energy, the characteristic final-state valence-band satellites (Kurtz et al., 1987) at 9.4 eV

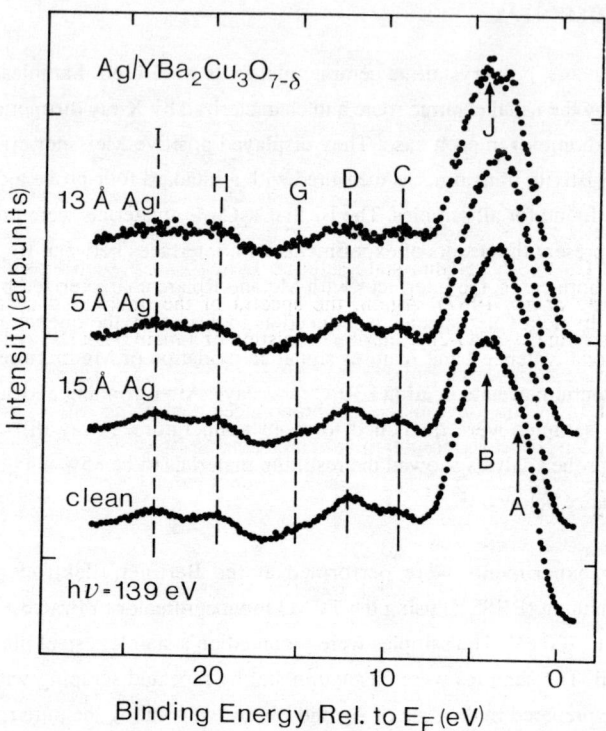

Fig. 1. Valence-band photoemission spectra of Ag/YBa$_2$Cu$_3$O$_{7-\delta}$ for increasing coverages of Ag. Note that the structure do not shift in energy with coverage. The two-hole final-state satellite (peak D) is present at all coverages.

(peak C) and 12.4 eV (peak D) mask a weak emission from Ba-5p states at about 16-eV BE (structure G). As is evident from Fig. 1, the spectra do not change in the studied coverage range up to $\simeq$ 13 Å of Ag. The persistence of the two-hole final-state Cu satellite at 12.4 eV (peak D); indicates that Cu(II) is preserved. Upon Ag deposition, the signals from the Y-4p and O-2s states at 24 eV (peak I) and at 20 eV (peak H) below E_F, respectively, neither change their energy positions nor their intensities relative to the valence-band emission (peak B); this indicates that the stoichiometry of the substrate is unchanged at the interface. The intensity of the

Ba-4d signal at about 90 eV below EF (not shown here) decreases monotonically with increasing Ag coverage following an exponential law, thereby suggesting a uniform coverage of the substrate in a layer-by-layer growth mode. A possible segregation of Ba to the surface, as reported for the Fe/$YBa_2Cu_3O_{7-\delta}$ interface (Gao et al., 1987), can be ruled out in the present case.

A still higher coverages (not shown here), the residual Ag-4d emission at a binding energy of $\simeq 5$ eV, first seen at 13-Å coverage (Fig. 1, feature J), becomes dominant, and the spectrum converges to the characteristic shape known for elemental Ag metal.

A totally different situation is observed for the interface of Rb with $YBa_2Cu_3O_{7-\delta}$, displayed in Fig. 2. The spectra were taken at the Cu-3p→3d resonance occuring at a photon energy of $\simeq 74$ eV, where the Cu two-hole final-state satellite D at 12.4 eV binding energy is resonantly enhanced (Kurtz et al., 1987). Again, the spectra of the uncovered ceramic material are dominated by strong Cu-3d/O-2p emission consisting of a main peak (B) at a binding energy of about 5 eV and a trailing-edge shoulder (A) at about 2.5 eV. The satellite structures C (at 9.4 eV) and D (at 12.4 eV) are obviously more pronounced than in the 139-eV spectra (see Fig. 2). With increasing Rb coverage monitored by the increasing Rb-4p PE emission intensity at about 15 eV (peak E), the following observations are made: (i) the 2.5-eV shoulder A decreases, (ii) the 12.4-eV satellite D is quenched at rather low coverages, and (iii) the whole spectrum shifts to higher binding energies for Rb coverages exceeding $\simeq 2$ Å.

Let us start with a discussion of observation (i): The shoulder A at a binding energy of 2.5 eV reflects mainly emission from Cu-3d/O-2p orbitals from atoms in the Cu-O chains (Massida et al.,1987; Redinger et al., 1987). Due to O-vacancies, the oxygen atoms at these sites are known to be more mobile than any other in the structure (Beech et al., 1987). Removal of oxygen from these positions is thought to be mainly responsible for the different oxygen-deficient stoichiometries with increasing δ in $YBa_2Cu_3O_{7-\delta}$ (Chaillout et al., 1987). The observed decrease in the intensity of shoulder A therefore reflects a removal of oxygen atoms from these chains through oxidation of Rb.

The emission from Rb oxides formed by this reaction at the interface coincides apparently with the main peak B at about 5-eV binding energy. This allows to draw some conclusions on the type of oxides formed. Assuming analogous chemical behavior for Rb and Cs oxides, we may use the results of a detailed PE study of Cs oxides for an interpretation of the present data (Su et al., 1982). Then, Rb compounds with low oxygen content, like Rb_2O or Rb_2O_2, are expected to show strong PE signals at a binding energy of $\simeq 3$ eV. Since no signal at this binding energy is observed in the present PE spectra upon Rb deposition, such Rb oxides are apparently not formed. On the other hand, compounds with higher oxygen content, like Rb_2O_3 or RbO_2, are expected to show strong PE signals at $\simeq$ 5-eV binding energy, which is consistent with the present observations. Furthermore, these oxides should also give rise to relatively weak features at binding energies of 8 to 10 eV. Indeed, an increase in PE intensity in the energy range of the

9.4-eV satellite (peak C) is observed with increasing Rb coverage.

The formation of these oxides requires a large amount of oxygen to be withdrawn from the ceramic superconductor substrate, causing strong effects on its chemical and physical properties.

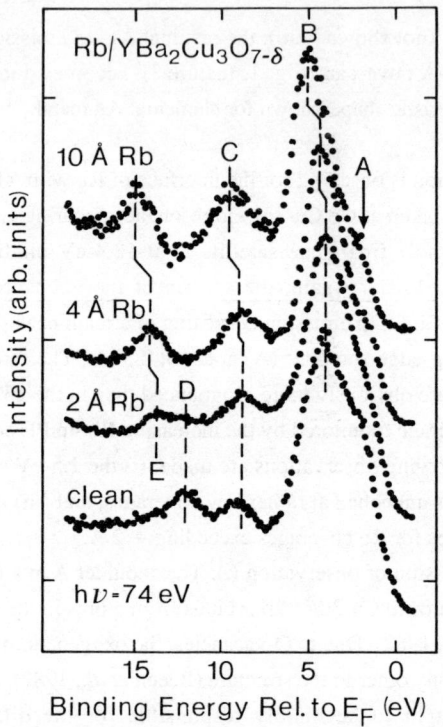

Fig. 2. Valence-band photoemission spectra of Rb/YBa$_2$Cu$_3$O$_{7-\delta}$ for increasing coverages of Rb taken at a substrate temperature of 180K. Note the shifts of all spectral features for Rb-coverages larger than 4 Å. The two hole final state satellite (peak D) is quenched by Rb exposure already at a coverage of 2-Å. Peak E is due to emission from the Rb-4p states.

Information on stoichiometry changes of the substrate adjacent to the interface is also provided by the observed quenching of the 12.4-eV satellite [observation (ii)] and the rigid shift of the whole spectrum to higher binding energy with increasing Rb coverage above $\simeq 4$ Å [observation (iii)]. The 12.4-eV satellite represents a shake-up structure characteristic for divalent Cu ions (Thuler et al., 1982). In monovalent Cu compounds, like e.g. Cu$_2$O or Cu metal, this structure is missing and a different, but weaker structure is observed at a binding energy of $\simeq 15$ eV (Thuler et al., 1982). Thus, the quenching of the intensity of the 12.4-eV satellite already at rather low Rb coverges strongly suggests a reduction of Cu(II) to oxygen and suggests that oxygen removal is consistent with the observed removal of involves both the Cu1 and Cu2 atoms in the YBa$_2$Cu$_3$O$_{7-\delta}$ structure. The corresponding 15-eV satellite due to monovalent Cu ions is probably masked in the spectra of Fig.2 by overlap with Rb-4p PE lines. The third observation in the spectra of Fig. 2 concerns the rigid shift of all spectral features to higher BE with increasing Rb coverage. This shift can also be related to a Rb-induced oxygen deficiency in the substrate adjacent to the interface. It affects both the valence-band spectra and the Ba-4d and Rb-3d core-level PE lines in an identical way, i.e. a rigid shift of the related PE signals to higher binding energies by $\simeq$ 1 eV at a Rb coverage of $\simeq$ 10 Å. Since the whole spectrum shifts in an identical way, an interpretation in terms of chemical reaction is ruled out.

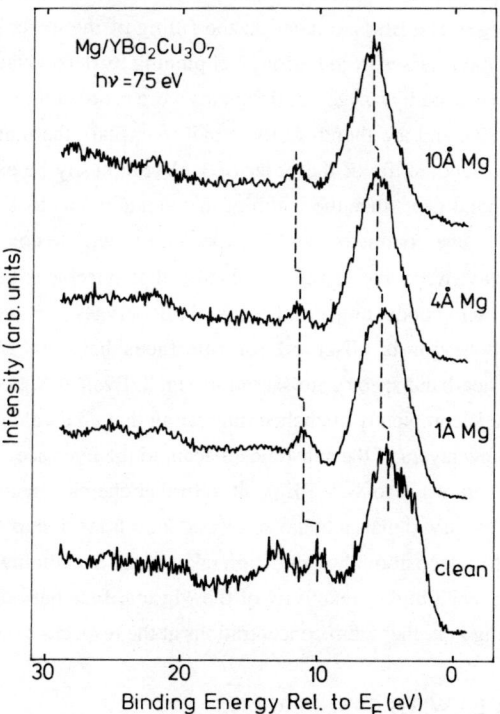

Fig. 3. Valence-band photoemission spectra of Mg/YBa$_2$Cu$_3$O$_{7-\delta}$ for increasing coverages of Mg taken at a substrate temperature of 150K.

Furthermore, since the same shift was observed for different photon energies as well as different photon fluxes, electrical charging of the samples can also be excluded as a possible origin. Instead, the shift may be consistently explained by a change in Fermi-level position relative to the valence-band edge at the interface. Such changes are well known to occur in metal/semiconductor interfaces through band-bending induced by defect states or virtual gap states upon metal deposition on a semiconductor substrate (Prietsch et al., 1988). In the present cases, the uncovered ceramic superconductor represents a semimetal, with two subbands separated by an energy gap of $\simeq 1.3$ eV, and the Fermi level close to the top of the lower band (Massida et al., 1987). Removal of oxygen leads to a metal-semiconductor transition in the substrate layers close to the interface, leading to an increase in the width of the band gap due to band narrowing (Lambin et al., 1988) and to a filling up of the lower band. As a consequence, E_F reaches the top of the lower band, giving rise to a situation similar to the one encountered in p-type semiconductors. A further removal of oxygen atoms through Rb will then lead to the formation of defect states, that in turn are expected to lead to a pinning of E_F at a position close to the middle of the band gap (Lindau and Kendelewicz, 1986).

Accordingly, two different mechanisms are actually expected to cause shifts of the PE signals to higher binding energy: The first is related to the filling of the lower valence band, while the second and dominant one arises from Fermi-level pinning by defect states. Changes of E_F due to a filling of the band caused by oxygen deficiencies were estimated to be of the order of 0.1 eV (Massida et al., 1987), and are therefore too small to explain the magnitude of the observed shifts. On the other hand, shifts of the order of 1 eV can easily be explained by Fermi-level pinning within the band gap, since the width of this gap amounts to 1.3 eV even in the case of the intrinsic high-T_c superconductors and is expected to increase considerably for the reacted interface. In this way, we come to the conclusion that interfaces between Rb and high-T_c superconductor turn semiconducting at rather low Rb coverages.

A rather similar behavior is observed for interfaces between Mg and $YBa_2Cu_3O_{7-\delta}$. Corresponding valence-band spectra are shown in Fig. 3. Even at Mg coverages of only $\simeq 1$ Å, the 12.4-eV satellite is completely quenched, indicating that all divalent Cu ions are chemically reduced in the first few layers of the substrate adjacent to the interfaces. Changes in the shape of the valence band structure at 2 to 8 eV BE point to further chemical reactions, and the shift of the whole spectrum by $\simeq 1$ eV signals a metal-semiconductor transition as for the interface with Rb. The fact that upon Mg deposition this transition takes place at even lower coverages than in the Rb case does not reveal a higher reactivity of the Mg interface but reflects the smaller atomic radius of Mg resulting in higher atom concentrations at the respective coverages.

INTERFACES WITH $Bi_5Sr_3Ca_2Cu_4O_y$

The interfaces of Ag and Mg with $Bi_5Sr_3Ca_2Cu_4O_y$ are characterized by closely similar properties as those with $YBa_2Cu_3O_{7-\delta}$, they are non-reactive in case of Ag and strongly reactive in case of Mg.

Fig. 4 shows valence-band PE spectra of the developing interfaces between Mg and $Bi_5Sr_3Ca_2Cu_4O_y$, taken at a photon energy of 135 eV and normalized to equal heights. At the bottom of the figure, the PE spectrum of the clean ceramic superconductor is displayed, which is quite similar to those reported for other Bi-based superconductors (Michel et al., 1988; Onellion et al., 1988). Strong emission from the valence band at a BE of about 5 eV is followed by a single, broad satellite at 10- to 13-eV BE. Measurements of the valence band at a photon energy of 75 eV (Cu-3p threshold) revealed a weak resonance at the high-BE part of this structure. Therefore, this part of the satellite is again attributed to the two-hole Cu-$3d^8$ final state. At binding energies of 18 and 21 eV, a weak doublet is observed and assigned to emission from Sr-4p and O-2s core levels, respectively. In this binding-energy range, the spectrum is dominated by a much stronger emission from Bi-5d core states at 26 eV ($5d_{5/2}$) and 29 eV ($5d_{3/2}$). The assignment of the 18-eV peak to emission from Sr-4p core levels is not equally straightforward, since the Sr-4p BE is known to be 21.2 eV for Sr metal and 21.8 eV for SrO (van Doveren and Verhoeven, 1980). However, the Sr-3d PE line of SrO is at 136.6 eV, while the observed BE in case of the ceramic superconductor is 132.8 eV (Michel *et al.*, 1988). This is the same BE difference (3.8 eV) as in case of the Sr-4p levels, a fact that strongly supports the given assignment.

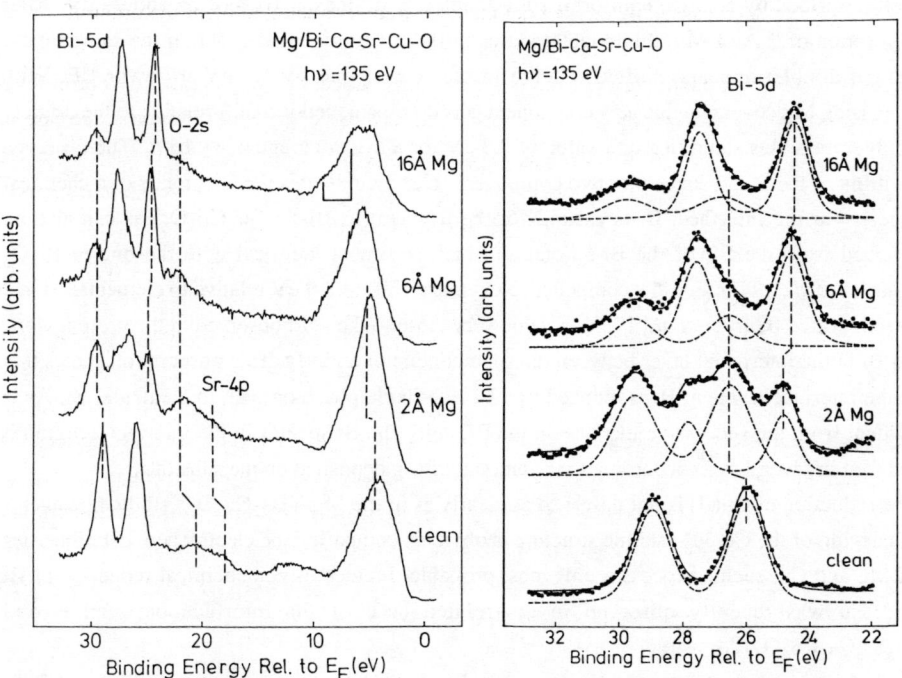

Fig. 4. Valence-band PE spectra of the developing interface between Mg and $Bi_5Sr_3Ca_2Cu_4O_y$, taken at a photon energy of 135 eV. The spectra are normalized to equal heights.

Fig. 5. Bi-5d core-level PE spectra for various increasing Mg exposures on $Bi_5Sr_3Ca_2Cu_4O_y$, together with the results of a least-squares fit analysis (solid line and dashed subspectra).

Upon deposition of 2 Å of Mg several spectral changes are observed that are quite similar to those found for $YBa_2Cu_3O_{7-\delta}$. The valence band narrows appreciably and its centroid shifts to higher BE, while the satellite loses intensity. Also, shifts of $\simeq 1$ eV to higher BE are observed for the Sr-4p, O-2s, and Bi-5d core levels, and in case of the Bi-5d core levels, a new component gets dominant, and both the satellite and the Sr-4p emission disappear, while a two-peaked structure at the high-BE side of the valence band (indicated by the bar diagram in Fig. 4 signals the formation of MgO (Fuggle, 1977). Our attempt to verify this assignment through measurements of the Mg core levels failed. Since the chemical shift of the Mg-2p line between MgO and Mg metal is known to be only 0.8 eV (Fuggle, 1977), a MgO-related structure could not be resolved in the relatively broad lines of the present Mg-2p PE spectra.

Fig. 5 displays the PE spectra of the Bi-5d core-levels in more detail, together with the results of a least-squares-fit spectral analysis based on Lorentzian lineshapes plus a linear background convoluted with a Gaussian resolution function. The spectrum of the pure superconductor is well described by a single spin-orbit-split doublet, with lines at BE's of 26 and 29 eV. After absorption of 2 Å of Mg, this doublet shifts by 0.8 eV to higher BE, and at the same time a second doublet appears, shifted relative to the former one by 1.7 eV to lower BE. With increasing Mg coverage, this new component gets dominant, and the shift relative to the original doublet increases saturating at a value of 2.3 eV for a Mg coverage of $\simeq 16$ Å. The observed splitting of the Bi-5d states into two components clearly indicates the occurence of a chemical reaction at the interface. Bi is assumed to be trivalent in $Bi_5Sr_3Ca_2Cu_4O_y$, as can also be inferred from the BE of the Bi-5d states, which is almost identical with the one of Bi_2O_3 (Gürtler et al., 1987). For this compound, a chemical shift of 1.9 eV relative to elemental Bi had been reported (Gürtler et al., 1987), a value very close to the shift observed in the present work for Bi in the interfacial layer between the superconductor and Mg. This observation shows that at the interface trivalent Bi is reduced to a Bi-rich oxide phase or even to elemental Bi. As is evident from the successive increase in the BE shift, the Bi species in the interfacial region is reduced in several steps according to the amount of Mg deposited on the substrate.

The reduction of Cu(II) is not observed as clearly as in the $Mg/YBa_2Cu_3O_{7-\delta}$ study, however a quenching of the $Cu-3d^8$ satellite structure is obvious; comparing the electrochemical properties of Mg and Cu, such a process seems most probable. Incidentally, a chemical reduction of Bi and Cu was recently observed in the related case of the interface between K and $Bi_5Sr_3Ca_2Cu_4O_y$ (Asensio et al.).

A further point worth considering concerns the observed attenuation of the PE signal from substrate atoms by the (Mg, MgO)-overlayer. We observe an exponential behavior of the PE intensity from Sr-core levels suggesting the formation of a rather homogeneous overlayer as in case of the $Rb/YBa_2Cu_3O_{7-\delta}$ interface. On the other hand, the Bi-related core-level PE lines are much less attenuated than those from Sr, as can be directly inferred from the spectra given in Fig. 4. This latter behavior indicates an accumulation of Bi close to the surface. The interface formed by deposition of Ag is again without an appreciable reaction. Fig. 6 displays valence band PE spectra for increasing coverages of Ag on $Bi_5Sr_3Ca_2Cu_4O_y$ taken again at a photon energy of 135 eV (at the Cooper minimum of Ag-4d). The spectra in Fig. 6 clearly show that the deposition of Ag has no effect on the shape of the valence band except for the seeming appearence of a weak peak at a BE of 24 eV that increases in relative intensity with increasing Ag coverage. However, this peak is also present in the spectrum of the uncovered substrate, and, in an absolute intensity scale, it even shrinks, but much less than the main component.

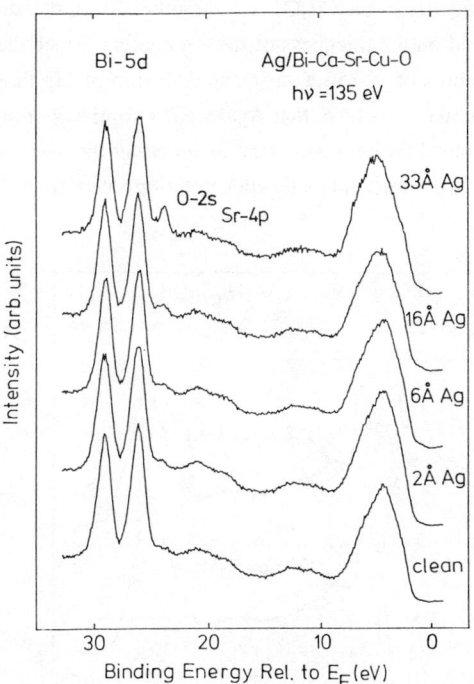

Fig. 6. Valence-band PE spectra of the formation of an interface between Ag and $Bi_5Sr_3Ca_2Cu_4O_y$, taken at a photon energy of 135 eV. Note that emission from Ag-4d core levels is strongly suppressed at this photon energy due to a Cooper minimum in the photoionization cross section.

On the basis of its BE, it is assigned to elemental Bi or Bi in a low oxidation state that is already present in the uncovered superconductor (remember that the sample is only $\simeq$ 85% single phase). Upon Ag deposition, this Bi species seems to segregate to the surface, causing the observed effect. Besides this minor reaction process, Ag forms a non-reactive overlayer. Thus, a thin Ag coating should be an effective means for passivating the superconductor substrate. We have checked this possibility by covering the clean superconductor substrate with 10 Å and 25 Å of Ag, respectively, and by subsequently depositing 5 Å of Mg. The results of this experiment are shown in Fig. 7. It is obvious that the deposition of Mg on top of a Ag overlayer leads again to strong chemical reactions with the substrate that manifest themselves in the same additional Bi-5d component as in Fig. 4. As estimated on the basis of the observed peak intensities, the strength of this reaction is about half as large as in the case of the direct interface between Mg and $Bi_5Sr_3Ca_2Cu_4O_y$. On the other hand, the strength of the reaction obviously does not depend on the thickness of the Ag coating. Since the measurements were made at sample temperatures of $\simeq$ 150 K, a strong diffusion of Mg though Ag may be ruled out. From these facts one may conclude, that Ag does not from a continuous overlayer for the coverages studied, but instead forms islands that do not cover the whole substrate surface; this would allow some of the Mg atoms to directly stick to the superconductor surface.

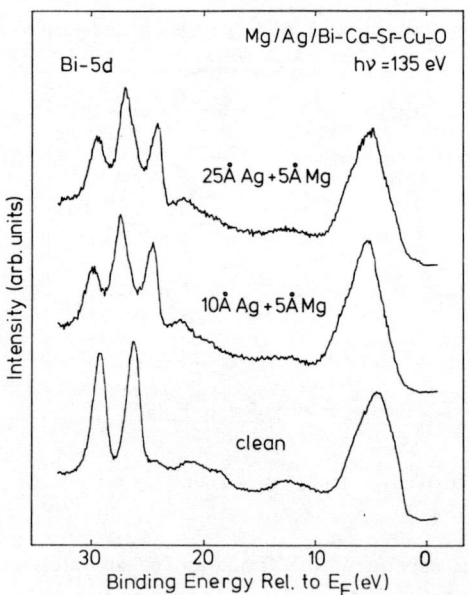

Fig. 7. Valence-band PE spectra of interfaces with $Bi_5Sr_3Ca_2Cu_4O_y$ formed by deposition of 5 Å of Mg on ceramic pellets pre-covered by 10 Å and 25 Å of Ag, respectively.

Possible causes for this behavior could be connected with a tendency of the Ag atoms to cluster, as observed for interfaces between Ag and amorphous carbon (Wertheim *et al.*, 1986), or with a shadowing effect during evaporation caused by the grainy surface morphology of the ceramic pellets (Meyer III et al., 1988). The fact, that the reaction is rather independent of the Ag coverage actually supports the latter alternative, since shadowing effects are expected to be almost independent of coverage in the studied range, whereas clustering should lead to an increase of the covered area with increasing Ag coverage.

SUMMARY

In this contribution we have presented the results of PE experiments using synchrotron radiation on interface formation between the high-T_c superconductors $YBa_2Cu_3O_{7-\delta}$ and $Bi_5Sr_3Ca_2Cu_4O_y$ and the metals Ag, Mg, and Rb. The interfaces with Mg and Rb were found to be reactive. For both types of superconductors, the overlayer is oxidized causing a removal of oxygen from the substrate layers adjacent to the interface. This causes a metal-semiconductor transition in the interfacial region of the superconductor. The interfaces with Ag have proven to be non-reactive, with the oxygen stoichiometry of the substrate being unchanged in the interfacial region. However, Ag does not form a continuous film on ceramic pellets of $Bi_5Sr_3Ca_2Cu_4O_y$, and is therefore not suitable for chemical passivation of the surface.

REFERENCES

Asensio, M.C., I. Alvarez, E.G. Michel, J.E. Ortega, R. Miranda, C. Laubschat, M. Domke, and G. Kaindl (to be published).

Beech, F., S. Miraglia, A. Santaro, and R.S. Roth (1987). *Phys. Rev. B*, 35, 8778.

Chaillout, C., M.A. Alario-Franco, J.J. Capponi, J. Chenavas, J.L. Hodean, and M. Marezio (1987). *Phys. Rev. B*, 36, 7118.

van Doveren, H. and J.A.Th. Verhoeven (1980). *J. Electr. Spectr. Relat. Phenom.*, 21, 265.

Fuggle, J.C. (1977). *Surf. Sci.*, 69, 581.

Gao, Y., T.J. Wagener, J.H. Weaver, B. Flandermeyer, and D.W. Capone II (1987). *Appl. Phys. Lett.*, 51, 1032.

Gao, Y., T.J. Wagener, J.H. Weaver, and D.W. Capone II (1988a). *Phys. Rev. B*, 37, 515.

Gao, Y., T.J. Wagener, C.M. Aldao, I.M. Vitomirov, J. H. Weaver, and D.W. Capone II (1988b). *J. Appl. Phys.*, 64, 1296.

Gürtler, K., K.H. Tan, G.M. Bancroft, and P.R. Norton (1987). *Phys. Rev. B*, 35, 6024.

Hill, D.M., H.M. Meyer III, J.H. Weaver, B. Flandermeyer, and D.W. Capone II (1987). *Phys. Rev. B*, 36, 3979.

Hill, D.M., Y. Gao, H.M. Meyer III, T.J. Wagener, J.H. Weaver, and D.W. Capone II (1988). *Phys. Rev. B*, 37, 511.

Kurtz, R.L., R. Stockbauer, D. Mueller, A. Shih, L.E. Toth, M. Osofsky, and S.A. Wolf (1987). *Phys. Rev. B*, 35, 8818.

Lambin, Ph., J.-P. Vigneron, and A.A. Lucas (1988). *Physica C*, 153-155, 1241.

Laubschat, C., M. Domke, M. Prietsch, C.T. Simmons, G. Kaindl, R. Miranda, E. Moran, F. Garcia, and M.A. Alario (1988a). *Physica C*, 153-155, 141.

Laubschat, C., M. Domke, M. Prietsch, T. Mandel, M. Bodenbach, G. Kaindl, H.J. Eickenbusch, R. Schoellhorn, R. Miranda, E. Moran, F. Garcia, and M.A. Alario (1988). *Europhys. Lett.*, 6, 555.

Lindau, I. and T. Kendelewicz (1986). *CRC Critical Review in Solid State and Material Sciences*, 13, 27.

Massida, S., Jaejun YU, A.J. Freeman, and D.D. Koelling (1987). *Phys. Lett. A*, 122, 198.

Meyer III, H.M., T.J. Wagener, D.M. Hill, Y. Gao, S.G. Anderson, S.D. Krahn, J.H. Weaver, B. Flandermeyer, and D.W. Capone II (1987a). *Appl. Phys. Lett.*, 51, 1118.

Meyer III, H.M., D.M. Hill, Steven G. Anderson, J.H. Weaver, and D.W. Capone II (1987b). *Appl. Phys. Lett.*, 51, 1750.

Meyer III, H.M., D.M. Hill, J.H. Weaver, D.L. Nelson, and K.C. Goretta (1988). *Appl. Phys. Lett.*, 53, 1004.

Michel, E.G., J. Alvarez, M.C. Asencio, R. Miranda, J. Ibanez, G. Peral, J.L. Vincent, F. Garcia, E. Moran, M.A. Alario Franco (1988). *Phys. Rev. B*, 38, 5146.

Onellion, M., Ming tang, Y. Chang, G. Margaritondo, J.M. Tarascon, P.A. Morris, W.A. Bonner, and N.G. Stoffel (1988). *Phys. Rev. B*, 38, 881.

Prietsch, M., M. Domke, C. Laubschat, and G. Kaindl (1988). *Phys. Rev. Lett.*, 60, 436.

Redinger, J., A.J. Freeman, Jaejun YU, and S. Massida (1987). *Phys. Lett. A*, 124, 469.

Su, C.-Y., I. Lindau, and W.E. Spicer (1982). *Chem. Phys. Lett.*, 87, 523.

Thuler, M.R., R.L. Benbow, and H. Hurych (1982). *Phys. Rev. B*, 36, 669.

Wagener, T.J., Y.Gao, I.M. Vitomirov, C.M. Aldao, J.J. Joyce, C. Capasso, J.H. Weaver, and D.W. Capone II (1988). *Phys. Rev. B*, 38, 232.

Wertheim, G.K., S.B. DiCenzo, D.N.E. Buchanan (1986). *Phys. Rev. B*, 33, 5384.

RESONANT PHOTOEMISSION NEAR O 1s THRESHOLD IN $YBa_2Cu_{2.7}Fe_{0.3}O_{6.9}$

D.D. Sarma[+], P. Sen[+], C. Carbone[+], R. Cimino[++], B. Dauth[+] and W. Gudat[+]

[+]*Institut für Festkörperforschung, Kernforschungsanlage Jülich
D-5170 Jülich, F.R.G.*
[++]*Bessy, Lentzeallee 100, Berlin 33 F.R.G.*

ABSTRACT

Superconducting samples of $YBa_2Cu_{2.7}Fe_{0.3}O_{6.9}$ ($T_c \sim 40K$) have been investigated in the photon energy range near the O 1s threshold. Valence band photoemission at hv = 524 eV and hv = 524 eV shows resonant enhancement of a $Cu(3d^8)$ related feature indicating interatomic resonance and considerable hybridization between Cu(3d) and O(2p) states. Comparing the photoemission results with the x-ray absorption spectra at O K-edge, we show that the 528 eV peak in XAS of these compounds cannot be related to stoichiometry induced holes at E_F, contrary to previous interpretation. We discuss various implications of our observations.

High energy spectroscopies like photoemission and x-ray absorption have been extensively applied to the study of the high-T_c cuprates in recent times. From these investigations, a general consensus has emerged that the excess holes which are induced by stoichiometry in the system (i.e. Sr or Ba in La_2CuO_4 or excess oxygen in $YBa_2Cu_3O_{6.5}$) primarily reside at the oxygen site, rather than the Cu side. Direct experimental evidence to support this interpretations comes from x-ray absorption (XA) studies at Cu L edges of these compounds; in the spectra the $2p^63d^8 \rightarrow 2p^53d^9$ transition is not observed for the high-T_c cuprates (Bianconi et al., 1987, 1988; Sarma et al., 1988), while such a transition is observed for formal Cu^{3+} compounds $NaCuO_2$ (Sarma et al., 1988). A peak at hv ~ 528 eV in the O K-edge excitation spectra (Yarmoff et al., 1987; Nücker et al., 1987, 1988; Kuiper et al., in press) in these compounds has also been associated with transitions into the pre-continuum and has thus been interpreted as an evidence for the holes residing at the oxygen sites. It is considered that a large intra-atomic Coulomb repulsion U within the Cu 3d electrons makes the two hole $3d^8$ configurations energetically unfavourable shifting the excess holes to the oxygen site. Various experimental (Fujimori et al., 1987; Takahashi, T. et al., 1987) and calculated (Sarma and Sreedhar et al., 1988; Chen et al., 1988) estimates of U range between 5 and 6 eV. Inspite of this large value of U, recently the validity of assigning the excess holes specifically to the oxygen sites has been questioned

(Takahashi and Zhang, 1988) on the basis of the expected large hybridization strength, t, between the Cu 3d and O 2p states. It is argued (Takahashi and Zhang, 1988) that a large hybridization will mix the Cu 3d and O 2p states and thus the distinction between a hole in the oxygen band and that in the copper band will be less meaningful and only in the limit of small Cu-O hybridization can such a distinction be retained. This argument, based on the magnitude of U/t ratio, is important, since in the regime of U/t > 1. the behaviour of the system is different for a very large value of this ratio (Bagaoka, 1966) compared to the situation with a moderate value of the ratio (Anderson *et al.*, 1987; Takahashi, Y., 1987). Moreover it has been shown (Zhang and Rice, 1988) that the two band Hubbard model for CuO square lattice can be reduced to a single band effective Hamiltonian in presence of sizable hybridization strength. Indeed, the hybridization strength between Cu 3d and O 2p states has been estimated to be in the range of ~2-3 eV by band structure (Mattheiss, 1987; Mattheiss and Hamman, 1987) and other (Zaanen *et al.*, 1988) calculations.

In our continued experimental efforts to understand the electronic structure of this highly interesting class of compounds, we have come across a direct experimental evidence for a large hybridization strength between Cu 3d and O 2p states in these systems. It manifests itself in an interatomic resonant enhancement of photoemission transition probability of Cu $3d^9 \rightarrow 3d^8$ related states in the valence band when the photon energy is tuned through the O 1s absorption threshold in a sample of $YBa_2Cu_{2.7}Fe_{0.3}O_{6.9}$. While we cannot estimate directly the magnitude of the hybridization strength from this experiment, it still points to an unusually large t consistent with previous estimates (Mattheiss, 1987; Mattheiss and Hamman, 1987; Zaanen *et al.*, 1988), in view of the fact that such a cross-resonance phenomenon has never been observed for any other compounds. We also discuss critically why the concept of excess holes specifically at the oxygen sites may still be valid in conformity with previous experimental observations (Bianconi *et al.*, 1987, 1988; Sarma *et al.*, 1988), in spite of this large hybridization strength. Comparing this resonance photoemission results with XAS data at O K-edge, we also indicate the possibility of a considerable reinterpretation of the existing O K-edge XAS results. At this stage, we should point out that the choice of $YBa_2Cu_{2.7}Fe_{0.3}O_{6.9}$ sample was entirely accidental. However we do not expect any drastic modifications in the electronic structure in the normal state of the Fe-doped sample and that of $YBa_2Cu_3O_{6.9}$. This expectation is supported by the very similar photoemission spectra of the valence band regions at all photon energies (unpublished results of this group) and by the very similar normal state resistivity behaviours (unpublished results of this group; Baggio-Saitovich *et al.*, 1988) between these two compounds. We also plan to investigate $YBa_2Cu_3O_{6.9}$ in future.

The sample of $YBa_2Cu_{2.7}Fe_{0.3}O_{6.9}$ was prepared in the standard way from a mixture of pure Y_2O_3, $BaCO_3$, CuO and Fe_2O_3 in the correct proportion. The photoemission experiments were performed at BESSY, Berlin, at the TGM3 and HE-TGM1 beamlines, while the x-ray absorption measurements in total yield technique (Kuiper *et al.*, in press) were recorded at the HE-TGM1 beamline. The resolution at the low photon energies (TGM3 beamline) was ~ 0.3 eV, while that around the O K threshold (HE-TGM1) was ~ 1.0 eV. In order to calibrate photon energies, the kinetic energy scale of the spectrometer was first calibrated with the Auger signals of oxygen in $YBa_2Cu_3O_{6.9}$ as well as of elemental Ni, Cu and Zn, these being independent of photon energies. Subsequently the photon energy was calibrated with known binding energies of valence band maximum and core levels in metallic samples. This procedure leads to uncertainties less than 0.5 eV in photon energies. The vacuum in the experimental chamber was maintained at $\leq 1 \times 10^{-10}$ mbar. The sample surface was scraped in-situ with a stainless steel blade in order to obtain a clean surface.

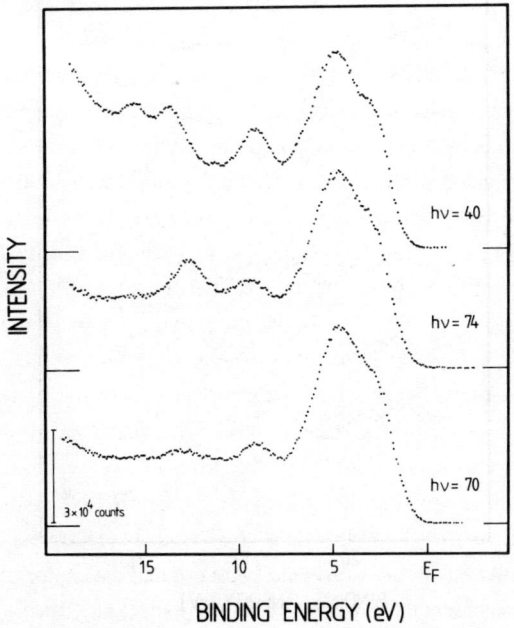

Fig.1. The valence band photoemission spectra of $YBa_2Cu_{2.7}Fe_{0.3}O_{6.9}$ with different photon energies. The spectrum with hv = 74 eV show the enhancement of the $3d^8$ related satellite at ~ 13 eV binding energy.

In Fig. 1. we show the spectra of $YBa_2Cu_{2.7}Fe_{0.3}O_{6.9}$ at hv = 40, 70 and 74 eV. The spectrum at 40 eV photon energy clearly shows the broad valence band emission, centered at ~4.8 eV below E_F, attributed primarily to the $d^9\underline{L}^1$ ($\underline{L}^1$ representing a hole in the ligand) features in the final state (Fujimori et al., 1987). At approximately 9 eV below E_F there is a weak peak, which is often present in these compounds and its origin is still unknown and controversial. The double peak structure at 13.6 eV and 15.4 eV is due to the spin orbit split Ba 5p states. The weak feature around 13 eV with hν = 70 eV is considerably enhanced in intensity with hν = 74 eV i.e. above the Cu 3p threshold; thus establishing the $3d^8$ satellite character of this feature (Fujimori et al., 1987; Takahashi, T. et al., 1987). The spectra shown in Fig. 1 are very similar to those obtained from $YBa_2Cu_3O_{6.9}$ with the same photon energies (unpublished results of this group).

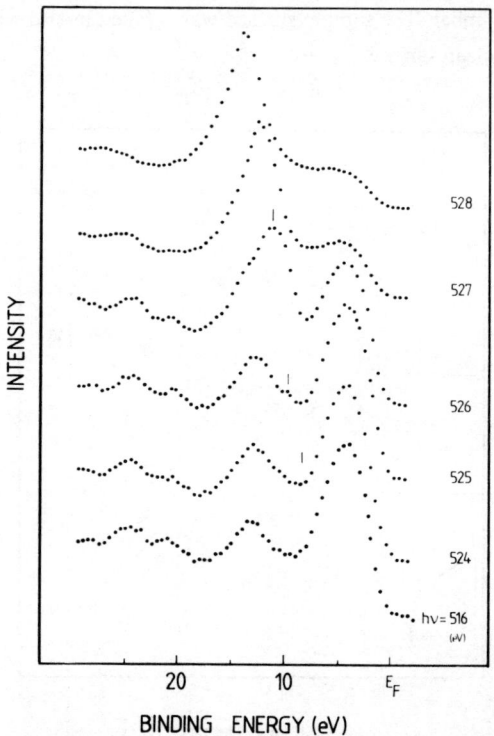

Fig. 2. Photoemission spectra in the valence band region of $YBa_2Cu_{2.7}Fe_{0.3}O_{6.9}$ with photon energies as shown in the figure. The expected energy of O KLL Auger signal is shown with a vertical tick for the spectra with hν = 524, 525 and 526 eV.

In Fig. 2 we show the spectra of the valence band region with various photon energies between 516 eV and 529 eV, i.e. in the region of O 1s excitation. Because of the considerably higher photon energies, as compared to the previous spectra, the cross sections for the various atomic states have changed strongly. The spectrum with hv = 516 eV shows the main peak due to Cu 3d states around 4 to 5 eV binding energy. The total Ba 5p cross-section is by a factor 20 weaker than that of Cu 3d. The weaker peak at ~ 13 eV below E_F is therefore predominantly due to the $3d^8$ satellite of Cu. On increasing the photon energy to 524 and 525 eV, we find a relative increase in the spectral intensity in this peak region. (Note that the spectra have to normalized to the same incident photon flux for quantitative evaluation given below; but here spectra are displayed with equal peak height). On further increase of the photon energy a distinct and intense peak develops. From the fact that the binding energy position of the latter peak shifts in synchronism with the photon energy we conclude that it is a constant kinetic energy feature.

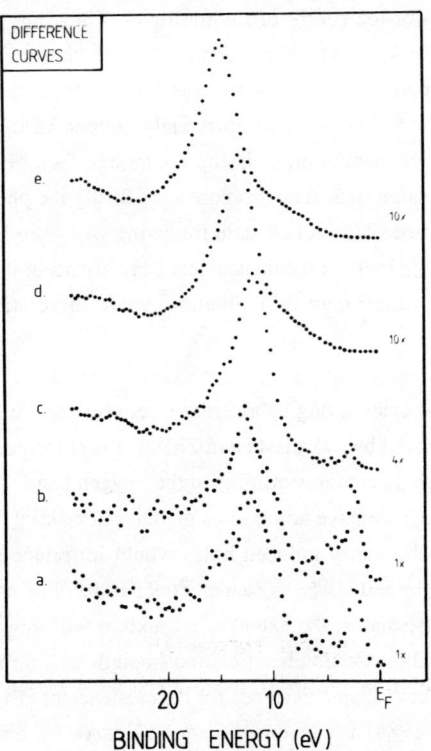

Fig. 3. The difference after subtracting the valence band spectra (hv = 516 eV) from the valence band spectra with photon energies: (a) 524 eV, (b) 525 eV, (c) 526 eV, (d) 527 eV, and (e) 528 eV. Intensity scaling factors are marked on the right of the figure.

Thus we attribute it to the oxygen KVV Auger transition in Fig. 2. We have marked with a tick the expected position of this Auger transition for the spectra recorded with hν= 524 eV to 526 eV. From this it is seen that the increase in the intensity around the $3d^8$ feature at these two photon energies cannot be attributed to the development of the oxygen Auger signal, but rather to a resonant enhancement of $3d^8$ states. This distinction can be seen more clearly in the difference spectra, shown in Fig. 3, between spectra recorded with hν = 524 to 528 eV and that with hν = 516 eV. While the most intense peak in Fig. 3 for hν ≥ 526 eV is clearly of constant kinetic energy, the increase in intensity for hν = 524 and 525 eV is of constant binding energy. The binding energy of the peak position of this enhanced feature (~13 eV) is exactly where the enhancement of the Cu $3d^8$ signal was observed previously for resonant photoemission at the Cu 3p threshold (Fig.1). The $3d^8$ configuration results in different terms with 1G being strongest giving essentially the centre of the multiplet. Thus, our experimental observation appears to be the first of its kind, where photoemission with photon energies at the threshold of the core level of one element (oxygen 1s in this case) gives rise to the enhancement of spectral features associated with another constituent (namely copper 3d in this case) of the sample. Consider the process of resonant photoemission i.e. an interface between different excitation channels resulting in the same final state, or more specifically the photoemission and the direct recombination of an intermediate excited state following with core hole absorption. It is them clear that such an interatomic resonant enhancement can only occur if the Cu 3d and O 2p states are strongly hybridized, rather than in a situation where these states primarily retain their separate identities.

While this observation indicates strong hybridization between the Cu 3d and O 2p states, it does not contradict (as conjectured by Takahashi and Zhang, 1987) the generally accepted notion of the excess hole in the high T_c cuprates residing at the oxygen band. While in the stochiometric compounds (e.g. La_2CuO_4) we have admixtures of states $|d^9>$, $|d^{10}\underline{L}^1>$ in the ground state due to hybridization, the stochiometry induced holes would introduce states with the following configurations: $|d^8>$, $|d^9\underline{L}^1>$ and $|d^{10}\underline{L}^2>$ (Zaanen et al., 1988). It is true that hybridization will lead to the mixing of these states; the extent of admixture will, however, depend on the bare energies of these states and the Coulomb correlation strength U. It turns out that these energetics are such that one finds spectroscopic evidence for the existence of $|d^9\underline{L}^1>$ states (Bianconi et al., 1987, 1988;Sarma et al., 1988) and no experimental evidence for a significant contribution of $|d^8>$ states (Sarma et al., 1988) in the ground state. Model calculations with the Anderson impurity Hamiltonian (Zaanen et al., 1988) also show negligible admixture of $|d^8>$ configuration in the ground state. It is in this sense that one refers to the stoichiometry induced holes residing in the ligand (i.e. $|d^9\underline{L}^1>$) rather than on copper (i.e. $|d^8>$), inspite of a large hybridization strength.

We have also measured x-ray absorption spectrum (XAS) of this compound in the region of the O K-edge and this is quite similar to that of the undoped sample of $YBa_2Cu_3O_{6.9}$ (Yarmoff *et al.*, 1987; Kuiper *et al.*, in press; unpublished results of this group). Given the previous results of the interatomic resonant photoemission (Fig. 2), it implies that the threshold for O K-edge XAS is at (or lower than) 524 eV, as resonant photoemission cannot take place without a preceeding resonant absorption process. Thus, we place 524 eV as an upper limit of the threshold. However, there is very little intensity in the spectrum at this energy (see for example Yarmoff *et al.*, 1987). This is not very suprising in view of the fact that photoemission experiments at all photon energies (see Fig. 1) exhibit hardly any density of states at the Fermi level of these compounds (Yarmoff *et al.*, 1987; Fujimori *et al.*, 1987; unpublished results of this group).

The placement of the threshold at 524 eV also indicates that the x-ray absorption feature at 528 eV cannot be attributed to the presence of stoichiometry induced oxygen holes <u>at the Fermi level</u> of these compounds, as been done till now (Yarmoff *et al.*, 1987; Kuiper *et al.*, in press). The peak at 528 eV is due to transitions into states that are well above the Fermi level and is most probably related to the density of states corresponding to the conduction band in the stoichiometric semiconducting phase. The above interpretation is also supported by our observation of qualitatively different behaviours in photoemission spectra with $h\nu = 524$ and 525 eV, compared to those with $h\nu \geq 526$ eV (Fig. 2). While $h\nu = 524$ and 525 exhibit a resonant enhancement of photoemission spectra, $h\nu \geq 526$ eV leads to the formtion of normal Auger signal of oxygen. Whether the intermediate state following resonant absorption decays via autoionization (and thus resonant enhancement of photoemission features) or via an ordinary Auger decay depends on the nature of the final state reached by the excited electron. If this final state is a delocalized one (in spite of the core hole potential), the decay process is dominated by the normal Auger channel. Comparing the photon energy (≥ 526 eV) for which the oxygen Auger signal appears with the x-ray absorption spectrum, we find that the peak in XAS around 528 eV must correspond to delocalized band states even in presence of the core-hole potential. It also appears that the density of states at the Fermi level, which leads to only negligible intensity XAS at the threshold, has a qualitatively different behaviour in that it leads to the autoionization as the main decay channel giving rise to the resonance observed (Fig. 2). We would like to point out that the systematic variation seen in the 528 eV XAS peak for $La_{2-x}Sr_xCuO_4$ and $YBa_2Cu_3O_{6+x}$ with increasing $h\nu$ (Nücker *et al.*, 1987, 1988; Kuiper *et al.*, in press), is most probably related to changes in the band structure (as will be discussed in detail in unpublished results of this group) and reflects changes in the density of states above E_F.

CONCLUSION

In conclusion, we have shown for the first time an interatomic resonant enhancement of 3d related features in the valence band of $YBa_2Cu_{2.7}Fe_{0.3}O_{6.9}$ for hv near the O 1s threshold, providing direct experimental evidence for strong hybridization between Cu 3d and O 2p states. This observation puts an upper limit for the threshold in O K-edge XAS at 524 eV, considerably below previous estimates. We further show that the peak at 528 eV in XAS is not related to the presence of stoichiometry induced holes at E_F of these compounds, as has been believed earlier.

ACKNOWLEDGEMENT

The authors are thankful to A.M. Bradshaw for the permission to use his beamline at BESSY, to W. Braun for experimental assistence and discussion and to U. Döbler for help with the programme for data-analysis. The authors thankfully acknowledge the cooperation of the BESSY staff. One of us (DDS) thanks Kernforschungsanlage Jülich for financial support during the course of this work.

* Permanent address: Solid State and Structural Chemistry Unit, Indian Institut of Science Bangalore - 560 012 India

REFERENCES

Anderson, P.W., (1987). *Science*, 235, 1196.

Bagaoka, Y., (1966). *Phys. Rev.*, 147, 39.

Baggio-Saitovich, E., I. Sonza Azevedo, R.B. Scorzelli, H. Saitovich, S.F. de Cunha, A.P. Guimaraes, P.R. Silva and A.Y. Takenchi, (1988). *Phys. Rev. B*, 37, 7967.

Beno, M.A., L. Soderholm, D.W. Capone II, D.H. Hinks, J.D. Jorgensen, I.K. Schuller, C.U. Segre, K. Zhang and J.D. Grace, (1987). *Appl. Phys. Lett.*, 51, 57.

Bianconi, A., A. Congiu Castellano, M. De Santis, P. Rudolf, P. Lagarde, A.M. Flank and A. Marcelli, (1987). *Solid State Commun.*, 63, 1009.

Bianconi, A. *et al.*, (1988). *Physica C*, 153-155, 115.

Chen, H., J. Callaway and P. K. Misra, (1988). *Phys. Rev. B*, 38, 19.

Fujimori, A., E. Takayama-Muromachi, Y. Uchida and B. Okai, (1987). *Phys. Rev. B*, 35, 8814.

Kuiper, P., G. Kruizinga, J. Ghijsen, M. Grioni, P.J.W. Weijs, F.H.M. de Groot, G.A. Sawatzky, H. Verweig, L.F. Feiner and H. Peterson, *Phys. Rev. B*, (in press)

Mattheiss, L.F., (1987). *Phys. Rev. Lett.*, 58, 1028.

Mattheiss, L.F. and D.R. Hamman, (1987). *Solid State Comm*, 63, 39.

Nücker, N., J. Fink, B. Renker, D. Ewert, C. Politis, P.J.W. Weijs and J.C. Fuggle, (1987). *Z. Phys. B*, 67, 9.

Nücker, N., J. Fink, J.C. Fuggle, P.J. Durham and W.M. Temmerman, (1988). *Phys. Rev. B*, 37, 7516.

Sarma, D.D., O. Strebel, C.T. Simmons, U. Neukirch, G. Kaindl, R. Hoppe and H.P. Müller, (1988). *Phys. Rev. B* , 37, 9784.

Sarma, D.D. and K. Sreedhar, (1988a). *Z. Phys. B*, 69, 529.

Yarmoff, J.A., D.R. Clarke, W. Drube, U.O. Karlsson, A. Taleb-Ibrahimi and F.J. Himpsel, (1987). *Phys. Rev. B*, 36, 3967.

Takahashi, T., *et al.*, (1987). *Phys. Rev. B*, 36, 5686.

Takahashi, Y., (1987). *Z. Phys. B*, 67, 50.

Takahashi, Y. and F. C. Zhang, (1988). *Z. Phys. B*, 69, 443.

Zaanen, J., O. Jepsen, O. Gunnarsson, A.T. Paxton, O.K. Andersen, and A. Svane, (1988). *Physica C*, 153-155, 163.

Zhang, F.C. and T.M. Rice, (1988). *Phys. Rev. B*, 37, 375.

Electron Spectroscopy Studies of High Temperature Superconductors: $Y_{1-x}Pr_xBa_2Cu_3O_{7-\delta}$

J.-S. Kang[1,2], J.W. Allen[1], B.-W. Lee[2], M.B. Maple[2], Z.-X. Shen[3], J.J. Yeh[3,4], W.P. Ellis[5], W.E. Spicer[3], and I Lindau[3]

[1] Dept. of Physics, University of Michigan, Ann Arbor, MI., 48109-1120, USA

[2] Dept. Of Physics and Institute for Pure and Applied Physical Sciences, University of California at San Diego, La Jolla, CA., 92093, USA

[3] Stanford Electronics Laboratories, Stanford, CA., 94305, USA

[4] AT&T Bell Laboratories, Murray Hill, NJ., 07974, USA

[5] Los Alamos National Laboratory, Los Alamos, NM., 87545, USA

ABSTRACT

We itemize our previous work on high T_c superconductors, and describe more fully the results of an electron spectroscopy study and impurity Anderson Hamiltonian analysis of the Pr 4f spectrum of $Y_{1-x}Pr_xBa_2Cu_3O_{7-\delta}$, a system in which superconductivity is quenched as x increases. It has been speculated that Pr has valence 4+, resulting in extra charge in the Cu-O planes, and causing T_c-suppression. We find that the Pr valence is close to 3+ for all x but that there is extensive Pr 4f hybridization with other valence band states. The Cu valence is essentially unchanged with x. From these findings, we speculate that Pr 4f hybridization with other valence band states has enabled Pr spin fluctuations to cause the T_c-suppression.

KEYWORDS

$Y_{1-x}Pr_xBa_2Cu_3O_{7-\delta}$; photoemission; BIS; 4f spectrum; quenched superconductivity; spin fluctuations; Anderson Hamiltonian.

INTRODUCTION

Previous Work

Previously (Shen *et al.*, 1987) we have studied the electronic structure of $La_{2-x}Sr_xCuO_4$ for x=0 and 0.2, and $YBa_2Cu_3O_{7-\delta}$ using resonant photoemission spectroscopy (RESPES) and x-ray photoemission spectroscopy (XPS). We analyzed these data in the impurity cluster approximation to the Anderson Hamiltonian to derive approximate values of Hamiltonian parameters and to discuss superexchange interactions. We pointed out that for these parameters Cu^{3+} is largely excluded from the ground state and that the non-metallic materials are charge transfer insulators (Sawatzky and Allen, 1984; Zaanen *et al.*, 1985) implying that oxygen holes are created by doping. An overview of this and related work has recently been given. (Shen et al., 1988a) We have also studied the dispersion of hole excitations in the archetype material NiO and found that a rigid-shift correction to band structure calculations is not valid. (Shih et al., 1988) Here we will concentrate on new results for $Y_{1-x}Pr_xBa_2Cu_3O_{7-\delta}$.

$Y_{1-x}Pr_xBa_2Cu_3O_{7-\delta}$

The superconductivity of the isostructural alloy system $Y_{1-x}Pr_xBa_2Cu_3O_{7-\delta}$ is quenched with increasing x. (Dalichaouch et al., 1987; Soderholm et al., 1987; Liang et al., 1987) The other $RBa_2Cu_3O_{7-\delta}$ compounds, where R is a rare earth or La, are superconductors with T_c near 90K except for Ce and Tb, for which the compounds do not form. (Yang et al., 1988) For $Y_{1-x}Pr_xBa_2Cu_3O_{7-\delta}$ it is found that the normal state electrical resistivity shows a transition from metallic to semiconducting behavior, with the monotonic suppression of T_c from $\sim$90K to $\sim$60K and $\sim$35K for x=0, 0.2 and 0.4, respectively, with the samples for x>0.6 being non-superconducting. The pressure dependence of these electrical properties is complex and interesting. (Neumeier et al., 1988) From magnetic susceptibility measurements an effective magnetic moment of 2.7 μ_B/Pr-ion was extracted, independent of x. (Dalichaouch et al., 1987) A possible intepretation of the magnetic moment is that the Pr valence has a fixed value of about 3.8, and that extra charge is then contributed to the Cu-O planes, filling the holes that are widely believed to be the superconducting carriers. This picture is supported by electron energy loss spectroscopy (EELS) measurements of the oxygen 1s→2p transition, which showed that a small pre-threshhold peak thought to indicate valence band holes is reduced, although not eliminated, for x=1 relative to x=0. (Nücker et al., 1988) In contrast, the lattice constants of the $RBa_2Cu_3O_{7-\delta}$ series strongly suggest that Pr is trivalent. (Yang et al., 1987) In fact, the Pr valence was concluded to be close to 3+ from X-ray absorption spectroscopy (XAS) measurements of the Pr L-edge (Horn et al., 1987; Alp et al., 1987; Lytle et al., 1988), although it was proposed that the T_c-quenching is due to charge transfer arising from a modest deviation of the valence from 3+ toward 4+. Other workers have subsequently come to the same conclusion from Pr M-edge studies. (Neukirch et al., 1988)

We have made an electron spectroscopy study of this system for x=0, 0.2, 0.4, 0.8 and 1.0, including XPS for various core levels, bremsstrahlung isochromat spectroscopy (BIS) for the conduction band, and synchrotron-excited RESPES for the Pr 4f and Cu 3d states of the valence band. Our Pr RESPES data were reported in a separate paper. (Kang et al., 1988) In agreement with the XAS results we concluded that the Pr valence is near 3+ for all x. However, the Pr 4f photoemission spectrum, which was found to be largely unchanged with x, shows strong hybridization with other valence band states. In this paper we summarize our BIS and XPS data and the results of efforts to quantify the hybridization by fitting the Pr 4f RESPES and BIS spectra using the impurity Anderson model.

EXPERIMENTAL DETAILS

The samples of $Y_{1-x}Pr_xBa_2Cu_3O_{7-\delta}$ with x=0, 0.2, 0.4, 0.6, 0.8, 1.0 were prepared as described by Dalichaouch et al. (1987). The samples with x=0, 0.2, 0.4, 0.6, and 0.8, were prepared in the same batch, while the sample with x=1.0 was prepared at a different time with a slightly different oxygen treatment. An impurity phase of $BaCuO_2$ in amounts less than 8 mol% is observed in the x-ray diffraction data, the amount of which decreases with increasing x until it is not observed for $x > 0.4$.

The synchrotron-excited photoemission spectroscopy (PES) was performed at the Stanford Synchrotron Radiation Laboratory (SSRL), using equipment and procedures described fully by Shen et al. (1987) and Kang et al. (1988). For all synchrotron-excited spectra presented below, normalization has been performed by recording the photoelectron yield from a beam flux monitor for each photoemission spectrum. The Fermi level of the system was determined from the valence-band spectrum of a gold sample evaporated onto a stainless-steel substrate *in situ*. Room temperature XPS and BIS spectra were obtained at the Xerox Palo Alto Research Center (PARC) using a Vacuum Generators (VG) ESCALAB spectrometer. The equipment and procedures are generally as described by Allen et al. (1987). XPS satellites due to the non-monochromatized Mg $K\alpha$ source were subtracted numerically. The base pressure during the BIS measurements rose to about 3×10^{-10} Torr, mostly due to the outgassing of the samples. The Cu 2p XPS spectrum was the same before and after BIS.

PHOTOEMISSION SPECTRA

Valence Band Cu RESPES and O 1s XPS

Panels (a) and (b) of Fig. 1 display the valence band spectra for x=0, 0.2, 0.4, 0.6, 0.8, and 1.0 at photon energies $h\nu$ = 70 eV and 75 eV, respectively, which are below and at the Cu 3d cross-section resonance at the Cu 3p edge. The two sets of spectra are scaled to have the same valence band maximum intensity in each set. As explained by Shen et al. (1987), interference from the Ba 4d core level excited by second order light has been removed. The Cu RESPES results of Fig. 3 are like our previous results for x=0. The emission of the d^8-like satellite located at -12.4 eV is clearly enhanced at $h\nu$ = 75 eV. An antiresonance of the states between -7 eV and E_F, which involves hybridized Cu 3d and O 2p states, is concealed by the scaling in the figure. We do not observe any correlation between x and the intensity of the d^8-like satellite relative to the -4.7 eV peak, which implies that there is no systematic change in the Cu 3d state due to the substitution of Pr.

Panel (c) of Fig. 1 shows the O 1s core level XPS spectra for x=0, 0.2, 0.4, 0.6, 0.8 and 1.0. For all x there are multiple peaks, with changing relative intensities. Multiple peaks are often found in ceramic samples, and the developing concensus (Kohiki et al., 1987; Takahashi et al., 1988; Shen et al., 1988b) is that only the 528 eV line is intrinsic. The decrease with increasing x in the intensity of the 531 eV line relative to the 528 eV line correlates with changes in the valence band spectra which show that with increasing x there is a decrease in the -4.7 eV maximum relative to the -2.4 eV shoulder. Since this change is essentially the same in both panels (a) and (b) of the figure, it does not involve primarily the Cu 3d states. We conclude that there is valence band emission in the -4.7 eV range from extrinsic oxygen, possibly water or gases such as CO_2 adsorbed in the grain boundaries, or the $BaCuO_2$ impurity phase. The higher binding energy O 1s peak decreases during BIS measurements.

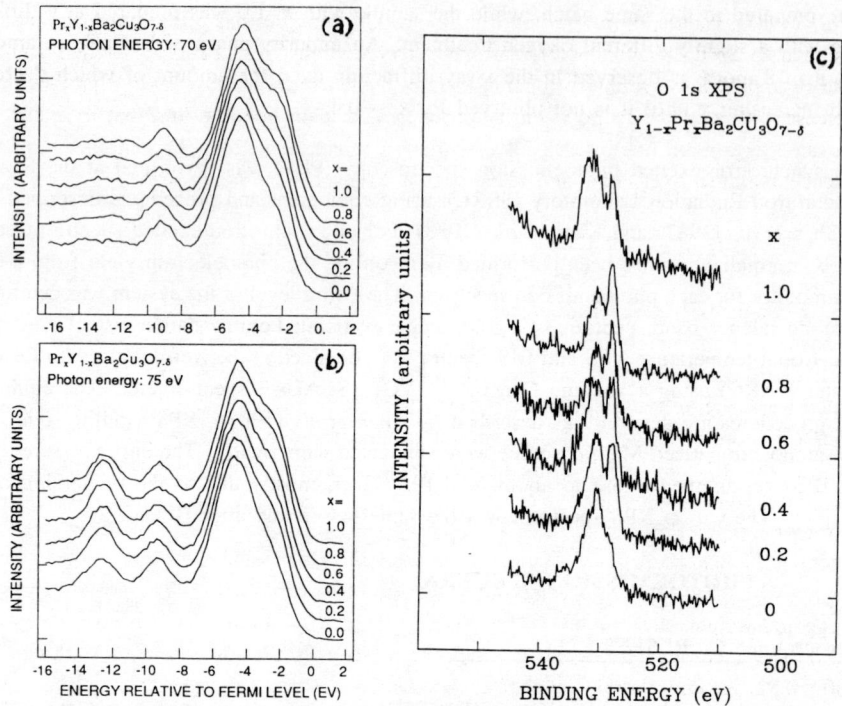

Fig. 1. The valence band spectra of $Y_{1-x}Pr_xBa_2Cu_3O_{7-\delta}$ at (a) $h\nu=70$ eV, below the resonance, and at (b) $h\nu=75$ eV, above the resonance. The interference from the Ba 4d core level excited by the second order light is removed, as described by Shen et al (1987).
(c) The O 1s core level XPS spectra of $Y_{1-x}Pr_xBa_2Cu_3O_{7-\delta}$.

Pr 4f RESPES

The Pr 4f emission is spread throughout the first 7 eV of the valence band and is not explicitly visible in Figs 1 (a,b). It has been determined by RESPES studies described in detail by Kang et al. (1988). It is important that the shapes of the extracted Pr 4f spectra for all x are essentiallly like that shown in Fig. 3 for x=1. We infer that Pr does not occur in combination with the extrinsic oxygen described above.

Pr 3d and Cu 2p XPS

The Pr 3d and Cu 2p core levels consist of spin-orbit split peaks which overlap. The Pr $3d_{5/2}$ states occur between 933 eV and 935 eV for several Pr materials, and the Cu $2p_{3/2}$ peak is observed at 933.7 eV in $YBa_2Cu_3O_{7-\delta}$. Since the two core levels have nearly identical spin-orbit splittings, the Pr 3d emission is expected under the Cu peaks. As discussed by Shen et al. (1987), each spin-orbit component of the Cu spectrum has a main line and a satellite assigned to $2p^5d^{10}L$ and $2p^5d^9$ final state configurations, respectively, where L denotes a hole of appropriate symmetry relative to filled ligand O 2p states. Taking into account the increase with x of the Pr 3d intensity under the Cu main line, we find that the ratio of the Cu satellite to main line intensity is essentially unchanged with x. This implies that the Cu valence averaged over chain and plane sites does not change much with x.

BIS SPECTRA

Fig. 2 compares BIS spectra for x=0, 0.2, 0.4, 0.6, 0.8, and 1.0. All the spectra are scaled to have the same magnitudes at the narrow peak at 15 eV. The spectrum for x=0 is very similar to those in the literature (van der Marel et al., 1987) in having a peak at $\sim$ 2 eV, which we assign to the Cu d^{10} final state, a wide band starting at $\sim$ 7 eV above E_F arising from a mixture of O 3s, Cu 4s, Ba 5d and Y 4d states, and a narrow peak at 15 eV due to the Ba $4f^0 \rightarrow 4f^1$ transitions. From this figure, it is clear that the spectral weight near E_F is essentially zero for all x, and that features near 2 eV and 7 eV increase as x increases. We have extracted the Pr spectra by subtracting the spectrum for x=0 from that for each x, after matching the magnitudes of the peaks corresponding to the Ba $4f^0 \rightarrow 4f^1$ transitions, assuming that highly atomic-like states are not affected by the Pr substitution. The areas under the Pr 4f spectral weights are roughly proportional to the Pr concentration x, with the exception of x=1.0. The area ratios are $A(x=0.2) : A(0.4) : A(0.6) : A(0.8) : A(1.0) = 0.19 : 0.38 : 0.60 : 0.80 : 0.82$. As with the oxygen 1s XPS spectra, we speculate that the different behavior for

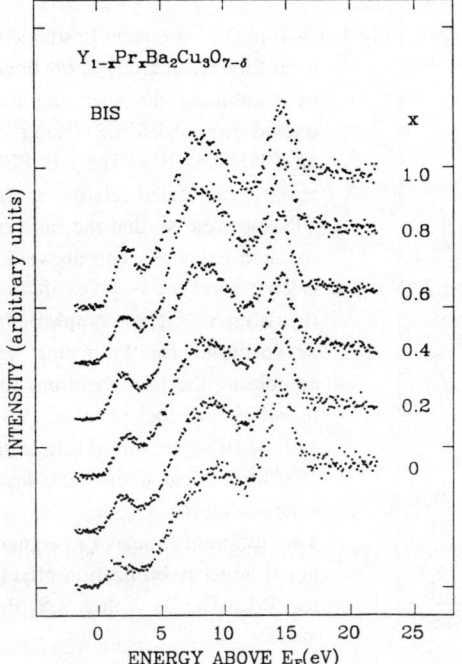

Fig. 2. The BIS spectra of $Y_{1-x}Pr_xBa_2Cu_3O_{7-\delta}$. The spectra are scaled to have the same magnitudes at the Ba $4f^0 \rightarrow 4f^1$ peak at 15 eV.

x=1.0 reflects the slight difference in its preparation. The extracted Pr spectral weight above E_F is shown in Fig. 3 for x=1. There are two features centered at 2 eV and at 7 eV, with complicated lineshapes and with essentially zero spectral weight at E_F. The spectra for other x are very similar, except that the intensity of the 2 eV feature decreases slightly relative to that of the 7 eV feature as x increases.

Pr 4f SPECTRUM

Qualitative Assignments

The top panel of Fig. 3 shows the complete Pr 4f spectral weight distribution for x=1.0, obtained by combining the spectra extracted from RESPES and BIS. The RESPES spectrum is scaled relative to the BIS spectrum so that the weight distribution has 2/14 of its area below E_F, corresponding to the Pr^{3+} state. For comparison, the bottom panel of the figure shows the analogous Pr 4f spectrum for Pr metal, obtained by combining BIS (Lang *et al.*, 1981) and PES data for $h\nu=80$ eV. (Wieliczka *et al.*, 1984) The spectrum is essentially that expected for a single-valent Pr^{3+} ion. Thus the vertical bars show the weights and positions of the final state multiplets of $4f^1$ and $4f^3$ configurations arising in a calculation of the $4f^2 \rightarrow 4f^1$ and $4f^2 \rightarrow 4f^3$ PES and BIS transitions using the method of coefficients of fractional parentage. (Lang *et al.*, 1981) The large 4f Coulomb interaction U_{ff} is shown by the 5 eV separation of the lowest PES ionization and BIS affinity lines. The PES spectrum also shows a small peak near E_F, which arises from hybridization between the 4f state and the conduction band, and can be

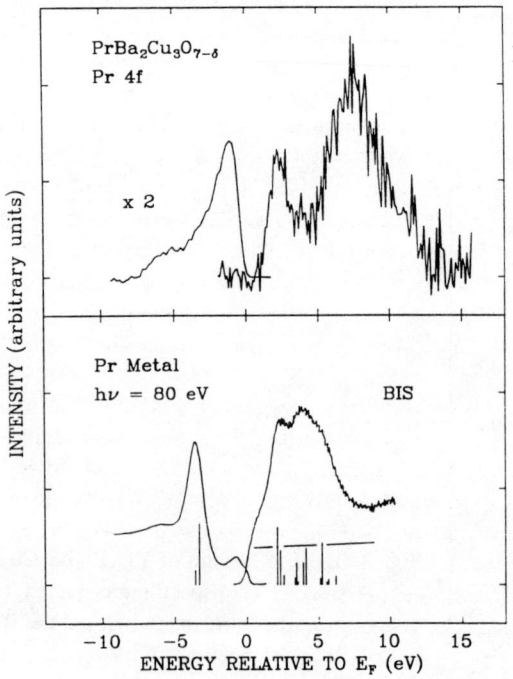

Fig.3. (a) Top: The complete Pr 4f spectrum for $PrBa_2Cu_3O_{7-\delta}$, obtained by combining the spectrum extracted from RESPES (Kang *et al*, 1988) and BIS. The RESPES spectrum is scaled relative to the BIS spectrum so that the ratio of the area below E_F and above E_F is 2:12.

(b) Bottom: The complete Pr 4f spectrum for Pr metal, by combining the PES spectrum for $h\nu=80$ eV (Wielickzka *et al*, 1984) and the BIS spectrum (Lang *et al*, 1981). Vertical lines show final state mulplets.

The difference near E_F signals much larger hybridization effects for $PrBa_2Cu_3O_{7-\delta}$ than for Pr-metal.

fit using the Anderson impurity model. (Gunnarsson and Schönhammer, 1983, 1985, 1987) More complex PES lineshapes in essentially trivalent Pr intermetallic compounds (Parks *et al.*, 1984; Sampathkumaran *et al.*, 1985) arise from hybridization to a structured conduction band and can also be fit with this model. (Gunnarsson and Schönhammer, 1985, 1987)

Considering this previous work on hybridization effects in Pr and its compounds, we interpret the PrBa$_2$Cu$_3$O$_{7-\delta}$ 4f spectrum as essentially trivalent, but with the BIS/PES peaks near E$_F$ signaling much larger hybridization effects than for Pr metal. We assign the 7eV BIS feature as the 4f$^2 \rightarrow$4f^3 transition. This interpretation implies a shift of the 4f^3 states by about 3 eV to higher energies, relative to the BIS spectrum of Pr metal, and in addition, a decrease by approximately 20% in the overall multiplet splitting, to about 4 eV at FWHM, from the metal value of about 5 eV. Such changes are known in compounds of the light lanthanides with transition metals (Laubschat *et al.*, 1987), and are most likely due to hybridization effects and to a larger 4f Coulomb interaction, as found in cerium oxides. (Nakano *et al.*, 1987)

Impurity Anderson Hamiltonian Analysis

We have attempted to quantify our interpretation by fitting the Pr 4f spectrum using the impurity Anderson Hamiltonian. This Hamiltonian models the Pr 4f state as a 14-fold degenerate local orbital, characterized by its binding energy ϵ_f relative to E$_F$, and its Coulomb interaction U$_{ff}$. The orbital is hybridized to a continuum density of states, with strength characterized by a parameter Δ_{av}. The Gunnarsson/Schönhammer (1985, 1987) calculation for Ce has been adapted approximately to Pr as described by them, with the 4f spin-orbit and multiplet splittings included only in simulating the BIS final states. Our fitting effort has been only partially successful. If we assume hybridization to a metallic density of states, the weights of the BIS and PES peaks near E$_F$ relative to those far from E$_F$ can be reproduced rather well with only a modest departure from trivalence. However, the theoretical spectra perforce show BIS and PES peaks meeting at E$_F$, whereas our experimental peaks lie away from E$_F$ somewhat. On the other hand, if we assume hybridization to a continuum with a small gap at E$_F$, we can produce theoretical 4f spectra which also display a gap, but to obtain a BIS feature near E$_F$ of adequate size, we must use such a large hybridization that the valence is driven nearly 50% away from f^2 (toward f^3, since f^1 is excluded with a filled valence band) and the PES spectrum also develops extra features not seen experimentally. A compromise scenario appears to be impossible within the framework of the impurity Anderson model.

The solid lines of panels (a) through (c) of Fig. 4 show, respectively, the 4f BIS, 4f PES, and 3d XPS spectra of our metallic mixed valence model. The inset of (b) shows the continuum density of states, with the important part below E$_F$ essentially the same as the Pr 4f off-resonance spectrum shown, where the Cu 3d emission dominates, but with an increased intensity between −2 eV and E$_F$ to account for the contribution from oxygen p-states. We note that interatomic spacings from Pr to both Cu and O permit strong hybridization, and Δ_{av} = 0.16 eV, intermediate between smaller values for α-like Ce intermetallics (Allen *et al.*, 1987) and larger values for cerium oxides. (Nakano *et al.*, 1987) The other model parameters

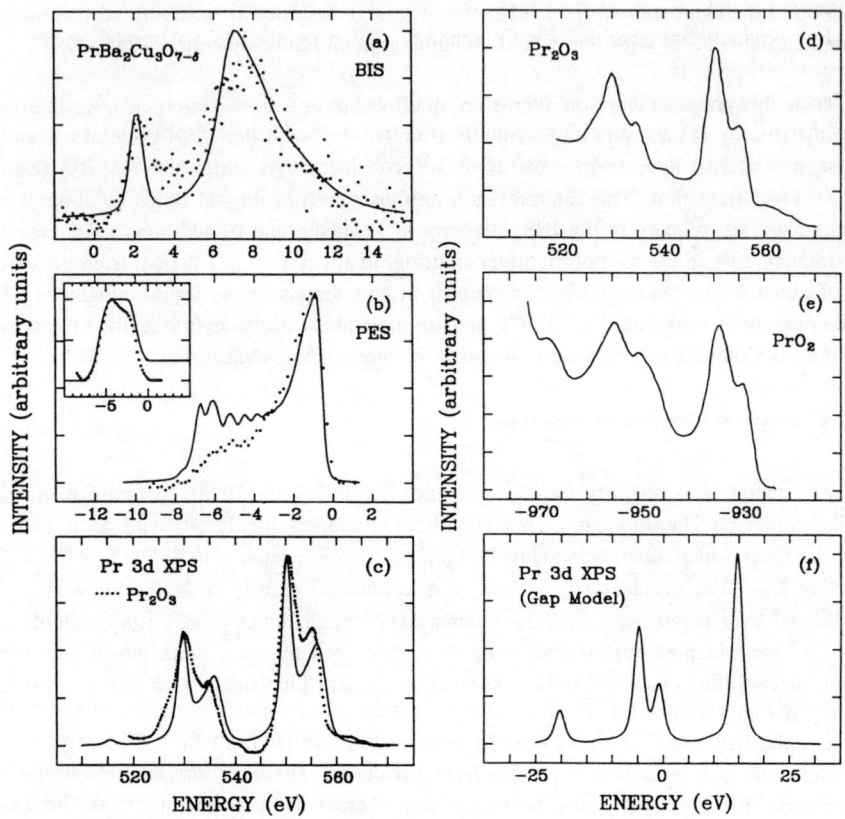

Fig. 4. The solid line and the dots in the inset of the panel (b) show, respectively, the continuum density of states and the Pr 4f off-resonance spectrum of $PrBa_2Cu_3O_{7-\delta}$. The solid lines in the panels (a), (b), and (c) show, respectively, the calculated 4f BIS, 4f PES, and 3d XPS spectra of our metallic mixed valence model. The theoretical PES and BIS spectra have been shifted away from E_F by 0.6 eV and 2 eV, respectively, and superposed on experimetnal spectra (in dots) of $PrBa_2Cu_3O_{7-\delta}$. The dots in the panel (c) is the 3d XPS spectrum of Pr_2O_3 (Burroughs, et al, 1976), with its inelastic background and non-monochromatic-source satellites (incompletely) removed.

Panels (d) and (e) show the 3d XPS spectra of Pr_2O_3 (Burroughs et al., 1976) on a kinetic energy scale, and PrO_2 (Bianconi et al., 1988) on a binding energy scale, respectively. Panel (f) shows the calculated 3d spectrum for our gapped-continuum model.

are $\epsilon_f = E(f^2 \to f^1) = -3$ eV, $U_{ff} = 8.45$ eV, and the 4f-3d Coulomb interaction $U_{fc} = 12.1$ eV. The theoretical PES and BIS spectra have been shifted away from E_F by 0.6 eV and 2 eV, respectively, and superposed on experimetnal spectra, producing the appearance of fits as good as typically are obtained for Ce spectra. (Allen *et al.*, 1987) Although the Pr 3d spectrum is obscured in $PrBa_2Cu_3O_{7-\delta}$ by the Cu 2p spectrum, as described above, we can guess that it should resemble that of Pr_2O_3, and be different from that of PrO_2, from the finding that the Pr L-edge XAS spectrum is very much like that of Pr_2O_3 and very much different from that of PrO_2. (Alp *et al.*, 1988) Panels (d) and (e) show the 3d XPS spectra of Pr_2O_3 (Burroughs *et al.*, 1976) and PrO_2 (Bianconi *et al.*, 1988), respectively. The Pr_2O_3 spectrum, with its inelastic background and non-monochromatic-source satellites at 540 eV and 561 eV (incompletely) removed numerically, is also shown in panel (c), and it is indeed much like that of our theory curve. Panel (f) shows the calculated 3d spectrum for our gapped-continuum model and it is obvious that it looks much different from that of Pr_2O_3 due to the strong f^2/f^3 valence mixing. It is not merely coincidence that it resembles the PrO_2 spectrum, which displays the effects of strong f^1/f^2 valence mixing, analogous to the nearly equal f^0/f^1 valence mixing in CeO_2. (Nakano *et al.*, 1987)

The metallic continuum model reproduces the various 3d and 4f spectral signatures generally thought to measure the Pr 4f valence and hybridization strength. The calculated ground state is 4.2% f^1, 89.9% f^2 and 5.9% f^3, giving 2.02 f-electrons. The seriousness or significance of the problem that we cannot reproduce the experimental 4f gap around E_F is not clear. Of great concern is the possibility that the gap is an experimental artifact associated somehow with the use of ceramic samples, or of surface oxygen loss, or a charging effect induced in some unknown way. For example, Arko *et al.*(1988) reported recently that single crystals of $YBa_2Cu_3O_{7-\delta}$ cleaved at 20K, where oxygen diffusion is very slow, show much larger photoemission intensity at E_F than we find in our room temperature measurements of ceramic samples for x=0. Fujimori *et al.* (1988) suggested that oxygen-vacancy disorder in these low carrier density materials can produce a Coulomb gap at E_F, which might be simulated by a shift. Our present inclination is to give credence to our characterization of the Pr valence and hybridization strength, while working to resolve the experimental uncertainties concerning the gap around E_F.

DISCUSSION

A suitable model for $Y_{1-x}Pr_xBa_2Cu_3O_{7-\delta}$ must account for the magnitude of the Pr magnetic moment and the change from a superconductor to a semiconductor with increasing x. We have found that the Pr valence is close to 3+ for all x, consistent with XAS results and with lattice parameter measurements. In addition, the extracted Pr 4f spectral lineshape indicates much hybridization to other valence band states. We have previously discussed (Kang *et al.*, 1988) various alternatives for explaining the Pr moment and concluded that its reduced value below that expected for Pr^{3+} must be attributed to the large hybridization but that the exact mechanism is not clear. Our findings also cast doubt on the possibility that the transformation from a superconductor to a semiconductor occurs merely by the charge-transfer mechanism. It is natural to speculate instead that the extensive hybridization between Pr 4f and other valence band states may have disrupted some features of the electronic or magnetic structure of the x=0 material which is essential for superconductivity. This mechanism would apply particularly to the rare earths Ce, Pr and Tb for which the 4f states have small 3+ ionization energies (Lang et al., 1981) degenerate with the other valence band states, a situation favoring strong hybridization.

Expanding our previous discussion (Kang et al., 1988), we suggest here that it is the Pr magnetic degrees of freedom that are most important. Much of the theoretical thinking within correlated electron models for the electronic structure of the Cu-O planes leads to the conclusion that the electrical behavior of holes in the Cu-O planes is determined in one way (Anderson, P.W. 1987) or another(Birgeneau et al., 1988; Hirsch et al., 1988; Stechel and Jennison, 1988) by Cu-O and/or Cu-Cu magnetic interactions. From an insulator point of view, it is expected that hybridization between Pr 4f states and the 2p states of the O atoms in the adjacent Cu-O planes leads to superexchange interactions between Pr and Cu or O moments which would have the potential to profoundly alter the behavior that would occur in their absence. Alternatively, from the view of the metallic model, there occur Kondo spin fluctuations with an associated energy scale of T_K. Although our model calculation is very unrealistic for T_K in neglecting spin-orbit and multiplet splittings, it is suggestive that the value of T_K obtained is about 125 K, the same order of magnitude as T_c. $T_c \sim T_K$ is the condition for maximum suppression of T_c by a magnetic impurity in a BCS superconductor with singlet pairing (Müller-Hartmann and Zittartz, 1970) and may have generic significance in the present case.

ACKNOWLEDGEMENTS

It is a pleasure to thank D. Fenner and the Xerox Palo Alto Research Center for allowing us to use their equipment to perform BIS and XPS measurements. We are deeply indebted to O. Gunnarsson for allowing us to use his computer program and for his help in checking that it runs correctly. The Stanford Synchrotron Radiation Laboratory (SSRL) and the work at Los Alamos National Laboratory are supported by the U.S. Department of Energy. We thank the staff at SSRL for skillful technical assistance. Research support by the U.S. National Science Foundation through Low Temperature Physics Grants Nos. DMR-87-21654 (J.W.A., J.-S.K.) and DMR-87-21455 (M.B.M., and B.W.L.), and through the National Science Foundation-Materials Research Laboratory Program at the Center for Materials Research at Stanford University, is gratefully acknowledged, as is also support at Stanford University from Air Force Contract AFSOFR-87-0389.

REFERENCES

Allen, J.W., S.-J. Oh, O. Gunnarsson, K. Schönhammer, M.B. Maple, M.S. Torikachvili, and I. Lindau (1987). Adv. in Phys. **35**, 275-316.

Alp, E.E., G.K. Shenoy, L. Soderholm, G.L. Goodman, D.G. Hinks, and B.W. Veal (1988). Mat. Res. Soc. Symp. Proc. **99**, 177-182.

Anderson, P.W. (1987). Science **235**, 1196-1198.

Arko, A.J., R.S. List, Z. Fisk, S.-W. Cheong, J.D. Thompson, J.A. O'Rourke, C.G. Olson, A.-B. Yang, T.-W. Pi, J.E. Schriver, and N.D. Shinn (1988). J. Less. Common. Metals **24 & 25**, to be published.

Bianconi, A., A. Kotani, K. Okada, R. Giorgi, A. Gargano, A. Marcelli, and T. Miyahara (1988). Phys. Rev. **B 38**, 3433-3437.

Birgeneau, R.J., M.A. Kastner, and A. Aharony (1988). Z. Phys. **B 71**, 57–62.

Burroughs, P., A. Hamnett, A. F. Orchard, and G. Thornton (1976). J. Chem. Soc., Dalton Trans. **17**, 1686-1698.

Dalichaouch, Y.D., M.S. Torikachvili, E.A. Early, B.-W. Lee, C.L. Seaman, K.N. Yang, H. Zou, and M. B. Maple, (1987). *Solid State Commun.* **65**, 1001-1006.

Fujimori, A., K. Kawakami, and N. Tsuda (1988). Phys. Rev. **B 38**, 7889-7892.

Gerken, F. (1982). *Ph. D. Thesis*, University of Hamburg, Germany.

Gunnarsson, O., and K. Schönhammer (1983). Phys. Rev. **B 28**, 4315-4341.

Gunnarsson, O., and K. Schönhammer (1985). Phys. Rev. **B 31**, 4815-4834.

Gunnarsson, O., and K. Schönhammer (1987). In: *Handbook on the Physics and Chemistry of Rare Earths* (K.A. Gschneidner, L. Eyring, and S. Hüfner, ed.), Vol. **10**, Chap. 64, pp. 103-163. North-Holland, Amsterdam.

Hirsch, J.E., S. Tang, E. Loh, and D.J. Scalapino (1988). Phys. Rev. Lett. **60**, 1668-1671.

Horn, S., J. Cai, S.A. Shaheen, Y. Jeon, M. Croft, C.L. Chang and M.L. denBoer (1987). Phys. Rev. **B 36**, 3895-3898.

Kang. J.-S., J.W. Allen, Z.-X. Shen, W. Ellis, J.J. Yeh, B.W. Lee, M.B. Maple, W.E. Spicer, and I. Lindau (1988). J. Less. Common. Metals **148 & 149**, to be published.

Kohiki. S., T. Hamada, and T. Wada (1987). Phys. Rev. **B 36**, 2290-2293.

Kohiki. S., T. Hamada, and T. Wada (1987). Phys. Rev. **B 36**, 2290-2293.

Laubschat, C., W. Grents, and G. Kaindl (1987). Phys. Rev. **B 36**, 8233-8236.

Liang, J.K., X.T. Xu, S.S. Xie, G.H. Rao, X.Y. Shao, and Z.G. Duan (1987). Z. Phys. **B 69**, 137-140.

Lytle, F., R. Greegor, E. Marques, E. Larson, J. Wong, and C. Violet (1988). preprint.

Muller-Hartmann, E. and J. Zittartz (1970). Z. Physik **234**, 58-69.

Nakano, T., A. Kotani, and J.C. Parlebas (1987). J. phys. Soc. Japan. **56**, 2201-2210.

Neukirch, N., C.T. Simmons, P. Sladeczek, C. Laubschat, O. Strebel, G. Kaindl, and D.D. Sarma (1988). Europhysics Lett. **5**, 567-

Neumeier, J.J., M.B. Maple, and M.S. Torikachvili (1988). to be published.

Nücker. N., J. Fink, J.C. Fuggle, P.J. Durham, and W.M. Temmerman (1988). Physica C **119**, 153-159.

Parks, R.D., S. Raaen, M.L. denBoer, Y.-S. Chang, and G.P. Williams (1984). Phys. Rev. Lett. **52**, 2176-2179.

Sampathkumaran, E.V., G. Kaindl, C. Laubschat, W. Krone, and G. Wortmann (1985). Phys. Rev. **B 31**, 3185-3187.

Sawatzky, G.A. and J.W. Allen (1985). Phys. Rev. Lett. **53**, 2339-2342.

Shen, Z.-X., J.W. Allen, J.-J. Yeh, J.-S. Kang, W. Ellis, W.E. Spicer, I. Lindau, M.B. Maple, Y.D. Dalichaouch, M.S. Torikachvili, J.Z. Sun, and T.H. Geballe (1987). Phys. Rev. **B 36**, 8414-8428.

Shen. Z.-X., P.A.P. Linberg, W.E. Spicer, I. Lindau, and J.W. Allen (1988a). *AVS Conf. Proc. AVS, 1988, Atlanta.*

Shen. Z.-X., P.A.P. Linberg, B.O. Wells, D.B. Mitzi, I. Lindau, W.E. Spicer, and A. Kapitulinik (1988b). to be published.

Shih, C.K., Z.-X. Shen, P.A.P. Linderberg, J.Hwang, I. Lindau, W.E. Spicer, and J.W. Allen (1988). to be published.

Soderholm, L., K. Zhang, D.G. Hinks, M.A. Beno, J.D. Jorgensen, C.U. Segre, and I.K. Schuller (1987). *Nature* **328**, 604-605.

Stechel, E.B. and D.R. Jennison (1988). Phys. Rev. **B 38**, 4632-4659.

Takahashi. T., F. Maeda, H. Katayama-Yoshida, Y. Okabe, T. Suzuki, A. Fujimori, S. Hosoya, S. Shamoto, and M. Sato (1988). Phys. Rev. **B 37**, 9788-9791.

van der Marel, D., J. van Elst, G.A. Sawatzky, and D. Heitmann (1988). Phys. Rev. **B 37**, 5136-5141.

Wieliczka, D.M., C. G. Olson, and D.W. Lynch (1984). Phys. Rev. Lett. **52**, 2180-2182.

Yang, K.N., Y. Dalichaouch, J.M. Ferreira, R.R. Hake, B.W. Lee, J.J. Neumeier, M.S. Torikachvili, H. Zhou, and M.B. Maple (1987). Jap. J. of Appl. Phys. supplement **26-3**, 1037-1038.

Yang, K.N., B.W. Lee, M.B. Maple, and S.S. Laderman (1988). to be published in Applied Physics.

Zaanen, J., G.A. Sawatzky, and J.W. Allen, (1985). Phys. Rev. Lett. **55**, 418-421.

ELECTRONIC STRUCTURE STUDIES OF Bi-Sr-Ca-Cu-O SINGLE CRYSTAL SUPERCONDUCTOR

Zhen Xiang Liu, Dian Hong Shen, Kan Xie, Shang Xue Qi, Hong Wei Xu and Nai Juan Wu

Laboratory for Surface Physics and Institute of Physics,
Chinese Academy of Science, Beijing, China

ABSTRACT

The valence band electronic structure of the Bi-Sr-Ca-Cu-O single crystal compound was studied by XPS and UPS, the changes of the valence band electronic structure with sputtering and annealing were examined. The density of states at the E_F edge is low, there are three features near the E_F centered around 2, 3.5 and 5.5 eV which arise primarily from Cu3d-O2p derived states. Their relative intensity changes with annealing and sputtering. The controversial features of ~ 9.7 and 531 eV in photoemission spectra are also investigated.

The electronic structure of high T_c superconductors are one key factor explaining the superconducting mechanism. There have been a number of electronic structure studies on Y(Gd or Eu)-Ba-Cu-O and La-Sr-Cu-O compounds (Samsavar *et al.*, 1988, Van der Marel *et al.*, 1988, Frommer, 1988, Ming Tang *et al.*, 1988). In general, these compounds contain Cu cations, an alkaline earth, and rare earth elements or yttrium. Recently, a new member of the high T_c superconductors, Bi-Sr-Ca-Cu-O compound has been prepared (Chu *et al.*, 1988, Michel *et al.*, 1987), which not only has high transition temperature (above 100K), but also does not contain rare earth elements or yttrium. To get more insight into its electronic structure, we performed XPS and UPS photoemission experiments on a cleaved surface of Bi-Sr-Ca-Cu-O single crystal compound with transition temperature at about 86K. We observed the changes of the UPS and XPS valence band spectra of the Bi-Sr-Ca-Cu-O superconductor with sputtering and annealing at different temperature. The XPS and UPS experiments were carried out under the pressure of 10^{-9}-10^{-10} torr on VGESCAlab5, AlK$_\alpha$, HeI and HeII are used as exciting source for XPS and UPS measurements respectively. The sample is sputtered with an Ar$^+$ ion beam of 2 kV and 5μA.

Fig. 1 shows the XPS valence band spectra of the Bi-Sr-Ca-Cu-O single crystal compound. For comparison, the valence band spectra of CuO, SrO and Bi_2O_3 are also shown in this figure.

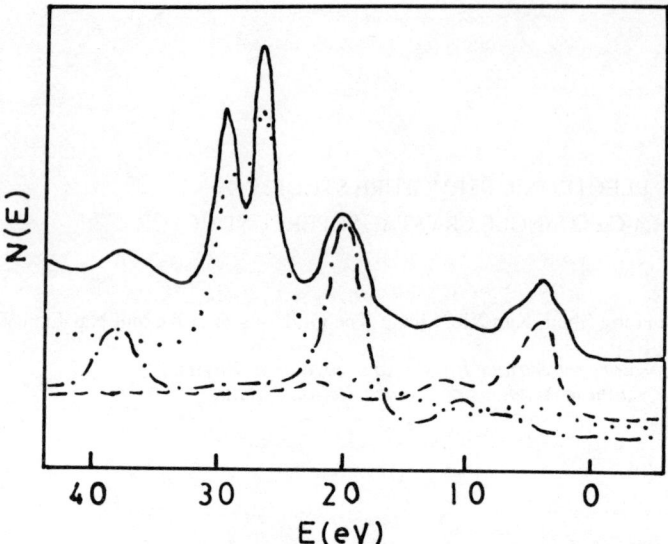

Fig. 1. XPS valence band spectra taken with Al K_α source for: ——— Bi-Sr-Ca-Cu-O compound after sputtering 0.5 min.; ---- CuO; —·—·— SrO; ······ Bi_2O_3.

Owing to the intensity of the valence band spectra near E_F is weak for SrO and Bi_2O_3, the valence band spetctra are not in the same scale. The spectra of Bi-Sr-Ca-Cu-O compound exhibits several core level peaks, 38, 29.4, 26.3 and 20 eV below E_F, comparing with the spectra of pure oxides, it is evident that, the first peak is due to the Sr4s, the second and third peaks arise from Bi 5d orbitals, the fourth peak is related to the Sr 4p levels. We can also see a shallower and broader feature centered around 10-11 eV below E_F, which may corresponds to a peak at 9.5 eV below E_F for Y-Ba-Cu-O and La-Sr-Cu-O compounds (Samsavar et al., 1988, Stoffel et al., 1988). This feature can not be attributed to any of the Y-Ba-Bi-Sr-Ca-Cu derived states basing on the known binding energies and calculated band structure of these materials. Therefore, there is a considerable controversy about this feature. Some authors attribute this feature to the contamination of carbon or water; some others identified this feature as an intrinsic electronic feature related to oxygen multihole states (Samsavar et al., 1988, Stoffel et al., 1988). We have carefully studied this feature by UPS, which comprises two peaks centred around 9.7 and 11.8 eV below the Fermi level as shown in Fig. 2 and 3. This feature will be discussed in some detail latter.

The electronic structure near the E_F is close related with the superconducting mechanism, so people are more interested in the valence spectra in this region. There are several features at about 2, 3.5 and 5.5 eV below the E_F for Bi-Sr-Ca-Cu-O compound, as shown in Fig.1, comparing with the valence band spectra of CuO, SrO and Bi_2O_3, it can be noted that the valence band feature near the E_F, arises primarily from the Cu3d-O2p derived states, little parts arises from Bi6p-O2p and Sr5s-O2p derived states. Or these features arising from the hybridization between Cu 3d, Bi 6p, Sr 5s, and the various type of O 2p orbitals.

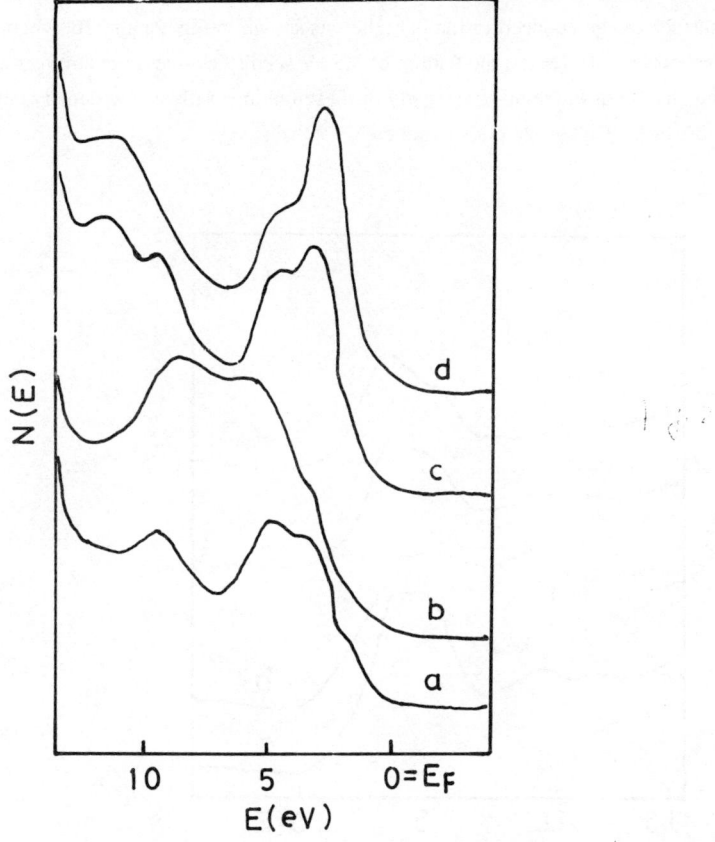

Fig. 2. HeI UPS spectra of Bi-Sr-Ca-Cu-O compound after various treatments. a) sputtering 0.5 min. b) annealing at 300°C for 25 min. c) annealing at 400°C for 25 min. d) sputtering 2 min., and then annealing at 500°C for 30 min.

In order to understand the valence band structure of Bi-Sr-Ca-Cu-O compound, we studied the valence band spectra using UPS, since its energy resolution is much better than that of XPS. Fig. 2 and Fig. 3 show the HeI (21.2 eV) and HeII (40.8 eV) UPS spectra of Bi-Sr-Ca-Cu-O compound after various treatment, respectively. Three points should be noted from these valence band spectra: 1) the spectral intensity near the E_F edge is small. 2) All these spectra near E_F exhibit three features around 2, 3.5 and 5.5 eV, the valence band spectra taken on Y-Ba-Cu-O and La-Cu-O compounds exhibit only two features in the same energy region, 2.5

and 4.5 eV below E_F (Steiner *et al.*, 1987). 3) The relative intensity of the three features changes obviously with various treatments. Comparing with the surface oxygen concentration obtained by quantitative XPS analysis, we noted that the feature of 5.5 and 2 eV seems more stronger when the surface oxygen concentration is higher, as shown in Fig. 2a and 2b; when the surface oxygen concentration decreases, the feature of 3.5 eV seems become more stronger as shown in Fig. 2d. The change of the relative intensity of these features reflects the density change of the occupied state of the Cu3d-O2p derived states.

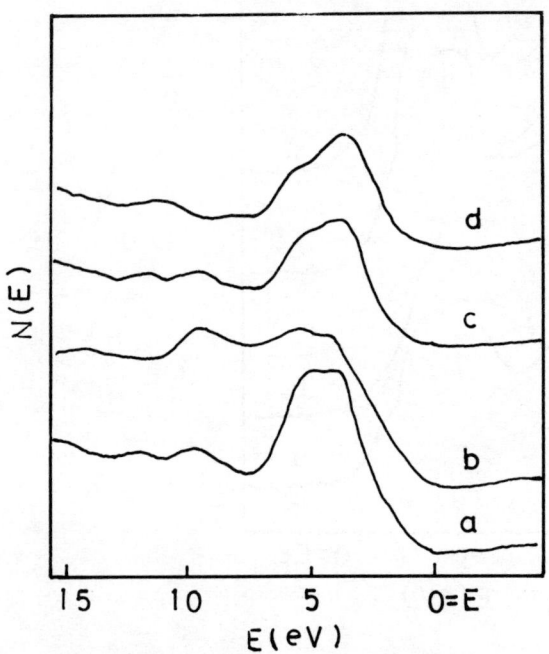

Fig. 3. HeII UPS spectra of Bi-Sr-Ca-Cu-O compound after various treatments. a-d) the experimental conditions are the same as in Fig. 2.

We can also see two features of ~ 9.7 and 11.8 eV in the HeI and HeII UPS spectra in Fig. 2 and 3, their relative intensity also sensitive to the various treatments. To get more insight into these features. We have carefully studied the relation between these features and O 1s as well as C 1s XPS spectra of Bi-Sr-Ca-Cu-O compound. Fig. 4 shows the O 1s spectra of Bi-Sr-Ca-Cu-O compound before and after various treatments. We observed a double peak structure of O 1s spectra centered around 528.8 and 531 eV, and the high binding energy O 1s (531 eV) is strong for the original surface, the 531 eV O 1s spectra decreases after sputtering (see Fig. 4a) and increase again after annealing at 300°C for 20 min. (Fig. 4b). However, further annealing at the temperature above 400°C, this peak decreases or disappears (Fig. 4c and 4d). Comparing Fig.2 and 3 with Fig. 4, we find that the feature, ~ 9.7 eV seems closed related with the high binding energy O 1s spectra, combining with the observation of C 1s XPS spectra, we suppose that both of the features of ~ 9.7 and 531 eV arise primarily from $CO_3^=$, especially when the intensity of these features are strong. Since the Sr^{2+} and Ba^{2+} cations in oxides or compounds near the surface region are contaminated easily by CO_2 in air forming $SrCO_3$- and $BaCO_3$-like complex.

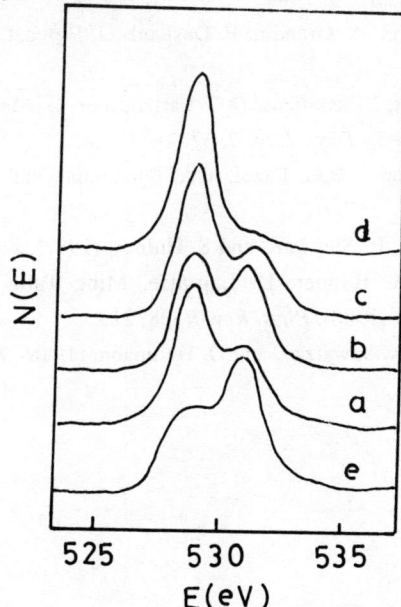

Fig. 4. O 1s XPS spectra of Bi-Sr-Ca-Cu-O compound before and after various treatments. a-d) the experimental conditions are the same as in Fig. 2; e) original surface.

The feature at 11.8 eV exihibits the intrinsic feature of copper-based oxides, especially when the surface oxygen concentration decreases, this feature seems become strong as shown in Fig.2d and Fig.3d

CONCLUSION

In summary, the valence band electronic structure of the Bi-Sr-Ca-Cu-O single crystal compound was studied using XPS and UPS. There are three features near the E_F centered around 2, 3.5 and 5.5 eV arise primarily from the hybrid states of the Cu 3d and O 2p states, their relative intensity changes with sputtering and annealing. The feature of ~ 9.7 eV and high binding energy O 1s spectra seems arise mainly from $CO_3^=$ caused by the contamination of CO_2

in air. More work is in progress.

REFERENCES

Chu, C.W., J.Bechtold, L. Gao, P.H. Hor, Z.J. Huang, R.L. Meng, Y.Y. Sun, Y.Q. Wang, and Y.Y. Xwe, (1988). *Phys. Rev. Lett.*, 60, 941.

Frommer, M.H., (1988). *Phys. Rev. B*, 38, 2444.

Michel, C., M. Hervieu, M.M. Borel, A. Grandin, F. Deslands, J. Provost, and B. Raveau, (1987). *Z. Phys. B*, 68, 421.

Ming Tang, Y. Chang, M. Onellion, J. Seuntjens, D.C. Larbalestier, G. Margaritondo, N.G. Stoffel, and J.M. Tarascon, (1988). *Phys. Rev. B*, 37, 1611.

Samsavar, A., T. Miller, T.C. Chiang, B.G. Pazol, T.A. Fnedmann, and D.M. Ginsberg, (1988). *Phys. Rev. B*, 37, 5164.

Steiner, P., V. Kinsinger, I. Sander, B. Siegwart, and S. Hufner, (1987). *Z. Phys. B*, 67, 19.

Stoffel, N.G., P.A. Morris, W.A. Bonner, D. Lagraffe, Ming Tang, Y. Chang, G. Margaritondo, and M. Onellion, (1988). *Phys. Rev. B,* 38, 213.

Van Der Marel, D., J. Van Elp, G.A. Sawatzky, and D. Heitmann, (1988). *Phys. Rev. B*, 37, 5136.

X-RAY SPECTROSCOPY

OXYGEN-DEPENDENT ELECTRONIC STRUCTURE IN POWDER AND SINGLE CRYSTAL OF $YBa_2Cu_3O_{7-\partial}$ OBSERVED IN-SITU BY X-RAY ABSORPTION SPECTROSCOPY

H.Tolentino, E.Dartyge, A.Fontaine, G.Tourillon
LURE Bât.209d 91405 Orsay - France

T.Gourieux, G.Krill
Lab.de Phys.des Solides UA155 CNRS,BP239 54506 Vandeuvre-les-Nancy - France

M.Maurer, M-F.Ravet
Lab. CNRS-Saint Gobain, 54703 Pont-a-Mousson - France

ABSTRACT

By using X-ray Absorption Spectroscopy (XAS) in dispersive mode, we have studied the evolution of the Cu K edge upon changing the **in situ** oxygen stoichiometry of $Y_1Ba_2Cu_3O_{7-\partial}$ powder and single crystal compounds. Upon the initial increase of ∂ starting with the highly oxygenated superconducting phase we observed directly the immediate transformation of the $|3d^9(\underline{L})>$ into the $|3d^9>$ configuration. The $|3d^9(\underline{L})>$ configuration is caused by p holes with z symmetry, as evidenced by the orientation-dependent difference signal. Furthermore, even before the compound becomes semiconducting, in the range of ∂ from 0.2 to 0.3, a dramatic amount of monovalent copper is produced. This means that the hole concentration in that range is greater than that given by the chemical formula. When the compound reaches a poorly oxygenated semiconducting phase a large amount of monovalent copper, i.e. $|3d^{10}>$ configuration is present.

KEYWORDS

High T_c Superconductors; XAS; X-ray absorption spectroscopy

The $Y_1Ba_2Cu_3O_{7-\partial}$ superconductor offers the unique possibility of varying continuously the electronic properties from a non-magnetic superconducting metal to an antiferromagnetic insulator, by only changing the oxygen stoichiometry ∂ (Müller and Olsen,1988). The orthorhombic to tetragonal transformation around $\partial \approx 0.6$ (Jorgensen et al., 1987; Rossat-Mignod et al., 1988) implies the breakdown of the long range ordering of the oxygen along chains in the Cu(1) plane. Both the number and the ordering (Hodeau et al., 1988; Werder et al., 1988; Tokumoto et al., 1987; Cava et al., 1987) of oxygen vacancies in the Cu(1) plane change throughout the oxygen loading. The other stricking feature when passing into the metallic state is the drastic reduction of the Cu(2)-O(1) distance from 2.47 to 2.30 Å (the O(1) site being between Cu(1) and Cu(2) sites along the z axis) (Beech et al., 1987). In addition, the unusually short distance Cu(1)-O(1) for a fourfold coordination is encountered only in formally trivalent compounds, like $NaCuO_2$.

High energy spectroscopies, including XPS Cu 2p and O 1s levels (Nücker et al., 1987; Gourieux et al., 1988; Fuggle et al., 1988; Fujimori et al., 1987; Marel et al., 1988), XAS at Cu K (Baudelet et al., 1987; Horn et al., 1988; Kosugi et al., 1988) and L_3 edges (Bianconi et al,1987 and 1988; Nücker et al., 1988) and O K edge (Nücker et al., 1988; Kuiper et al., 1988) and resonant photoemission (Thiry et al., 1988), have been extensively applied in order to study the electronic ground state of these compounds. These former investigations have lead to the following clear-cut conclusions:

i) The dedoped system has features of a Mott-Hubbard semiconductor with a gap of charge transfer type (Zaanen et al., 1985) like the antiferromagnetic oxides of 3d elements. The high correlation $U_{dd}/W > 1$ yields a ground state made of an empty narrow d-like band separated from a broad oxygen p-type full band located just below the Fermi level (Birgeneau et al., 1988; Eskes and Sawatzky, 1988). Additional complexities come from the mixing of chains and sheets and the change in oxygen stoichiometry which is used as a doping process.

ii) the charge compensation under doping is not given by a $|3d^8>$ copper configuration but it occurs primarily by hole injection in the O 2p band.

iii) The Cu L_3 spectroscopy points out the anisotropy of the $|3d^9(\underline{L})>$ configuration ($|3d^9(\underline{L})>$ meaning the ocurrence of a hole in the narrow d-like band and a hole ($\underline{L}$) injected in the oxygen p-band). The difference between the high energy tail-like x,y-polarized and the line-type z-polarized additional final states should reflect both the degree of hybridization of $|3d^9>$ and $|3d^{10}\underline{L}>$ configurations ($\underline{L}$ for charge transfer) and the localized/delocalized nature of the doping-induced holes ($\underline{L}$). Due to shorter Cu-O bond lengths the hybridization is larger for the chains, meanwhile holes in chains are expected to be localized. This view is well supported by the results on formally trivalent insulators such as $NaCuO_2$ and $LaLi_{1/2}Cu_{1/2}O_4$.

THE COPPER K-EDGE OF SOME SIGNIFCANT STANDARDS COMPOUNDS

In order to provide a frame to the analysis of our results on the Cu K edge spectroscopy of the $Y_1Ba_2Cu_3O_{7-\partial}$ compound, let us summarize some of the main features of well characterized standard materials. Figure 1 shows the copper K edge spectra of a) Cu_2O, b) CuO and c) $NaCuO_2$ whose formal oxidation states are Cu(I), Cu(II) and Cu(III), respectively (througout this paper all the energies refer to the first inflection point of the absorption spectrum of the metallic copper, i.e. E_o=8976.8eV). Differences between them (d) $CuO-Cu_2O$ and e) $NaCuO_2-Cu_2O$) are also displayed. The structure at 2.5 eV present in the Cu(I) spectrum reflects the dipolar allowed transition from the 1s state to the p_x and p_y degenerated empty states.

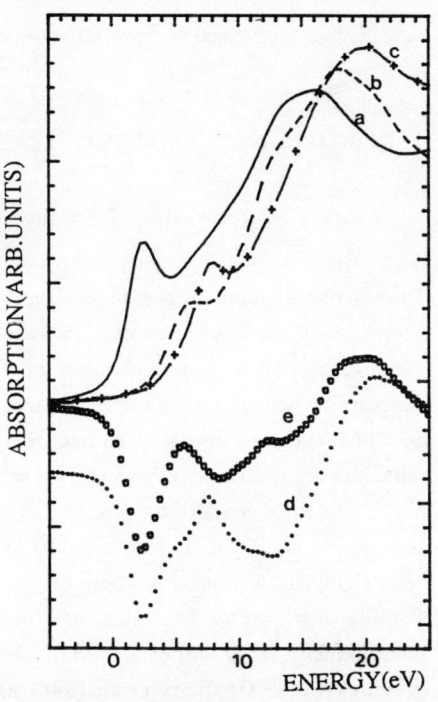

Fig.1: Copper K-edge XAS spectra of a) Cu_2O, b) CuO and c) $NaCuO_2$. The differences between the monovalent spectrum and the others (d,e) are plotted, showing that the $|3d^{10}>$ configuration is very well separated from the other features.

According to a well-documented and systematic survey of XAS in many copper compounds (Smith et al.,1985), this structure is always observed in linearly coordinated copper systems and its position comes from the $|3d^{10}>$ configuration in the ground state. As can be seen, this structure is well separated from the others appearing in the spectra of the other Cu compounds with unfilled d-band.

The energy shift between these valencies reflects the difference between the final state configuration of these cations undergoing the X-ray absorption and is essentially due to the Coulomb interaction between the core hole $\underline{c}$ and the electrons in the d copper band (U_{cd}). In addition, holes in the d band enables the charge transfer between the atom and the ligand, the so-called shake down in the photoabsorption process. This situation exists in the case of Cu(II) where the final state is a mixture of $|3d^9>$ and $|3d^{10}\underline{L}>$ configurations. The correlation between the electrons in the **d** band (U_{dd}) has to be included as soon as more than one hole exists in that band. Such a complexity does not exist in the present case.

IN-SITU UPTAKE AND REMOVAL OF OXYGEN UPON ANNEALING

So far, to our knowledge, all the investigations on $Y_1Ba_2Cu_3O_{7-\partial}$ compounds have been carried out with samples whose stoichiometry was controlled stepwise by quenching. The aim of our present work is to investigate the Cu valence admixture by XAS upon changing continuously the oxygen content in the same sample. The Dispersive X-ray Absorption Spectrometer at LURE (Tolentino et al.,1987) enables these **in situ** time-resolved experiments with an energy resolution of about 1 eV at 9 keV provided by an asymmetrically cut Si (311) curved crystal. The stability of the whole system yields a great sensitivity and reliability in the difference XAS spectra between two states corresponding to different values of ∂. Thus, the kinetics of the process can be followed and subtle variations of the electronic structure versus ∂ will not be masked by sample dependent effects. The stoichiometry ∂ has been changed by controlling both the oxygen partial pressure and the temperature with a stability of $\partial T \approx \pm 1°C$. The values of ∂ have been determined from the thermogravimetry measurements of Kishio et al.(1987). The data for 1 atmosphere oxygen pressure in our previous work (Gourieux et al.,1988) agrees with these measurements. These **in-situ** experiments have been performed on both powder and single crystal samples, starting in the high oxygenated superconducting phase.

Powder Sample: In the first step a well loaded sample ($\partial \approx 0.05$) with a narrow superconducting transition at 92 K was progressively heated from room temperature (RT) up to 600 °C under 1 atmosphere of O_2 (i.e. up to $\partial \approx 0.20$). In this range the sample remains in the 90 K phase. The two XAS spectra, corresponding to the sample at RT and 450°C, are shown in figure 2-a and 2-b, respectively. The difference between them (fig.2-d) displays two maxima at about 4.5 and 10.2 eV, which increase progressively with temperature. This difference compares very well with the derivative of the XAS spectrum from a $Y_1Ba_2Cu_3O_{6.95}$ sample (fig.3) and is due to a shift of the spectrum to lower energies.

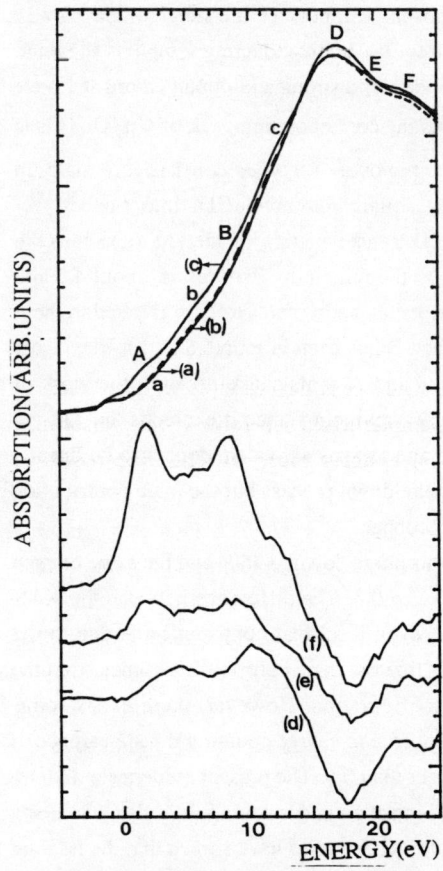

Fig.2: In situ kinetics of oxygen removal from $Y_1Ba_2Cu_3O_{7-\partial}$ powder sample: under 1 atm of O_2 a) at RT and b) at 450°C; and c) at 600°C under 10^{-4} Torr. Difference between XAS spectra under different ∂ values: d) sample annealed from room temperature to 450°C under 1 atm of O_2; oxygen partial pressure varying from 1 atm to 10^{-4} Torr for sample at e) 450°C and at f) 600°C.

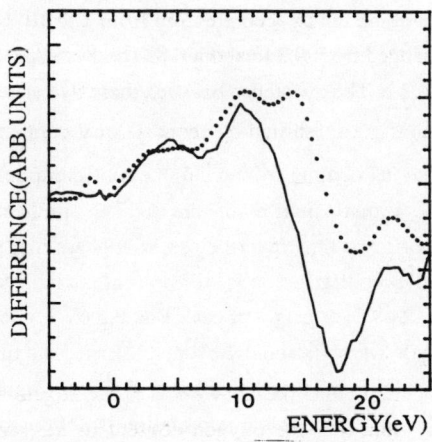

Fig.3: Derivative of the copper K-edge XAS spectrum of $Y_1Ba_2Cu_3O_{6.95}$ (....). The difference signal (___) is a derivative-like one and corresponds to a shift of about 1.2 eV between the two spectra.

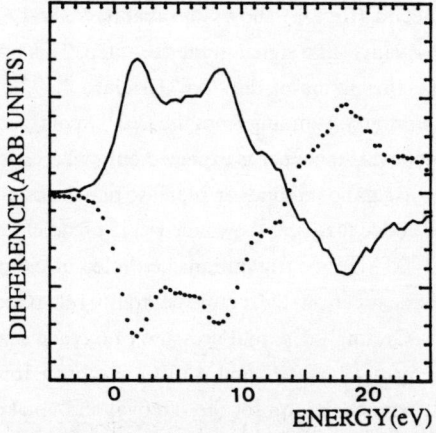

Fig.4: Difference signal corresponding to the oxygen removal (___) and uptake(....) showing that the process is fully reversible.

The most straightforward explanation is that, when ∂ increases, a small fraction of the holes disappears, corresponding to a transformation of the high energy $|3d^9(\underline{L})\rangle$ configuration into the $|3d^9\rangle$ one. A simulation of this difference spectra reveals a shift $\partial E \approx 1.2$ eV, which is consistent with the difference ($\partial E \approx 1.5$ eV) evaluated by Cu 2p XPS and L_3 XAS. By changing the oxygen pressure from 1 to 10^{-4} atmosphere at 600 °C, the stoichiometry ∂ was changed from 0.2 to about 0.8. The XAS spectrum for the semiconducting sample is shown in fig.2-c. The difference between these two states evidences a strong additional feature at 1.6 eV (**A**) (fig.2-f), shifted by about -0.8 eV compared to the corresponding peak of Cu_2O. This is a signal coming from a $|3d^{10}>$ configuration. The removal of oxygen consists essentially in the transformation of the $|3d^9>$, induced by square planar Cu(II), into the $|3d^{10}>$ configuration. Nevertheless, it appears that partial transformation of the $|3d^9(\underline{L})>$ into the $|3d^9>$ configuration is still present, as indicated by the remaining features at about 4.5 and 10.2 eV. The origin of peak **B** at 8.6 eV is unclear. Its intensity seems to be correlated to the **A** peak, which is about 7 eV apart. Based on this fact, **B** has been interpreted as a π-like direct transition in the chains for a $|3d^9>$ configuration and **A** as its shake-down in the work of Kosugi et al.. This interpretation is not completely convincing since it excludes the $|3d^{10}>$ configuration necessary for charge compensation and assigns a divalent copper to be linearly coordinated. Peak **A** could be partially due to a shake-down process but the main contribution is coming from the existence of 'real' monovalent copper.

After, the sample was cooled under oxygen atmosphere down to 450° and the same oxygen cycling was accomplished with ∂ varing from 0.1 to 0.3. The difference between the XAS spectra (fig.2-e) shows the feature **A** at 1.6 eV, even if a great contribution is due to the derivative-like signal in the first step. This proves that the main contribution comes from the transformation of the $|3d^9(\underline{L})>$ into the $|3d^9>$ configuration. However, there exists some contribution coming from the $|3d^{10}>$ configuration. Due to charge counting the appeareance of monovalent copper is expected only when ∂ is larger than 0.5. The present evidence of $|3d^{10}>$ configuration implies an increase rate of the **p**-hole injection under doping. It is also interesting to notice that over the whole range of stoichiometry, XPS studies have shown that the fraction of $|3d^9(\underline{L})>$ configuration, i.e. holes in the **p** oxygen band, is significantly larger than the prediction from the simple neutrality rule (Gourieux et al., 1988).

Cycling the partial pressure of oxygen during annealing the superconductive fine powder ($\approx 5\mu m$) at 600°C and 450°C has been found totally reversible, as shown by the XAS difference spectra for the removal and uptake kinetics (fig.4). In Table I we summarize the main features appearing in the spectra.

Single Crystal: Before performing the **in-situ** measurements on the $Y_1Ba_2Cu_3O_{7-\partial}$ single crystal, orientation dependent spectra were measured for angles ß ranging from 0° to 60°, where ß is tha angle between the electric field polarisation ($\hat{e}$) and the x-y plane. The sample was in a superconducting phase.

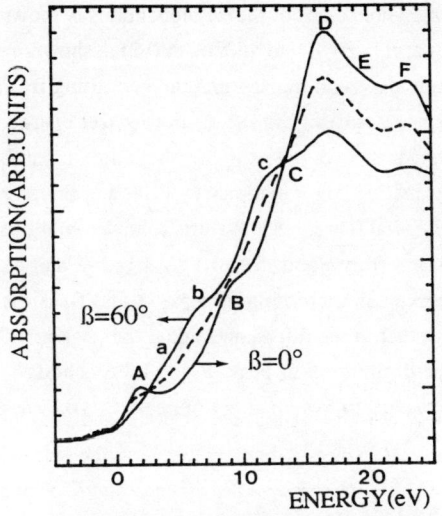

Fig.5: Orientation-dependence of copper K-edge XAS spectrum of a single crystal of $Y_1Ba_2Cu_3O_{7-\partial}$ for ß=0°, 36° and 60°, where ß is the angle between the electric field polarisation (ê) and the x-y plane.

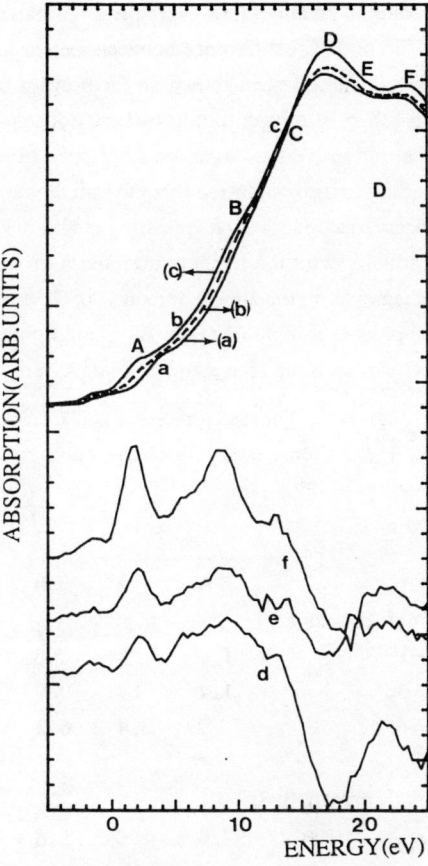

Fig.6: In situ kinetics of oxygen removal from $Y_1Ba_2Cu_3O_{7-\partial}$ single crystal: under 1 atm of O_2 a) at RT and b) at 700°C; and c) under 1 atm of N_2 at 700°C. Difference between XAS spectra under different ∂ values: sample annealed d) from RT to 450°C and e) from 450°C to 700°C under 1 atm of O_2 and f) partial pressure varying gradually from 1 atm of O_2 to 1 atm of N_2 at 700°C.

We display in fig.5 the spectra for ß=0° (ê//x-y), ß=36° and for ß=60° (where the main contribution comes from ê//z-axis). In the first two lines of Table I we display all the main features of these spectra. We label with capital letters the features appearing mainly in the ê//x-y orientation and with small letters the one appearing mainly in the ê//z orientation.

The **in-situ** kinetics of oxygen uptake and removal was performed for the ß=0° orientation, i.e., the electric field ê in the x-y plane of the single crystal. The sample was heated from RT to 700° under oxygen atmosphere and then, keeping this temperature, the atmosphere was gradually diluted by replacing oxygen by nitrogen until a final replacement leading to an almost complet nitrogen atmosphere. Three steps of the whole kinetics is shown in figure 6. The difference between the spectrum at RT ($\partial\approx 0$) and 450°C ($\partial\approx 0.2$) is shown in fig.6-d. As has been shown, in the powder sample the main transformation is coming from the filling of oxygen **p**-holes whose consequence is a shift of the spectrum to lower energy. The derivative-like signal has not been observed in that range of ∂ for the single crystal, leading to the conclusion that the initial hole elimination has **z** symmetry. Instead of this, we observed a contribution appearing at about 3.7 eV, that increases and shifts to lower energies gradually with the temperature, and a broad signal from about 5 to 13 eV (fig.6-d and e). When cycling the atmosphere at 700°C, we observed an increasing of peaks **A** and **B**, as for the powder sample (fig.6-f). We should point out that in the difference signal the peak at 1.9 eV is due to the increasing of peak **A** and to the diminution of peak **a** (at 3.8 eV) and the other one, at 8.9 eV, is coming essentially from a shift to lower energy of peak **B** (10.4 eV - 9.5 eV).

TABLE I

	A	a	b	B	c	C	D	E	F
Single crystal									
ß=0°	**1.6**	-	5.6	**8.9**	-	**13.1**	**16.3**	**19.7**	**23.6**
ß=36°	**1.4**	3.4	5.9	8.6	11.7	12.9	**16.5**	19.5	**23.4**
ß=60°	1.2	**3.4**	6.2	8.2	**11.4**	13.3	**17.1**	19.6	**23.5**
Single crystal (ß=0°)									
$O_{6.2}$	**2.0**	4.0	5.8	**9.5**	-	13.1	**16.6**	19.0	**23.6**
$O_{6.6}$	**2.4**	3.7	6.2	**10.2**	-	-	**16.3**	-	**23.6**
$O_{6.8}$	**2.8**	**3.8**	6.1	**10.4**	-	13.3	**16.3**	-	**23.7**
$O_{6.9}$	-	3.9	6.6	**10.4**	-	13.0	**16.6**	-	**23.8**
Powder									
$O_{6.2}$	**1.5**	-	5.7	8.5	11.9	13.0	**16.2**	19.4	**22.9**
$O_{6.7}$	1.6	**4.1**	5.6	8.4	**11.4**	-	**16.2**	-	**23.0**
$O_{6.9}$	1.1	3.6	5.7	-	**11.6**	-	**16.5**	-	**23.0**

GENERAL DISCUSSION.

In a depleted sample ($\partial \approx 1$), the added neutral oxygen atom nests in between two Cu(I) atoms, which act as an electron reservoir and provide two electrons to this oxygen. The net result is the transformation of $|3d^{10}\rangle$ into mainly $|3d^9\rangle$ configuration, i.e. a d-like hole is created. For an almost oxygen-full lattice ($\partial \approx 0$), the neutral oxygen atom fills an isolated oxygen vacancy in between two tri-coordinated Cu(1), which have already **d**-like holes. New holes are injected now in the upper oxygen **p** band. The net result is the transformation of the $|3d^9\rangle$ into the $|3d^9(\underline{L})\rangle$ configuration. In the intermediate ∂ range, holes are created both in the **d** copper band and in the **p** oxygen band.

The $|3d^{10}\rangle$ configuration, which is localized on Cu(1) sites, exists well beyond the tetragonal–orthorhombic transition. More precisely, in the range from $\partial \approx 0.2$ to $\partial \approx 0.3$. This implies that the hole concentration in the oxygen **p** band is higher than that given by the chemical formula. Indeed, the initial oxygen uptake ($\partial > 0.75$), which drains electrons from the Cu(1) reservoir, does not change the Neel temperature. On the contrary, when the hole creation in the oxygen **p** band starts ($\partial < 0.75$) the antiferromagnetic order is strongly affected and the Neel temperature decreases, vanishing at $\partial \approx 0.4$ (Rossat-Mignod et al.,1988).

Whether **p** holes are spread over sheets and chains is no more an open question for well oxygenated samples. It has been clearly demonstrated that $p_{x,y}$ holes sit at 929.5 eV and p_z holes are found at lower energy (928 eV) (Nücker et al.,1988). Let us notice that p_z holes do not exist in the Bi phase. For intermediate ∂, where the compound is insulator or metallic but with low Tc ($\approx 50K$), the ∂-dependence of the hole repartition should be different between the sheet and the chain oxygen **p** band. One way to consider this problem is to assume a rigid band model focusing on the similarities with the layered chalcogenides of the intercalation chemistry (McKinnon and Selwyn, 1987): lowering the Fermi level by doping or shifting up rigidly the oxygen **p** bands, keeping the Fermi level fixed to account for screening (Friedel, 1954), should implement **p** holes with **x-y** symmetry first. A small amount in the sheets should disturb the antiferromagnetic order of the undoped insulator. Our K-edge measurements are well correlated with this possibility which implies that, if we restrict the holes of the chains to be of the **z** symmetry, the chains are doped last of all (or dedoped first of all in removal of oxygen process). This fact is also supported by Cu L_3 spectroscopy which does not exhibit **z**-oriented $|3d^9(\underline{L})\rangle$ configuration for samples with $\partial > 0.45$ (Bianconi et al.,1988). The energy shift under doping found in this study (1.2 eV) is close to 1.5 eV measured in Cu L_3 spectroscopy. The real answer to such a question should come from the

∂-dependence of the polarisation-dependent measurements of the O K edge on single crystals. A starting clue to solve that question is given by the determinant -because bulk-sensitive- EELS experiment (Nücker et al., 1988) which observed a rapid initial decrease of the p_z holes under doping whereas the $p_{x,y}$ holes decreases more slowly. If the answer confirms this possible sequence, one can infer that the 2 steps transition to the 90 K superconductivity is connected to the "metallisation of the chains" which should transforms the single sheet 50 K superconductivity into the slab (3 layers) 90 K superconductivity.

Schematically we can say that doping by rising the oxygen content from 6 to 7 per unit formula leads to the progressive injection of **d** holes plus, with two successive delays, holes in the sheets and then in the chains. The only ascertainties are that monovalent copper is existing above 6.5 and that holes do not exist in the plane below 6.25. At the end (y≈7) the ratio 2:1 for holes in sheets and holes in chains comes from polarization argument (Cu L_{III} and O K edges). If this argument is not correct revised values should be given.

This sequence can found support in the analysis of Birgeneau (Birgeneau et al., 1988) of the Mattheiss' band structure (Mattheiss, 1987) or Guo (Guo et al., 1988; Chen et al., 1988) calculations which classify the oxygen band in three sets for single $(CuO_2)^{2-}$ sheet. The deeper one is the p_σ bonding one. The non bonding p_z is intermediate, whereas the in plane p_π orbitals pointing towards the empty centroid of four negative charge is less stabilized. Thence we expect **p** holes in a sheet to take place in this non bonding in plane orbitals which lay at the highest energy.

For the chain the two oxygens of the square structure should play a different hole. All the orbitals of the capping oxygen are bonding because the BaO distances (2.76Å) are close to that one found in pure BaO and p_z points towards Cu_1. Orbitals of the additional oxygen included under doping are stabilized either by Cu(1) or by the four Baryum. This stabilisation is however weak since the Ba atoms are farther (2.92Å) than they are in BaO (2.76Å).Thence it is possible to envisage two upper **p** bands for the chains.

Tokura (Tokura et al.,1988) derived an opposite conclusion from the investigation of substituted 123 compounds. Doping with Ca(II) on the Y(III) site or La(III) on the Ba(II) site they can decoupled doping and chemical composition, because the oxygen content can be maintained at the desired level. They found the insulator–metal transition after an initial doping which is quite appreciable (0.17 per Cu for $Y(Ba_{1-y}La_y)_2Cu_3O_7$).This large threshold lead them to claim that chains house holes in the initial doping and sheets in the end, with the argument that holes in one dimensional system should not percolate before 100%. According to their analysis, once the metallic phase is obtained, the additional holes are injected in the sheets. At the end the holes are equally distributed between chains and sheets. This sequence does not fit the respective energy of the hole levels found by the oxygen K-edge spectroscopy (Kuiper et al., 1988) which detects the shallowest hole with the **x,y** symmetry. The only way to reconcile these speculations is to envisage a more complex scenario which distinguishes two types of holes in the chains because of the distorted squared coordination around Cu(1), one with the **x-y** symmetry and the other one **z**-polarized, this one appearing the last of all to metallize the chains. A possible sequence of hole injection is displayed in figure 7. One hole is injected in the copper d-band and the other one is distributed among the oxygen p-bands having **x-y** symmetry in the chains (1), **x-y** symmetry in the sheets (2) and last of all having **z**-symmetry in the chains (3).

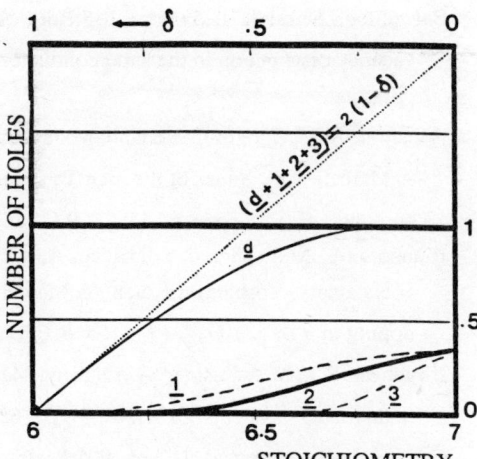

Fig.7: A possible sequence of hole injection: One **d**-hole in the copper and another one distributed among the oxygen p-bands having **x-y** symmetry in the chains (1), **x-y** symmetry in the sheets (2) and **z**-symmetry in the chains (3).

ACKNOWLEDGEMENTS:

We would like to thank A.Bianconi, J.Y.Henry, J.Rossat-Mignod, A.M.Flank, P.Lagarde et Y.Petroff for very helpful discussions and for showing their unpublished results. One of us (HT) is supported as a PhD Fellow by a grant of CAPES from the Brazilian Government. This work has been supported by the Stimulation Program of the European Economic Community (EEC) under contract STP-0040-1 (CD) and the MRES France.

REFERENCES

Baudelet, F., G.Collin, E.Dartyge, A.Fontaine, J.P.Kappler, G.Krill, J.P.Itie, J.Jegoudez, M.Maurer, P.Monod, A.Revcolevschi, H.Tolentino, G.Tourillon, M.Verdaguer, X-ray absorption spectroscopy of the 90K superconductor $YBa_2Cu_3O_{7-\delta}$, *Z.Phys.B*, **69**, 141 (1987)

Beech, F., S.Miraglia, A.Santoro, R.S.Roth, Neutron study of the crystal structure and vacancy distribution in the superconductor $YBa_2Cu_3O_{7-\partial}$, *Phys.Rev.B*, 35, 16, 8778 (1987)

Bianconi, A., A.Congiu-Castellano, M.de Santis, P. Rudolf, P. Lagarde, A.M.Flank, A.Marcelli, $L_{2,3}$ Xanes of the high Tc superconductor $YBa_2Cu_3O_{7-\partial}$ with variable oxygen content, *Sol. St. Comm.* 63,11, 1009 (1987)

Bianconi, A., M.de Santis, A.di Cicco, A.M.Flank, A.Fontaine, P.Lagarde, H.Katayama-Yoshida, A.Kotani, A.Marcelli, Symmetry of the $3d^9$ ligand hole induced by doping in $YBa_2Cu_3O_{7-\partial}$, *Phys.Rev.B*, 38,7,(1988)

Birgeneau R.J., M.A.Kastner, A.Aharony, Magnetic frustation model for superconductivity in planar CuO_2 systems, *Z.Phys.B*, 71, 57-62 (1988)

Cava, R.J., B.Batlogg, C.H.Chen, E.A.Rietmann, S.M.Zahurak, D.Werder, Single-phase 60K bulk superconductor in annealed $YBa_2Cu_3O_{7-\partial}$ ($0.3<\partial<0.4$) with correlated oxygen vacancies in the Cu-O chains, *Phys.Rev. B*, 36, 10, 5719 (1987)

Chen, G., W.A.Goddard III, The magnon pairing mechanism of superconductivity in cuprate ceramics, *Science* 239, 899 (1988)

Eskes, H., G.A.Sawatzky, Tendency towards local spin compensation of holes in high Tc copper compounds, *Phys.Rev.Let.*, 61, 12, 1415-1418 (1988)

Friedel, J.,Electronic structure of primary solid solutions in metals, *Adv. Phys.* 3, 446 (1954)

Fuggle, J.C., P.J.W.Weijs, R.Schoorl, G.A.Sawatzky, J.Fink, N.Nücker, P.J.Durham, W.M.Temmerman, Valence bands and electron correlation in the high Tc superconductors, *Phys.Rev.B*, 37, 123 (1988)

Fujimori, A., E.Takayama-Muromachi,Y.Uchida, B.Gkai, Spectroscopic evidence for strongly correlated electronic states in La-Sr-Cu-O and Y-Ba-Cu-O oxides, *Phys.Rev.B*, 35, 6, 8814, (1987)

Gourieux, T., G.Krill, M.Maurer, M-F.Ravet, A.Menny, H.Tolentino, A.Fontaine, Oxygen stoichiometry dependence of the electronic structure of $YBa_2Cu_3O_{7-\partial}$ with ∂ ($0<\partial<0.7$): possibility of a highly correlated mixed-valent state, *Phys.Rev.B*, 37,7516 (1988)

Guo, Y., J.M.Langlois, W.A.Goddard III, Electronic structure and valence-bond band structure of cuprate superconducting materials, *Science* 239, 896 (1988)

Hodeau, J.L., P.Bordet, J.J.Caponni, C.Chaillout, M.Marezio, Oxygen vacancy ordering twinning and copper substitution in $YBa_2Cu_3O_{7-\partial}$, *Proceedings of the International Conference on High Temperature Superconductors and Materials and Mechanisms of Superconductivity*, Interlaken, Switzerland, p.582 (1988)

Horn, S., K.Reilly, Z.Fisk, R.S.Kwok, J.D.Thompson, H.A.Borges, C.L. Chang, M.L.denBoer, X-ray spectroscopy of $EuBa_2(Cu_{1-y}Zn_y)_3O_{7-x}$, *Phys.Rev.B*, 38, 4, 2930-2933 (1988)

Jorgensen, J.D., M.A.Beno, D.G.Hinks, L.Soderholm, K.J.Volin, R.L.Hitterman, J.D.Grace, I.K.Schuller, C.U.Segre, K.Zhang, M.Q.Kleefisch, Oxygen ordering and the orthorhombic-to-tetragonal phase transition in $YBa_2Cu_3O_{7-x}$, *Phys.Rev.B*, 36, 7, 3608 (1987)

Kishio, K., J.Shimoyana, T.Hasegawa, K.Kitazawa, K.Fueki, Determination of oxygen nonstoichiometry in a high Tc superconductor $YBa_2Cu_3O_{7-\partial}$, *Jap.J.Appl. Phys.*, 26 L1228 (1987)

Kosugi, N., H.Kondoh, H.Tajima, H. Kuroda, The 1s-4p X-ray absorption near edge structure of $(La_{1-x}Sr_x)CuO_4$, $YBa_2Cu_3O_y$ and related compounds, *subm. to Phys.Rev.B*, (1988)

Kuiper, P., G.Kruizinga, J.Ghijsen, M.Grioni, P.J.W.Weijs, F.M.F.de Groot, G.A.Sawatzky, H.Verweij, L.F.Feiner, H.Peterson, X-ray absorption study of the O 2p hole concentration dependence on O stoichiometry in $YBa_2Cu_3O_x$, *Phys.Rev.B*, 37, 10, 6483 (1988)

Marel, D.van der, J.van Elp, G.A.Sawatzky, D.Heitmann, X-ray photoemission, bremsstrahlung isochromat, Auger-electron, and optical spectroscopy studies of Y-Ba-Cu-O thin films, *Phys.Rev. B* , 37, 5136 (1988)

Mattheiss, L.F., Electronic band properties and superconductivity in $La_{2-y}X_yCuO_4$, *Phys.Rev.Lett.* 58,1028, (1987)

McKinnon,W.R., L.S.Selwin, Ionic and electronic contributions to the Li chemical potential in $Li_xRu_zMo_{6-z}Se_8$, *Phys.Rev.B*, 35, 13, 7275 (1987)

Müller, J. and J.L.Olsen editors (North Holland), *Proceedings of the International Conference on High Temperature Superconductors and Materials and Mechanisms of Superconductivity*, Interlaken, Switzerland, Physica C 153-155 -(1988)

Nücker, N., J.Fink, B.Renker, D.Ewert, C.Politis, P.J.W.Weijs, J.C.Fuggle, Experimental electronic structure studies of $La_{2-x}Sr_xCuO_4$, *Z.Phys.B*, 67,9-14 (1987).

Nücker,N., H.Romberg, X.X.Xi, J.Fink, B.Gegenheimer, Z.X.Zhao, On the symmetry of holes in high Tc superconductors, *subm. to Phys.Rev.B*, (1988)

Rossat-Mignod, J., P.Burlet, M.J.Jurgens, L.P.Regnault, J.Y.Henry, C.Ayache, L.Forro, C.Vettier, H.Noel, M.Potel, P.Gougeon,J.C.Levet, Antiferromagnetic ordering and phase diagram of $YBa_2Cu_3O_{6+x}$, invited paper at the International Conference on Magnetism, (July 25-29, 1988)

Smith, T.A., J.E.Penner-Hahn, M.A.Berding, S.Doniach, K.O.Hodgson, Polarized X-ray absorption edge spectroscopy of single-crystal copper(II) complexes, *J.Am.Chem.Soc.*, <u>107</u>, 5945-5955 (1985)

Thiry, P., G.Rossi, Y.Petroff, A.Revcolevschi, J.Jegoudez, Observation of strong electron correlations in $YBa_2Cu_3O_{7-\partial}$ by $h\nu$ -dependent photoelectron spectroscopy, *Europhys. Let.* <u>5</u>, 55 (1988)

Tokumoto, M., H.Ihara, T.Matsubara, M.Hirabayashi, N.Terada, H. Oyanagi, K.Murata, Y.Kimura, Evidence of critical oxygen concentration at y=6.7 approximately 6.8 for 90K superconductivity in $YBa_2Cu_3O_{7-\partial}$, *Jap.J.App.Phys.*, <u>26</u>, L1565 -8 (1987)

Tokura, Y., J.B. Torrance, T.C.Huang, A.I.Nazzal, Broader perspective on the high temperature superconducting $YBa_2Cu_3O_{7-\partial}$ system: the real role of the oxygen content, *Phys.Rev.B.* <u>38</u>, 10, 7156 (1988).

Tolentino, H., E.Dartyge, A.Fontaine, G.Tourillon, X-ray absorption spectroscopy in dispersive mode for synchrotron radiation: optical considerations, *J.App.Crys.* <u>21</u>, 15 (1988)

Werder, D.J., C.H.Chen, R.J.Cava, B.Batlogg, Diffraction evidence for oxygen-vacancy ordering in annealed $YBa_2Cu_3O_{7-\partial}$ ($0.3<\partial<0.4$) superconductors, *Phys.Rev. B*, <u>37</u>, 4, 2317 (1988)

Zaanen, J., G.A.Sawatzky, J.W.Allen, Band gaps and electronic structure of transition-metal compounds, *Phys.Rev.Let.* <u>55</u>, 4, 418-421 (1985)

ON THE VARIATION OF THE Cu K-EDGE XANES
OF $YBa_2Cu_3O_{7-x}$ WITH OXYGEN CONCENTRATION

S. Della Longa*, M. De Simone, C. Li+, M. Pompa and A. Bianconi*

*Gruppo di Fisica, Dipartimento di Medicina Sperimentale, Università dell' Aquila,
Collemaggio, 67100 L'Aquila
Dipartimento di Fisica, Università degli Studi di Roma "La Sapienza" I-00185 Roma, Italy

ABSTRACT

The variation of the Cu K-edge x-ray absorption near edge structure (XANES) determined by the structural changes that accompany the loss of oxygen going from the high T_c superconductor $YBa_2Cu_3O_7$ to the antiferromagnetic insulator $YBa_2Cu_3O_6$ has been calculated by multiple scattering theory. The elongation of the Cu(2)-O(4) distance (between the Cu(2) in the CuO_2 plane and its apical oxygen) induces the shift of the main peak in polarized XANES spectra **E//c** toward lower energy. The formation of the oxygen vacancies induces changes in the **E⊥c** spectra for the Cu(1) ion in the linear chains. The results are in agreement with the experimental changes of the Cu XANES spectra induced by oxygen vacancies.

KEYWORDS

High T_c superconductivity, XANES, oxygen deficency, stuctural changes;

INTRODUCTION

A large interest has been addressed in Cu K-edge XANES spectroscopy of high T_c superconductors because of the local character of this experimental probe. The x-ray absorption spectroscopy has been applied in particular to the study of $YBa_2Cu_3O_{7-x}$ that changes from a superconductor metal to an antiferromagnetic insulator decreasing the oxygen content. (Iwazumi,et al., 1988; Oyanagi, et al., 1987; Tranquada, et al., 1988; Tolentino, et al., 1989; Baudelet, et al., 1987; Alp, et al., 1989, Bianconi et al., 1987). The Cu K-edge XANES probes the local structure of a cluster of atoms of finite size around the absorbing Cu ion. In a previous work (Garg, et al., 1988) an analysis of the XANES of $YBa_2Cu_3O_7$ was made using the multiple scattering approach. It was found that to get an an agreement with the experimental data it was necessary to calculate the XANES of large clusters of about 35 atoms within a sphere of 5Å radius as shown in Fig. 1.

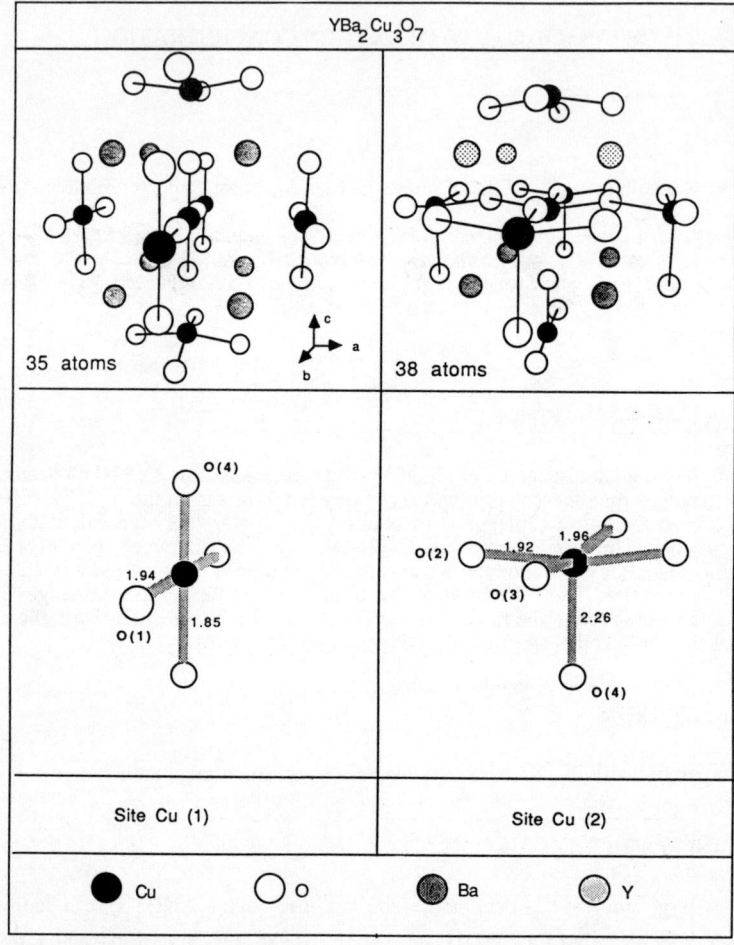

Fig. 1. Structure of the clusters of neighbor atoms surrounding the Cu(1) and Cu(2) sites in $YBa_2Cu_3O_7$.

The orthorhombic crystalline structure of $YBa_2Cu_3O_7$ changes to the tetragonal structure of $YBa_2Cu_3O_6$ (see for example Cava, et al., 1988). There are two types of Cu sites, named Cu(1) and Cu(2), in the crystalline cell. Therefore two clusters of about 35 atoms, relevant for the Cu K-edge XANES, within a sphere of 5Å radius around each of the two central Cu ions for $YBa_2Cu_3O_7$ ($YBa_2Cu_3O_6$), are shown in Fig.1 (Fig.2). We have used the coordinates of the atoms according to the crystallographic data of Capponi, *et al.*, (1987) for $YBa_2Cu_3O_7$ at 30K and of Roth, *et al.*, (1987) for $YBa_2Cu_3O_6$ at room temperature.

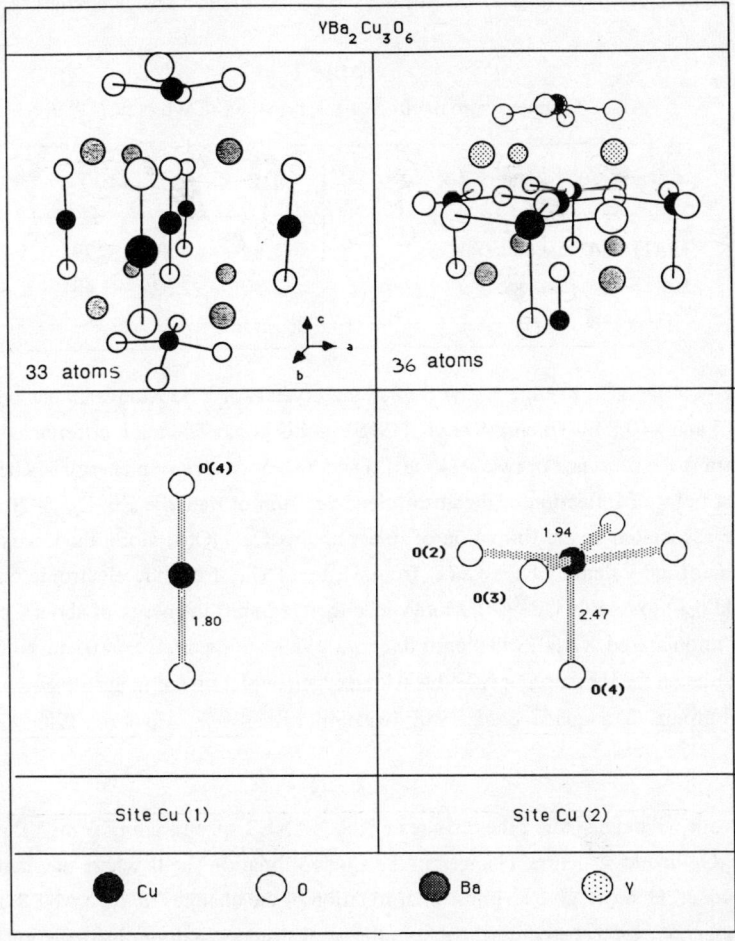

Fig. 2. Structure of the clusters of neighbor atoms surroundings the Cu(1) and Cu(2) sites in $YBa_2Cu_3O_6$.

Starting from the oxygen rich phase $YBa_2Cu_3O_7$ the following structural changes are observed in the Cu coordination spheres: i) the Cu ion in site Cu(1) changes from a square plane configuration with 4 oxygens in the **bc** plane (Fig. 1) to a linear coordination (Fig. 2) in $YBa_2Cu_3O_6$; ii) the Cu ion in site Cu(2) has a fivefold square pyramid configuration where the Cu(2)-apical oxygen O(4) distance changes from 2.26 Å to 2.47 Å in $YBa_2Cu_3O_6$.

The axial distances of Cu(1)-O(4) and Cu(2)-O(4), measured by neutron diffraction experiments (Cava *et.al.*, 1988), are shown in table 1. Moreover the different Cu(2)-O(2) and Cu(2)-O(3) nearest neighbor distances (Fig. 1) become the same in tetragonal $YBa_2Cu_3O_6$ (Fig. 2).

Table 1
Axial distances from neutron diffraction data (Cava *et al.*, 1989)

oxygen stoichiometry 7-x	0.0	0.2	0.5	1.0
Cu(1)-O(4) distance (Å)	1.84	1.84	1.79	1.79
Cu(2)-O(4) distance (Å)	2.30	2.32	2.43	2.47

The $E \perp c$ polarized Cu K-edge XANES of single crystals of $YBa_2Cu_3O_{7-x}$ has been reported for x=0.15 and x=0.8 by Tolentino *et.al.*, (1989) in this book. The main difference between the two spectra is the presence of two peaks at 1.6 and 8.6 eV (the zero in energy has been choosen at the first point of inflection of the absorption spectrum of metallic Cu, E_0=8976 eV). These peaks are associated to the formation of linear chains Cu(1)O(4) along the **c** axis, and to the change of formal valence of the Cu(1) from +2 to +1 (i.e. from the electronic configuration mainly $|3d^9\underline{L}>$ to mainly $|3d^{10}>$). Moreover only a red-shift in energy of about 1 eV has been found in unpolarized XANES (Tolentino, *et.al.*, 1989) going from x=0.05 to x= 0.15. These recent results on the variation of XANES with oxygen depletion are in agreement with findings of several groups (Tranquada, *et al.*, 1988; Iwazumi, *et al.*, 1988; Alp *et al.*, 1989; Heald *et al.*, 1989).

In this work we have studied the changes of the XANES spectra going from $YBa_2Cu_3O_7$ to $YBa_2Cu_3O_6$ due to structural changes in the Cu coordination shells where the main structural changes occur. Here we give a simple interpretation of the changes in the XANES observed by (Tolentino, *et.al.*, 1989) based on changes in the local structure induced by depletion of oxygen.

RESULTS

The XANES calculation has been performed with the multiple scattering (MS) approach using the program of Durham, *et al.*, (1982), starting from a set of muffin-tin (MT) potentials whose effect is described in terms of phase shifts, in the same way of the KKR procedure for band calculations. The MS calculations has been utilized to interpret a large variety of materials, from the transition metal oxides, to molecular complexes and biological molecules (Bianconi, *et al.*, 1988) and recently it was applied to the study of the distortions of Cu sites (Onori, *et.al.*, 1988)

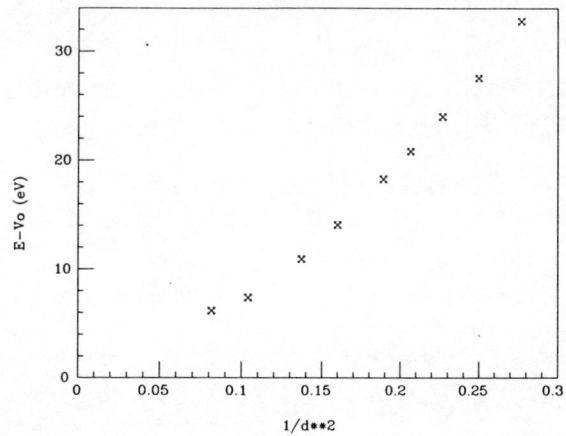

Fig. 3. Energy of the strongest peak in Cu K edge XANES due to the multiple scattering resonance determined by the first shell of oxygens for a CuO_2 linear cluster as a function of $1/d^2$ where d is the Cu-O distance.

This approach has some intrinsic limits due to the MT approximation of the potential and to the break-down of the one electron picture for highly correlated systems. Moreover, it was observed in several cases (Garg, *et al.*, 1988) that an energy scale expansion is necessary to take into account the exchange interaction between the excited electron and the valence electrons. Last, the calculation is made by considering elastic scattering and by using an energy independent pseudopotential and a correction is required to take into account the finite life-time of the photoelectron due to inelastic scattering. All these limitations have to be considered to get agreement with experimental data.

We have used a criterium of not overlapping MT spheres and neutrality of the sites, respecting the Mattheis prescriptions (Mattheis, 1964). For the calculation of the potential a larger cluster of about 100 atoms has been considered. The MT radii of Cu, O, Ba and Y, were firstly tested by varying R_{MT}(Cu) independently, and consequently varying the other R_{MT} in order to have as larger as possible, but not overlapping, MT spheres.

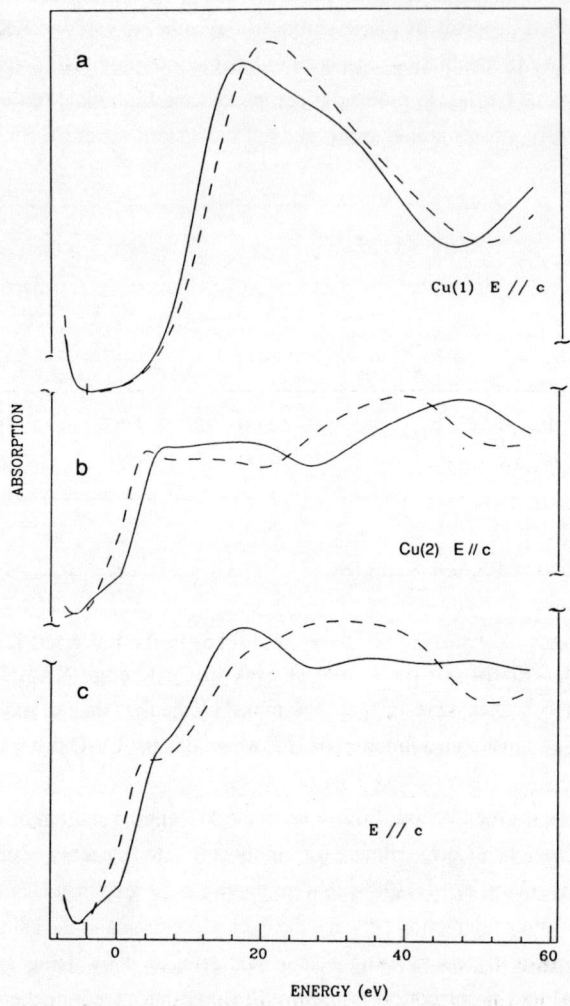

Fig. 4. Calculated polarized Cu K-edge XANES for one oxygen shell clusters with the electric field **E** parallel to the **c** axis, solid line YBa$_2$Cu$_3$O$_7$, dashed line YBa$_2$Cu$_3$O$_6$. From top to bottom:
panel a) Cu(1) site: Cu(1)O(4)$_2$O(1)$_2$ cluster of YBa$_2$Cu$_3$O$_7$ and Cu(1)O(4)$_2$ cluster of YBa$_2$Cu$_3$O$_6$,
panel b) Cu(2) site: Cu(2)O(2)$_2$O(3)$_2$O(4)$_1$ cluster for, and for YBa$_2$Cu$_3$O$_6$. *Panel c)* Weighted sum of the two Cu sites in the unit cell.

Looking for the convergence of the MT potentials inside the interstitial volume, a final choice for the set of MT radii was taken. The value of the potential in the interstitial volume is the intesphere constant V_0. Since to different sites of Cu correspond different first shell distances, also the sets of R_{MT} choosen are different for site 1 and site 2; see also table 2. We discuss separately the results of the calculations for two polarizations.

Table 2
Muffin tin radii R_{MT} (Å) and intersphere constant V_0 (Ryd)

	$YBa_2Cu_3O_7$		$YBa_2Cu_3O_6$	
	Cu(1)	Cu(2)	Cu(1)	Cu(2)
R_{MT} (Å)	1.00	1.03	0.96	1.03
V_0 (Ryd)	-2.00	-1.76	-2.15	-1.76

a) The $E \parallel c$ polarized XANES spectra.

The main changes observed by Heald, *et al.*, (1988) in the polarized $E \parallel c$ XANES spectrum going from $YBa_2Cu_3O_7$ to $YBa_2Cu_3O_6$ are: i) a red-shift of the energy of the absorption edge (features at 7.3 eV and at 14.6 eV, measured relative to the Cu metal K-edge, for $YBa_2Cu_3O_7$) and ii) a lowering of the intensity of the main absorption peak (at 17.1 eV, measured relative to the Cu metal K-edge, for $YBa_2Cu_3O_7$). The crystallographic data show that in the direction of the c axis, the main structural changes are: i) the contraction of the axial distance Cu(1)-O(4) and ii) the elongation of Cu(2)-O(4) distance, as shown in Fig. 1. We have studied the effects of these changes of the Cu-O distances along the c axis on the XANES.

It is well established that the bond length variations induce an energy shift of the XANES peaks according with the rule $(E_c-E_0)d^2$ = constant (in which E_0 is a reference energy and d the interatomic distance) (Bianconi, *et al.*, 1983). This rule is expected to be valid for small bond length variation $\Delta d/d$. In order to test this rule we have calculated the XANES for both linear and square plane coordinated copper ions by varying the Cu-O distance in the range 1.85<d<3.5 Å. We have used as reference energy the intersphere constant V_0 and the results of these calculations are shown in Fig. 3. There is an important deviation from the linear trend for d >2.6 Å therefore we expect a linear red-shift versus decreasing $1/d^2$ (increasing bond distance) in the range of Cu-O distances 1.8<d<2.5 Å.

In Fig. 4 the calculated **E//c** XANES of the $Cu(1)O(4)_2O(1)_2$ and $Cu(2)O(2)_2O(3)_2O(4)_1$ clusters for $YBa_2Cu_3O_7$, $Cu(1)O(4)_2$ and $Cu(2)O(2)_2O(3)_2O(4)$ for $YBa_2Cu_3O_6$ are shown.
By comparing the variation of the spectrum of $Cu(2)O(2)_2O(3)_2O(4)$ going from $YBa_2Cu_3O_7$ to $YBa_2Cu_3O_6$ in panel b we observe the energy red shift induced by the elongation of the $Cu(2)O(4)$ distance by 0.21 Å. The small contraction of the $Cu(1)O(4)$ distance and the change of the overall shape of the **E//c** spectrum of Cu(1) site (panel a) due to the change from square plane to linear coordination induce a blue shift.

The experimental polarized spectra are determined by the weigthed sum 1:2 of the spectra of Cu(1) and Cu(2) site. The sum of the two spectra is shown in Fig. 4, panel c. The weighted sum of the two sites, is in good agreement with the experiment of Heald, *et al.*, (1988), concerning the red-shift of about 3 eV both in the calculated XANES and in the experiment. Moreover the calculated spectra predict the lowering of intensity of the main absorption peak (about 9% from theory, versus 5% from experiment). The calculated energy interval between the shoulder on the rising edge and the main peak in calculation is about 12 eV, and it is consistent with experimental. We don't observe the peak at 14.6 eV therefore we assign this peak to a contribution of further shells.

b) The **E ⊥ c** polarized XANES spectra.

The data of Tolentino, *et al.*, (1988) show the formation of two peaks named A and B at 1.6 and 8.6 eV respectively and the lowering of the main peak D at 16.5 eV going from $YBa_2Cu_3O_7$ to $YBa_2Cu_3O_6$. As discussed previuosly the main structural changes in the directions parallel to the ab plane are the missing of O(1) ions in the Cu(1) site of $YBa_2Cu_3O_6$. We see from Fig. 1 and 2 that the Cu(2) clusters of the two crystalline structures are very similar up to the 3th shell. Therefore in the **E//ab** polarized Cu XANES spectra the most prominent differences are expected to arise from the site Cu(1).

In Fig. 5 we report the results of the XANES calculations of the same clusters as in Fig. 4 for the **E ⊥ c** polarization. We see in panel b that the spectra for the Cu(2) sites exhibit only small difference between the two compounds that can be assigned to the orthorhombic to tetragonal transition, and to the effect of changing Cu(2)O(4) distance. The XANES calculation for the Cu(1) cluster in panel a shows a large change determined by the fact that there are no atoms in the ab plane in the Cu(1) site of $YBa_2Cu_3O_6$, therefore the spectrum **E ⊥ c** of Cu(1) in $YBa_2Cu_3O_6$ is determined mainly by the atomic muffin tin potential.

Summing the absorption of the two sites we obtain the spectrum in the bottom in Fig. 5 panel c, showing only the main absorption peak similar to experimental finding by Tolentino, *et al.* (1989), but without features A and B.

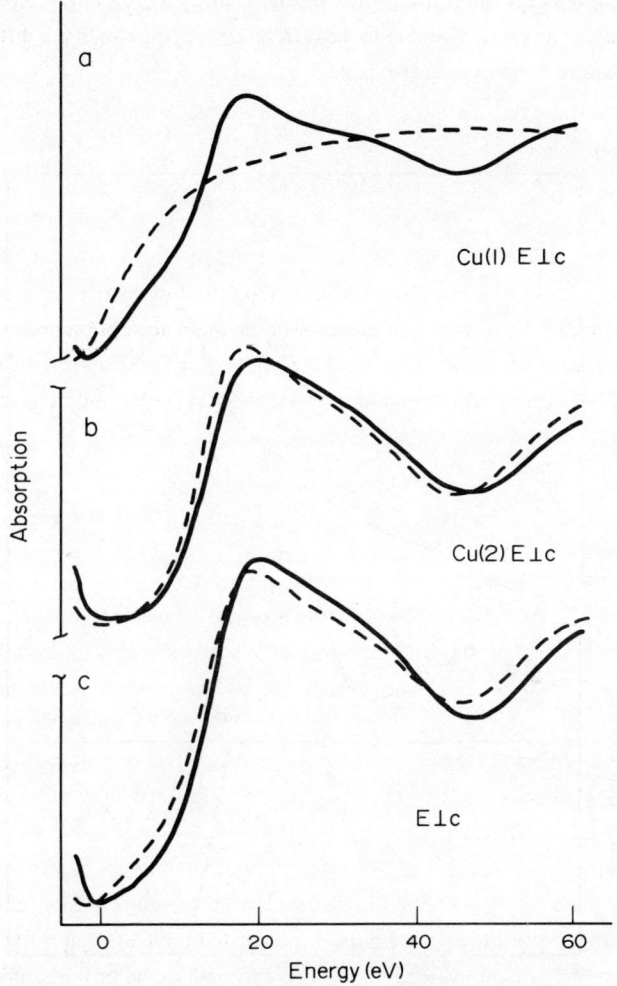

Fig. 5. Calculated polarized Cu K-edge XANES for one oxygen shell clusters with the electric field **E** perpendicular to the **c** axis, solid line $YBa_2Cu_3O_7$, dashed line $YBa_2Cu_3O_6$. From top to bottom: as in Fig.4 a) Cu(1) site, b) Cu(2) site, c) Weighted sum of the two Cu sites in the unit cell.

In Fig. 6 the calculation of the polarized $E \perp c$ XANES for the Cu(1) site for a larger cluster including the second shell of 8 Ba ions is shown. Two new features A and B appear in

$YBa_2Cu_3O_6$, at energy ≈ 1.5 and 10 eV. We have calculated the difference spectrum between the calculated spectra, shown in Fig. 6, to be compared with the experimental results of of Tolentino, *et al.,* (1989) i.e. curve f reported in Fig. 6 in their paper. We reproduce well the experimental features named A, C and D in the difference spectrum, while a difference of about 1.5 eV for the B feature remains. See also table 3.

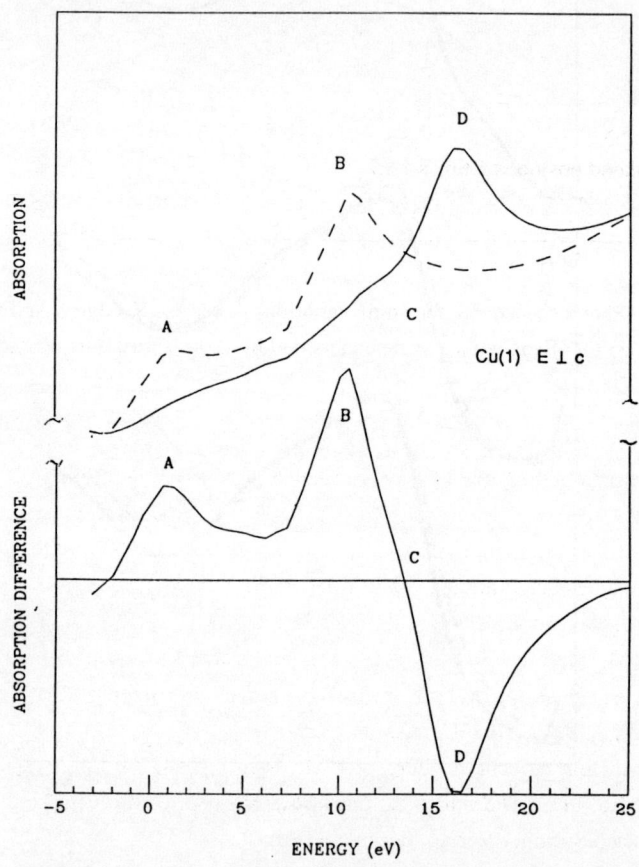

Fig. 6. Upper panel: calculated polarized Cu K-edge XANES of Cu(1) site for two shell clusters (including Ba ions) with the electric field **E** perpendicular to the **c** axis, solid line $YBa_2Cu_3O_7$, dashed line $YBa_2Cu_3O_6$.

Lower panel: Difference between the spectra of $YBa_2Cu_3O_6$ and $YBa_2Cu_3O_7$ in the upper part of the figure.

Table 3
Energy positions (eV) of difference $E \perp c$ XAS spectra

	A	B	C	D
experiment (Tolentino et al., 1989)				
x=0.05 -> x=0.15	4.4	10.0	13.6	18.5
x=0.15 -> x=0.3	1.6	8.6	12.5	17.6
x=0.2 -> x=0.8	1.6	8.6	13.6	17.6
calculated positions				
x=0 -> x=1	0.9	10.4	13.4	16.2
calculated positions x times 1.15				
x=0 -> x=1	1.03	11.9	15.4	18.6

In conclusion we have shown that the some variations of the Cu K-edge XANES spectra going from $YBa_2Cu_3O_7$ to $YBa_2Cu_3O_6$ can be understood in terms of structural changes.

REFERENCES

+ Current address: Department of Physics, Academia Sinica, Beijing, China

Alp, E.E., S.M. Mini, M. Ramanathan, B.W. Veal, L. Soderholm, G.L. Goodman, B. Dabrowski, G.K. Shennoy, A. Bommanavar and O.B. Hyun, (1989). Polarized x-ray absorption studies in oxide superconductors, *preprint, to be published*

Baudelet, F., G. Collin, E. Dartyge, A. Fontaine, J.P. Kappler, G. Krill., J.P. Itie, J. Jegoudez, M. Maurer, P. Monod, A. Revcolevschi, H.Tolentino, G. Tourillon, M. Verdaguer, (1987). X-ray absorption spectroscopy of the 90K superconductor $YBa_2Cu_3O_{7-\delta}$, Z. *Physik B Condensed Matter*, 69, 141.

Bianconi, A., M. Dell' Ariccia, A. Gargano, C.R. Natoli, (1983). Bond length determination using XANES, in *EXAFS and Near Edge Structure* edited by A. Bianconi, L. Incoccia and S. Stipcich Springer Verlag, Berlin p.57-61.

Bianconi, A., A. Congiu Castellano, M. De Santis, C. Politis, A. Marcelli, S. Mobilio, A. Savoia, (1987). Lack of delocalized Cu p States at the Fermi level in high T_c superconductor $YBa_2Cu_3O_{\sim7}$ by XANES Spectroscopy, Z. *Physik B Condensed Matter*, 67, 307-312.

Capponi, J.J., C. Chaillout, A.W. Hewat, P. Lejay, M. Marezio, N. Nguyen, B. Raveau, J.L. Soubeyroux, J.L. Tholence, and R. Tournier, (1987). Structure of the 100K superconductor $Ba_2YCu_3O_7$ between (5-300) K by neutron powder diffraction, *Europhysics Lett.*, 3, 1301-1307.

Cava, R.J., B. Batlogg, S.A. Sunshine, T. Siegrist, R.M. Fleming, K.Rabe, L.F. Schneemeyer, D.W. Murphy, R.B. van Dover, P.K.Gallagher, S.H. Glarum, S. Nakahara, R.C. Farrow, J.J. Krajewski, S. M. Zahurak, J.V. Waszczak, J.H. Marshall, P. Marsh, L.W. Rupp, Jr. W.F. Peck and E.A. Rietman, (1988). Studies of oxygen deficient $YBa_2Cu_3O_{7-\delta}$ and superconductivity Bi(Pb)-Sr-Ca-Cu-O, *Physica C*, 153-154, 560-565.

Durham, P.J., J.B. Pendry, and C.H. Hodges, (1982). Calculation of x-ray absorption near-edge structure, XANES, *Comp. Phys. Comm.*, 25, 193-205.

Garg, K.B., A. Bianconi, S. Della Longa, A. Clozza, and M. De Santis, A. Marcelli, (1987), Multiple scattering analysis of K-edge x-ray absorption near edge spectrum of $YBa_2Cu_3O_7$, *Phys. Rev. B*, 38, 244-251.

Heald, S.M., J.M. Tranquada, A.R. Moodenbaugh, and Y.Xu ,(1988). Orientation dependent x-ray absorption near-edge studies of high-T_c superconductors, *Phys. Rev. B*, 38, 761-764.

Iwazumi,T., I. Nakai, M. Izumi, H. Oyanagi, H. Sawada, H. Ikeda, Y. Saito, Y.Abe, K. Takita, R. Yoshizaki, (1988). Study on Cu valency of high T_c superconductor $YBa_2Cu_3O_{7-y}$ by high temperature x-ray absorption spectroscopy, *Solid State Commun.*, 65, 213-217.

Mattheis, L.F., Energy bands for the iron transition series, (1964). *Phys. Rev. A*, 134, 970-973.

Onori, G., A. Santucci, A. Scafati, M. Belli, S. Della Longa, A. Bianconi, L. Palladino (1988). Cu K-edge XANES of Cu(II) ions in aqueous solution: a measure of the axial ligand distances, *Chem. Phys. Lett.*, 149, 289-294.

Oyanagi, H., H. Ihara, T. Matsubara, T. Matsushita, M. Hirabayashi, M. Tokumoto, K. Murata, N. Terada, K. Senzaki, T. Yao, H. Iwasaki, and Y. Kimura, (1987). Local Structure in Orthorhombic and Tetragonal $Ba_2YCu_3O_{7-y}$. The role of oxygen vacancies for high T_c superconductivity, *Jpn. Journ. Appl. Phys.*, 26, L1233-L1236.

Roth, G., B. Renker, G. Heger, M. Hervieu, B. Domenges, and B. Raveau, On the structure of non-superconducting $Ba_2YCu_3O_{6+\epsilon}$ (1987).*Z.Phys. B- Condensed Matter*, 69, 53-59.

Tolentino, H., E. Dartyge, A. Fontaine, G. Tourillon, T. Gourieux, G. Krill, M. Maurer, and M.F. Ravet, (1989). Oxygen dependent electronic structure in powder and single crystal of $YBa_2Cu_3O_{7-\delta}$ observed in-situ by x-ray absorption spectroscopy, *in "High T_c Superconductors: Electronic Structure"* edited by A. Bianconi and A. Marcelli, this book.

Tranquada, J.M., S.M. Heald, A.R. Moodenbaugh, and Y. Xu, (1988), Mixed valency, hole concentration, and T_c in $YBa_2Cu_3O_{6+x}$, *Phys. Rev. B*, 38, 8893-8899.

ON THE TEMPERATURE DEPENDENCE OF THE Cu-O STRUCTURE IN EuBa$_2$Cu$_3$O$_{7-\delta}$

Jürgen RÖHLER and Udo MUREK

II. Physikalisches Institut, Universität zu Köln,
Zülpicherstr. 77, 5000 Köln 41, Fed. Rep. Germany

ABSTRACT

We have have investigated the temperature dependence (83 - 300 K) of the Cu-O bonds in EuBa$_2$Cu$_3$O$_{7-\delta}$ through the extended absorption finestructure (EXAFS) beyond the Cu 1s (K) absorption threshold. The data were recorded from bulk polycrystalline material in the fluorescence mode simultaneously together with the electrical resistance. The temperature change of the relative mean square displacement, $\Delta\sigma^2(T)$, of the Cu-O pairs exhibits a kink at $T \leq 110$ K and a singularity at 250 K. The kink at $T \leq 110$ K is due to the decomposition of the nonstoichiometric sample into oxygen rich and oxygen lean products. The singularity at 250 K is attributed to a displacive phase transition. We have extracted the temperature dependence of the Cu(1)-O(1) bondlength from the beats (jumps and minima in the total phase shift and amplitude, respectively) in the filtered absorption fine structure. The Cu(1)-O(1) bond stretches by about + 0.01 Å as the temperature decreases from 300 K to 90 K.

KEYWORDS

X-ray absorption, EXAFS, Superconductivity

INTRODUCTION

All known high-T_c superconducting cuprates have in common square-planar Cu-O bond configurations in the a-b plane of their crystallographic cells. The square-planar Cu-O bond configurations are mostly corner linked and form a two-dimensional network, which is believed to be of central importance for the mechanism of high-T_c superconductivity. "Capping" oxygen atoms link the square-planar Cu-O bond configurations with adjacent hypoelectronic metal-oxygen clusters (i.e. Cu$^{I, III}$-O, Bi$^{III, V}$-O, Tl$^{I, III}$-O). Pyramides therefore describe more exactly the Cu-O bond configurations, forming the superconducting layers in all members of the homologous series $RE_2Ba_4Cu_6O_{14-n}$ and in all various types of superconducting Bi- and Tl-cuprates. Each pyramide has one "long" (typically 2.2 - 2.5 Å) and four "normal" (typically 1.95 Å) Cu-O bonds. The length of the "long" bonds depends sensitively on the specific composition of the compound, the number of Cu-O layers, the concentration of oxygen vacancies, *and* on the temperature. On the other hand, the "normal" Cu-O bonds are more or less independent on the specific crystalline structure and on the chemical composition. The length of the "normal" bonds varies only slightly with temperature, also.

In this contribution we report on the thermal behavior (83 K - 300 K) of the pyramidal Cu-O bond configurations in superconducting $EuBa_2Cu_3O_{7-\delta}$ as extracted from the extended absorption fine structure (EXAFS) beyond the Cu - K (1s) absorption threshold. We discuss in particular the temperature dependence of the average Cu-O Debye-Waller factor, $\sigma^2(T)$, and the displacement of the "capping" oxygen atom O(1) with temperature.

EXPERIMENT

Sample Preparation and Characterization

The polycrystalline sample was prepared using the usual solid state reaction. Stoichiometric amounts of $BaCO_3$, Eu_2O_3 and CuO were finely ground, pressed into pellets, fired in air at 970°C for 12 hours and then were slowly cooled to ambient temperature. No further grinding procedures and no further heat or oxygen treatments were carried out. The structure of the sample was found to be orthorhombic, spacegroup P*mmm*, with the lattice parameters (at 300 K): a = 3.847 Å, b = 3.907 Å, c = 11.722 Å, and (at 80 K): a = 3.840 Å, b = 3.898 Å, c = 11.692 Å. These numbers compare well with data from the literature (cf. Table I).

Table I
Lattice parameters on superconducting $EuBa_2Cu_3O_{7-\delta}$ at 300 K

Composition	Lattice parameters			Ref.	Remarks
	a[Å]	b[Å]	c[Å]		
$EuBa_2Cu_3O_{7-\delta}$	3.840(2)	3.897(2)	11.722(4)	this work	see text
	3.857(2)	3.900(2)	11.749(5)	Hodorowicz et al.	once heated in air
$\delta \leq 0.1$	3.840(2)	3.900(2)	11.712(4)	and references	After heating in O_2
	3.869(2)	3.879(3)	11.693(6)	therein	
$\delta \approx 0.5$	3.820(3)	3.871(2)	11.576(5)	this work and Olsen et al.	minority phase see text

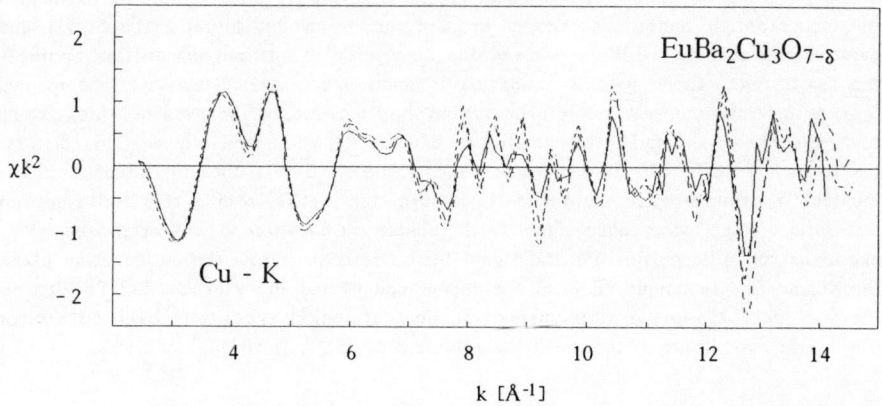

Fig. 1. EXAFS absorption coefficient $\chi k^3(k)$ of Cu in $EuBa_2Cu_3O_{7-\delta}$ recorded in the fluorescence mode at 83 K (----), 110 K (- - -), and at 253 K (——).

A grain (< 60 μm), picked up from this pellet, was subjected to a diffraction study at high pressures. The diffraction pattern, d-spacings and the lattice parameters are reported up to 600 kbar (Olsen *et al.*, 1988). The bulk modulus B and its pressure derivative B' are 1760(60) kbar and 1.4(3), respectively. Very surprisingly, the lattice parameters measured in this crystallite (at ambient pressure) turned out to be anomalously *small* (cf. Table I) and to deviate considerably from the diffraction result recorded from the pellet. The symmetry of the cells in the crystallite is still orthorhombic.

The c -axis increases as the oxygen defect, δ, decreases (see *e.g.* Schmahl *et al.*, 1988). Using c_0 = 11.71(2) Å and $\Delta c/\Delta\delta$ = - 0.2 [Å], we can attribute to the minority phase an average oxygen deficiency of about 0.5(1). Such a concentration of oxygen vacancies characterizes phases located just at the orthorhombic/tetragonal phase boundary.

The pellet (8 mm diameter, 2mm thick) was mounted on a pivoted sample holder placed in a He - bathcryostat. Electrical contacts to the sample were made by attaching platinum leads with silver epoxy to the back of the pellet. The temperature was measured with a Si-diode closely attached to the sample. The resistance was measured using a four point DC-technique (10 mA). Initially the sample exhibited a superconducting transition temperature of 101 K. The sample then was thermally cycled several times between 77 K and 300 K in He exchange gas (800 mbar). T_c decreased to 92.5 K (ΔT_c = 1.5 K) and then remained uninfluenced by further thermal treatments. The spectroscopic data shown here were recorded at increasing temperatures (0.1 K/min.), actually during the last cycle of a 36 h lasting thermal treatment.

Spectroscopic Measurement

The X-ray absorption spectra (Cu-K edge) were recorded at HASYLAB (DESY, Hamburg) using the focussing vacuum spectrometer EXAFS II. The content of harmonics in the monchromatic beam was reduced by a premirror and by detuning the Si 111 double crystal monochromator. We measured the absorption of the polycrystalline compact in the fluorescence mode with a cylindrical ionization chamber filled with 200 mbar Xe. Self-absorption processes of the primary and the fluorescence radiation in "thick" samples may non-linearily falsify the amplitude of the EXAFS oscillations. The distortions due to such damping effects were determined to be ≤ 0.5 % by the simultaneous measurement of the fluorescence yield and the transmitted signal from a thin foil ($5 mg/cm^2$). The ratio of the fluorescence signal to the scattering background was optimized with the use of narrow slits collimating the signal to the detector within the polarization plane of the beam. This technique reduced the acceptance of the detector by 70% but increased the S/N up to 2. Fig. 1 exhibts normalized raw data $\chi(k^2)$ at 83, 110, and 253 K. Notice the damped amplitude of the spectrum recorded at 253 K (full line).

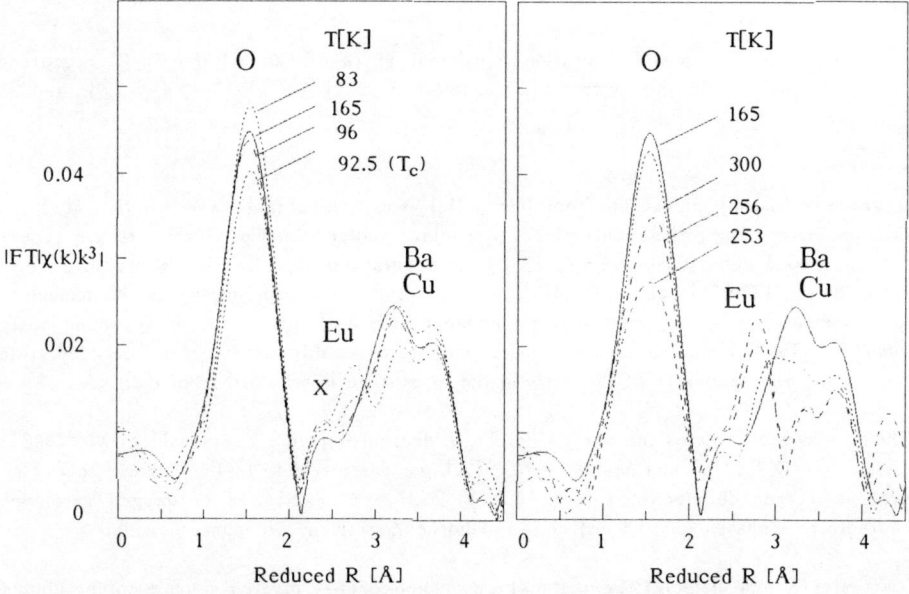

Fig. 2. Magnitudes of the Fourier transforms of $k^3 \chi(k)$ at the Cu K-edge in $EuBa_2Cu_3O_{7-\delta}$. The transform range is $k = 2.5 - 12.5$ $Å^{-1}$, weightened with a cosine type window centered at 7.5 $Å^{-1}$, optimizing the oxygen peak (0.8 - 2.1 $Å^{-1}$). Between 2.1 and 4 $Å^{-1}$ the labels indicate only the dominant contributions. X labels a second peak occurring in the Cu-Eu pair distribution function due to the jump in the Eu backscattering phase shift at $k = 8 Å^{-1}$ ("Ramsauer effect") *Left hand side:* 83 K, 92.5 K (T_c), 96 K, 165 K. *Right hand side:* 165 K, 253 K, 256 K 300 K.

RESULTS AND DISCUSSION

Temperature Dependence of the Cu-O Debye-Waller Factor, σ^2_{Cu-O} (T)

The data reduction was performed in the usual manner. Fig. 2 exhibits the magnitudes of the Fourier transforms, |FT(R)| of $k^3 \chi(k)$, at low temperatures. The choice of the window function and the weight of $\chi(k)$ optimizes the isolation of the nearest neighbour oxygen peaks. |FT(R)| is a sum of Gaussian-like pair distribution functions, having the widths σ_i, centered at the nearest neighor distances, R_i (reduced distances are not corrected for the phase shifts). Assuming, N_i, the number of nearest neighbors, and R_i do not

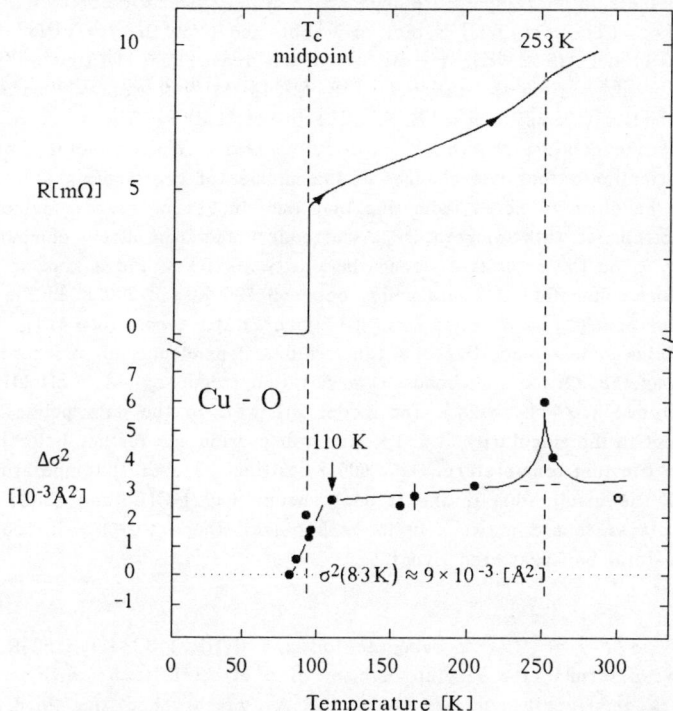

*Fig. 3. **(bottom)**:* The temperature dependence of the relative effective Cu-O Debye-Waller factor $\Delta\sigma^2 = \sigma^2(T) - \sigma^2(83K)$ in EuBa$_2$Cu$_3$O$_{7-\delta}$ recorded after the thermal treatment described in the text. The open circle (300 K) was measured before the thermal treatment. Notice the strong singularity at 250 K and the kink at 110 K. The dashed line connecting the data points between 200 K and 300 K indicates the slope of $\sigma^2(T)$ as obtained from an Einstein model with $\Theta_E = 600$ K. ***(top)**:* The temperature dependence of the electrical resistance was measured simultaneously together with the spectroscopic data. Notice the coincidence of the resistance anomaly at 250 K with the strong singularity of $\Delta\sigma^2$.

change appreciably with temperature, the amplitudes of the Gaussian-like pair distribution functions are inversely proportional to σ_i. The figure at the left hand side compares with each other $|FT(R)|$ obtained at $83 \text{ K} \leq T \leq 165 \text{ K}$, including the $|FT(R)|$ obtained at the midpoint (within 1 K) of the resistance jump. Fig. 2(right hand side) exhibits some of the $|FT(R)|$ obtained at $165 \leq T \leq 300 \text{ K}$. Primarily we focus on the analysis of the temperature dependence of the nearest neighbor Cu-O peaks at 0.8 - 2.2 Å (cf. Murek et al., 1988). Two anomalies occur: *i.* the heights of the maxima clearly decreases as T increases by only 13 K from 83 K to 96 K, and *ii.* the maxima at 253 and 256 K are drastically damped. An anomalous strong damping occurs also at the midpoint of the superconducting transition.

Fig. 3 exhibits the relative temperature dependence of σ^2_{Cu-O} and the electrical resistance, R(T). $\Delta\sigma^2(T) = \sigma^2(T) - \sigma^2(83 \text{ K})$ is plotted as obtained from the *log*-ratio of the Cu-O amplitudes, A(T) and A(T=83 K). A straight line was least squared fitted to ln (A(T, k^2)/ A(83 K, k^2)) = $\alpha k^2 + \beta$ in the range $k^2 = 8$ Å^{-2} to 74 Å^{-2}, *i.e.* well below the beat minimum in the Cu-O amplitude at 11 Å^{-1} (see further below). The slope $\alpha = -2\Delta\sigma^2(T)$ yields directly the relative change of the effective Debye-Waller factor. The intercept, β, yields information on possible changes of the number of coordinating atoms. β always turned out to be close to zero, indicating that the number of nearest neighbor oxygen atoms does not change with temperature. As already found from direct comparison of the $|FT(R)|$ in Fig. 2, $\Delta\sigma^2(T)$ exhibits a strong singularity at 253 K and a kink at $T \leq 110 \text{ K}$. The dashed line connecting the data points between 200 K and 300 K in Fig.3 (bottom) shows the slope of $\sigma^2(T)$ as obtained from $\sigma^2(T) = h/2\pi M\omega \times coth$ (h$\omega/4\pi k_B T$) with Θ_E = 600 K (Einstein model, modelling the temperature dependence of incoherent thermal vibrations along the axes of the bonds: ω = vibration frequency, M = effective mass of the pair of atoms, h$\omega/4\pi k_B = \Theta_E$). The model fits well to the data points at $110 \leq T \leq 300$ K, except to the singularity at 253 K (for a discussion see further below). The large number of the Einstein temperature, $\Theta_E = 600$ K, fitting the overall temperature behavior at $T \geq 110$ K, is mainly due to the strongly bound Cu(2)-O(2) and Cu(2)-O(3) pairs forming the planar-square network in the *a-b* plane. These Cu-O pairs contribute by ≈ 60 % to the total backscattered signal.

The Kink of $\Delta\sigma^2(T)$ at 110 K. Using $\Delta\sigma^2(T)$ and $|FT(R, T=83 \text{ K})| / |FT(R,T)| = \sigma(T)/\sigma(T=83 \text{ K})$, we determine the absolute number of σ^2 at 83 K to be $9(3) \times 10^{-3}$ Å^2. At $83 \leq T \leq 110$ K σ^2 strongly increases by 3×10^{-3} Å^2, *i.e.* by about one third of the zero point motions along the Cu-O bonds. Between 110 K and 300 K, however, the overall variation of σ^2 is only about 5×10^{-4} Å^2. It might be concluded that the Cu-O bonds soften at $T \leq 110$ K and that these data could be also fitted to the Einstein model. However, such a fit would yield an unreasonably small $\Theta_E < 90$ K. It is more likely that the Cu coordination spheres at $T \leq 110$ K are more structurally ordered or disordered than is predicted by fitting the high-temperature data to the (thermal) Einstein model. We emphasize that σ^2 comprises thermal disorder as well as disorder arising from static or dynamic displacements of the coordinating atoms: $\sigma^2 = \sigma^2_{thermal} + \sigma^2_{disorder}$. Evidence for ordering of the oxygen atoms in REBa$_2$Cu$_3$O$_{7-\delta}$ at about 110 K comes also from other EXAFS measurements (*e.g.* Zhang et al., 1988) and several other experiments.

According to the high resolution diffraction study of Horn et al. (1988) the orthorhombicity, $2(b-a)/(b+a)$, in $YBa_2Cu_3O_{7-\delta}$ increases at low temperatures, exhibits a kink at 120 K, passes over a maximum near T_c and then saturates. Several low frequency anelastic relaxation measurents in "123" compounds (e.g. Cannelli et al., 1988) reproducibly show maxima of the elastic energy dissipation at 110 K (and at 240 K) indicating the superconducting compound decomposes into various phases at low temperatures.

Khachaturyan and Morris,Jr. (1988) investigated theoretically the decomposition of nonstoichimetric "123" compounds into transient homologous phases and showed that oxygen deficient $YBa_2Cu_3O_{7-\delta}$ must be thermodynamically unstable at low temperatures. For thermodynamical reasons an oxygen deficient compound can attain stability through decomposition into a homologous series of transient ordered structures with oxygen vacancy concentrations δ = 1/3, 2/5, 3/7, 1/2. The oxygen vacancies order by a sequence of *completely empty* chains alternating with *completely filled* Cu-O chains. Partially filled chains are thermodynamically unstable. Experimental support for this model comes from several investigations of the superstructures in $YBa_2Cu_3O_{7-\delta}$. There is also microscopic evidence for such an ordering scheme: T-shaped Cu-O bond configurations (required for the formation of partially filled Cu-O chains) are extremely unstable and are not observed in other cuprates (cf. Renault et al., 1987). Following Kachaturyan and Morris, Jr. (1988), nonstoichimetric $EuBa_2Cu_3O_{7-\delta}$ decomposes into two metastable orthorhombic phases at cryogenic temperatures. The oxygen content of the one decomposition product evolves towards $EuBa_2Cu_3O_7$ (δ = 0) while the oxygen content of the other evolves towards $EuBa_2Cu_3O_6$ (δ = 1).

Such an ordering of transient homologous tetragonal and orthorhombic structures should be clearly observable in the Cu-O EXAFS. Variations of $\sigma^2(T)$ arise from the dramatic change of the Cu(2)-O(1) bondlength ("long" bond to the "capping" atom) by -0.18 Å ≙ -7.2% (Schmahl et al. 1988) as the oxygen lean phase transforms into to the oxygen rich phase. The variance, σ^2, of the experimentally observed Cu-O pair distribution function depends very sensitively on the position of the capping O(1) atom (see further below). Shortening of the Cu(2)-O(1) bond by only 0.03 Å (i.e. δ decreases) reduces the variance of the pair distribution function by 3×10^{-3} Å^2 and thus can account quantitatively for the decrease of σ^2 at T ≤ 110 K. In other words, σ^2_{Cu-O} decreases by vacancy ordering, enriching the oxygen rich orthorhombic phases with more oxygen and depleting the oxygen lean tetragonal phases. A quantitative anlaysis of this problem and further experiments are in progress.

The Singularity of σ^2 at 250 K. From the *log*-ratio method we find σ^2 increases by 3×10^{-3} Å^2 relative to the fit of the Einstein model (Θ_E = 600 K). The singularity of σ^2 coincides with a step-like increase of the electrical resistance. The decrease of the Cu-O peaks at these temperatures seemingly is correlated with an increase of the Cu-Eu peaks (cf. Fig. 2, right hand side). The amplitude functions, A(253 K, 256 K, k), turn out to exhibit bumps at k ≈ 8.5 Å^{-2} not observed at all other temperatures. These bumps were already realized in the linear *log*-ratio fits at 253 K and 256 K. Such bumps in metaloxygen amplitudes mostly are due to additional scattering paths from a distorted coordination shell. Therefore we attribute the observed singularity at 250 K to a displacive phase transition. For a detailed discussion of the phase transition in $YBa_2Cu_3O_{7-\delta}$ at 240 K see e.g. Canelli et al. (1988) and references therein.

The Anomalous Variation of the Cu(1) - O(1) (capping oxygen) with Temperature

The lengths of the 14 Cu-O bonds (per formula unit, $\delta=0$) deviate from each other. They are *discretely* distributed between 1.85 Å and 2.3 Å. 2 bonds are "very short" ($\approx$1.85 Å, Cu(1)-"capping" O(1)), 10 are "normal" ($\approx$1.95 Å, in the planes and along the chain) and 2 are "long" ($\approx$2.3 Å, Cu(2)-"capping" O(1)). The discrete distribution of slightly different distances causes beats in the Cu-O EXAFS. We use these beats to extract the temperature dependence of the bondlength differences (Martens et al., 1977). The temperature (300 K-4K) induced variation of the stiff bonds within the planes is < 0.004 Å. Thus the shifts of the beat minima (and jumps in the phase shifts) at $k_{beat} = \pi/(2\Delta R)$ yield directly the temperature induced variation of the Cu(1)-"capping" O(1) bondlength ("very short" bond) relative to the Cu(2)-O(2),O(3) bonds in the plane ("normal" bonds). The jumps in the phase shifts emerge clearly at $\approx$11 Å^{-1} and 6 Å^{-1}, respectively. The variation of R(Cu(1)-O(1)) (relative to the "normal" bonds) is plotted in Fig. 4. This bond stretches by +0.01 Å as the unit cell thermally contracts. More data points at T < T_c are needed to extract unambiguously a change of the slope at T_c. According to a recent diffraction study of monocrystalline $YBa_2Cu_3O_{6.93}$ (Simon et al., 1988), the positional parameter, z, of O(1) increases by +0.001 on cooling (300 K-100 K). The thermal contraction along the c-axis, however, nearly compensates for this positional change such that R(Cu(1)-O(1)) increases by +0.005(1), in coarse agreement with our result. Well known (see e.g. Miceli et. al., 1988) the Cu(1)-O(1) bondlength increases (at fixed T) as the

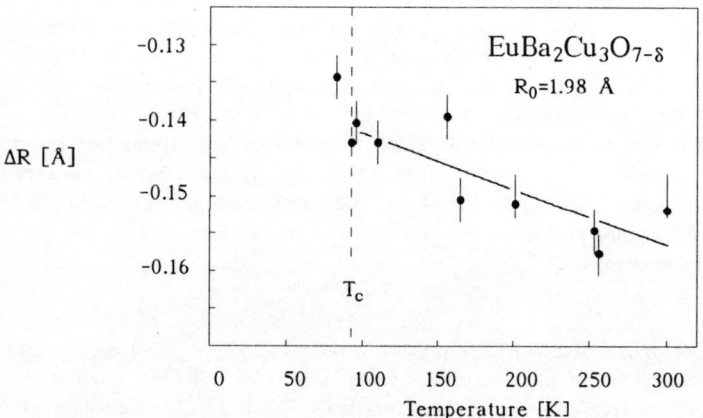

Fig. 4. Relative variation of the bondlength R(Cu(1)-O(1)) with temperature as extracted from beats in the EXAFS. ΔR is the difference of the very short ($\approx$1.8 Å) and the "normal" ($\approx$1.98 Å) bonds in $EuBa_2Cu_3O_{7-\delta}$. ΔR decreases at low temperatures, *i.e.* the Cu(1)-O(1) bond stretches (+0.01 Å, 300 $\rightarrow$ 100 K) as the unit cell contracts.

oxygen defect, δ, decreases. Thus cryogenic temperatures have the same effect on the position of the "capping" O(1) as a reduced umber of oxygen defects. From simple Coulomb repulsion arguments, this finding may be understood from the differential thermal expansion of the Ba-Ba and Ba-Y layers.

REBa$_2$Cu$_3$O$_{7-\delta}$ contracts anisotropically along the c-axis This is mainly due to the large polyhedron thermal expansion coefficient of the two Ba^{2+}- polyhedra differing from that of the one RE^{3+}-polyhedron. (see Fig. 5). When cooled to low temperatures the Ba^{2+}-Ba^{2+} pairs move more easily towards each other than the RE^{3+}-Ba^{2+} pairs. Thus Ba moves into the Ba—O(1) plane and stretches the O(1)-Cu(1) bond. One can imagine that filled chains weaken the O(1)-Cu(1) bondstrength and have a similar effect on the Ba-O(1) planes as thermal contraction.

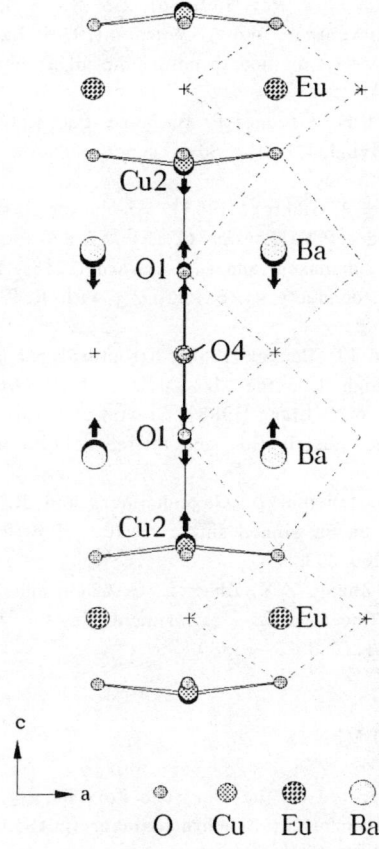

Fig. 5 The structure of EuBa$_2$Cu$_3$O$_{7-\delta}$. The black circles and the arrows indicate the local distortions of the Ba-O(1) planes and the anisotropic contraction of the Cu(1)-O(1)-Cu(2) bonds occuring on cooling from 300 K to cryogenic temperatures.

REFERENCES

Canelli B., R. Cantelli, and F. Cordero (1988). Ordering and diffusion at low temperature in Y-Ba—Cu-O by measurements of eleastic dissipation and modulus. *J. Modern Phys. B* (Proceedings of the Adriatico Research Conference on: "Towards the theoretical understanding of high T_c superconductors", Trieste, 26-29 july, 1988 (in press).

Hodorowicz, S.A., A. Lasocha, W. Lasocha, and H.A. Eick (1988). The high-temperature Eu_2O_3-$BaCO_3$-CuO—atmospheric oxygen phase diagram: Phase characterization in the 90 K superconducting region. *J. Solid State Chem.*, 75, 270 - 278.

Horn, P.M., D. T. Keane, G.A. Held, J.L. Jordan-Sweet, D.L. Kaiser, F. Holtzberg, and T.M. Rice (1987), Orthorhombic distortion at the superconducting transition in $YBa_2Cu_3O_{7-\delta}$: Evidence for aniostropic pairing. *Phys. Rev. Lett.*, 59, 2772-2775.

Kachaturyan, A.G., and J.W. Morris, Jr. (1988). Transient homologous structures in non-stoichiometric $YBa_2Cu_3O_{7-x}$. *Phys. Rev. Lett.*, 61, 215-218.

Martens G., P. Rabe, N. Schwentner, and A. Werner (1977). Extended x-ray - absorption fine-structure beats: A new method to determine differences in bond lengths. *Phys. Rev. Lett.*, 39, 1411 - 1414.

Miceli, P.F., M. Tarascon, L.H. Greene, P. Barboux, F.J. Rotella, and J.D. Jorgensen (1988). Role of the bond lenghts in the 90-K superconductor: A neutron powder-diffraction study of $YBa_2Cu_3Co_xO_{7-y}$.

Murek U., K. Keulerz, and J. Röhler (1988). Anomalies in the Cu-O structure of $EuBa_2Cu_3O_{7-\delta}$ at 110 K and 250 K. *Physica C, 153-155*, 270 - 271.

Olsen J.S., S. Steenstrup, I. Johannsen, and L. Gerward (1988). High pressure studies of the high temperature superconductors $RBa_2Cu_3O_{9-\delta}$ with R: Y, Eu and Ho up to 60 GPa. *Z. Phys.*, 72, 165 - 168.

Renault, A., J.K. Burdett and J.P. Pouget (1988). Structural changes with oxygen content and ordering of defects in high-T_c oxide $YBa_2Cu_3O_{6-x}$. *J. Sol. State Chem.*, 71, 587-590.

Schmahl W.W., E. Salje, and W.Y. Liang (1988). Crystal structure of 1-2-3 superconductor precursor material: Vacancy distribution and lattice relaxation. *Phil. Mag. Lett.*, 58, 173 - 181.

Simon, A., J. Köhler, H. Borrmann, B. Gegenheimer, and R. Kremer (1988). X-ray structural investigation of an untwinned single crystal of orthorhombic $YBa_2Cu_3O_{6.93}$. *J. Sol. State Chem.*, 77, 200-203.

Zhang, K., G.B. Bunker, G. Zhang, Z.X. Zhao, L. Q. Chen, and Y.Z. Huang (1988). Extended x-ray-absorption fine-structure experiment on the high-T_c superconductor $YBa_2Cu_3O_{7-\delta}$. *Phys. Rev. B*, 37, 3375 - 3380.

ACKNOWLEDGEMENTS

This work was partially supported by the Deutsche Forschungsgemeinschaft through SFB 125. The authors thank the Hamburger Synchrotronlabor (HASYLAB) at DESY in Hamburg for their hospitality.

CORRELATION BETWEEN T_C OF CUPRATE SUPERCONDUCTORS AND THE ENERGY SPLITTING BETWEEN IN-PLANE AND OUT-OF-PLANE POLARIZED Cu 2p->3d TRANSITION

A. Bianconi[*], P. Castrucci, A. Fabrizi

GNSM, Dipartimento di Fisica, Università degli Studi di Roma "La Sapienza" I-00185 Roma, Italy

A. M. Flank, P. Lagarde

LURE, CNRS-CEA-MEN, Bâtiment 209 D, Université Paris Sud, 91405 Orsay, France

S. Della Longa

Gruppo di Fisica, Dipartimento di Medicina Sperimentale, Università dell'Aquila, Collemaggio, 67100 L'Aquila

A. Marcelli

Laboratori Nazionali di Frascati, Istituto Nazionale di Fisica Nucleare, 00044 Frascati, Italy

Y. Endoh, H. Katayama-Yoshida

Department of physics Tohoku University, Sendai, 980 Japan.

Z.X. Zhao

Institute of Physics, Academia Sinica, P.O.Box 603, Beijing, P.R. China

ABSTRACT

High resolution polarized Cu L_3 x-ray absorption spectra of single crystals of three classes of high T_c superconductors $La_{2-x}Sr_xCuO_4$ (x=0.04 and x=0.1), $Bi_2CaSr_2Cu_2O_8$ and $YBa_2Cu_3O_{7-\delta}$ have been measured by using synchrotron radiation at Super ACO. We have investigated the energy splitting Δ between the in-plane and out-of-plane 2p->3d transition. We find that the splitting Δ is correlated with the critical temperature of the superconducting crystals for each class of superconductors.

KEYWORDS

High T_c superconductivity, x-ray absorption, electronic structure;

INTRODUCTION

A key point in the investigation of the electronic structure of high T_c superconductors is the determination of the changes of unoccupied states induced by doping in the antiferromagnetic insulating background. It is well established that the holes induced by doping are mainly in the oxygen derived valence band i.e. ligand holes giving the $3d^9\underline{L}$ configuration (Bianconi et al., 1987; De Santis et al., 1989; Kaindl et al., 1988; Fink et al., 1989; Bianconi et al., 1988a; and 1988b). The two-hole states were found by photonduced IR (Taliani et al., 1989) and by many other methods. The determination of the symmetry of the two holes is an open problem.

In this work we have addressed our interest to the variation of the unoccupied Cu d states induced by doping. The electronic configuration of the Cu ions in the antiferromagnetic insulating compounds is expected to be the Cu $3d^9$ configuration, where the d hole is in the Cu $3d_{x^2-y^2}$ orbital of b_1 molecular symmetry in the CuO_4 square plane cluster. This is expected for a perfect square plane but it is well known that a simple system of regular CuO_4 square planes does not exhibit high T_c superconductivity above 20K. The tilting of the elongated octahedron of the $La_{2-x}Sr_xCuO_4$ and the distortions of the CuO_5 square pyramids of $Bi_2CaSr_2Cu_2O_8$ and $YBa_2Cu_3O_{7-\delta}$ induce a mixture of $3d_{z^2-r^2}$ (a_1 molecular symmetry) or of the $3d_{xz}$ (e molecular symmetry) with the $3d_{x^2-y^2}$ in the ground state (Bacci, 1986), (Bacci, 1988). There are several experimental indications that the holes are not present only in the σ bond between the $3d_{x^2-y^2}$ and the $O(2p_x)$ or $O(2p_y)$ in the CuO_2 layers. The holes induced by doping in $YBa_2Cu_3O_7$ (T_c=80K system) are present in the apical oxygen O(4) of Cu(2) which is also in the linear Cu(1)O chains (Bianconi, et al., 1988b and 1988c). Evidence for d-d excitations in $YBa_2Cu_3O_7$ has been reported (Geserich, et al., 1988). The study of the polarization dependence of the "white line" in the Cu L_3-edge XAS of $Bi_2CaSr_2Cu_2O_8$ has indicated the presence of 10%-20% of Cu 3d holes oriented in the z direction (Bianconi, et al., 1988d).

Several theoretical models for high T_c superconductivity such as the three band model (Castellani et al., 1988a; and 1988b) the spin polaron pairing mechanism (Kamimura, 1988) the d-d excitation model (Weber, 1988a, and 1988b) (Jarrell et al., 1988) (Shelankov et al., 1988) and the Kanamori model (Nishino et al., 1988) the interlayer pairing (Askenazi, et al. 1988; Kuper et al., 1989) (Kurihara, 1988) require the presence of electronic states with two holes in two different orbitals.

Here we have investigated the polarization dependence of the white line due to $2p_{3/2} \rightarrow 3d$ transitions in antiferromagnetic insulating La_2CuO_4 and superconducting crystals $La_{2-x}Sr_xCuO_4$ (x=0.04 and x=0.1), $Bi_2CaSr_2Cu_2O_8$ and $YBa_2Cu_3O_{7-\delta}$ (δ= 4.5 and 0.2) in order to investigate the presence of the electronic transitions polarized in the z directions which probe the components of the orbitals of the d holes in the z direction.

EXPERIMENTAL

The L_3 x-ray absorption experiment has been carried out on the super-ACO storage ring of the synchrotron radiation facility LURE at Orsay. A double crystal $10\bar{1}0$ beryl monochromator with an energy resolution about 0.35 eV at 900 eV has been used. The absorption coefficient has been measured by total electron yield method. The thickness of the probed surface layer is about 100 or 200 Å. The crystals have been cleaved with the surface normal direction z in the direction perpendicular to the CuO_2 planes. The polarized spectra of the single crystals with several incidence angles between the electric field of the photon beam **E** and the sample surface normal z have been recorded by rotating the sample, using synchrotron radiation linearly polarized in the horizontal plane. The polarized XAS spectra ($\mathbf{E} \perp z$) (**E** parallel to the **xy** plane of the CuO_2 layers) and ($\mathbf{E} // z$) have been obtained by extrapolation of the spectra taken at four different incidence angles. In the case of $Bi_2Sr_2Ca_1Cu_2O_8$ and $YBa_2Cu_3O_{7-\delta}$ the z axis is parallel to the **c** axis while for $La_{2-x}Sr_xCuO_{4-y}$ the z axis is parallel to the **b** axis. The line-shape of the white line shows a broadening going from ($\mathbf{E} \perp z$) to ($\mathbf{E} // z$) as measured by studying the changes of the white line of a powder sample of CuO with the incidence angle. This instrumental effect has been taken into account in the data analysis and will be discussed in a longer paper in preparation. The energy resolution depends also on the working conditions of the electron beam in the storage ring, therefore variations of the width and of the line shape can be extracted only by comparing spectra taken in the same experimental conditions. For these reasons the spectrum of CuO reference sample has been recorded immediately after each spectrum.

We have investigated the following set of samples: a) several crystals cleaved from two $Bi_2Sr_2Ca_1Cu_2O_8$ large crystals which were prepared at the Tohoku University and at the Institute of Physics of Academia Sinica with T_c=85K, b) an insulating La_2CuO_{4-y} single crystal, c) a superconducting $La_{1.96}Sr_{0.04}CuO_{4-y}$ single crystal grown by one of us YKY at the Tohoku University, d) a large $La_{1.9}Sr_{0.1}CuO_{4-y}$ single crystal, T_c=12K, grown at NTT laboratories and with lattice constant a=5.365 Å and b=13.19 Å, e) two Y-Ba-Cu-O single crystals, one in the low T_c range, T_c =60K, and another in the high T_c range, T_c=82K.

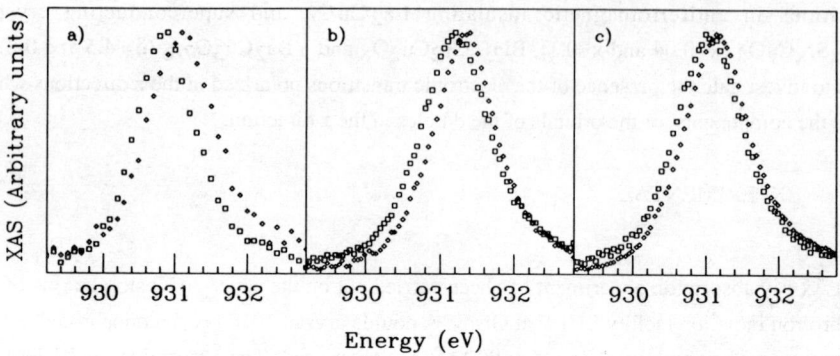

Fig. 1. Polarized L_3-edge XAS spectra measured with electrical field parallel ($\mathbf{E}$ //z) (squares) and perpendicular ($\mathbf{E} \perp z$) (rhombs) to the sample z axis, for : a) La_2CuO_{4-y}, b) $La_{1.96}Sr_{0.04}CuO_4$; and c) $La_{1.9}Sr_{0.1}CuO_4$

RESULTS

Fig.1 shows the white line due to Cu 2p->3d transitions of the polarized, $\mathbf{E}$ //z and $\mathbf{E} \perp z$, Cu L_3 XAS of the insulating La_2CuO_{4-y} and of superconducting $La_{1.96}Sr_{0.04}CuO_4$ and $La_{1.9}Sr_{0.1}CuO_4$. The white line maximum for the $\mathbf{E}$ //z polarization is always at lower energy than for the $\mathbf{E} \perp z$ polarization. The energy separation Δ between the maxima of the two polarized white lines has been obtained by curve fitting by taking into account the experimental artifact of the variation of the lineshape with the incidence angle. The energy separation Δ decreases from 230 ± 50 meV to 80 ± 50 meV going from La_2CuO_{4-y} to $La_{1.9}Sr_{0.1}CuO_4$. The curves in Fig.1 are normalized to the white line maximum to show the energy shift. The intensity of the $\mathbf{E}$ //z white line is about 25% of the intensity of the $\mathbf{E} \perp z$ white line. The error in the measure if the relative intensity is about 15-20%. The presence of a component of the white line for $\mathbf{E}$//z polarization indicates that there are Cu d holes having a component of their orbital in the z direction $\underline{d_z}$ ($\underline{d_z}$ is introduced here to indicate one of the 3d orbital: $3d_{z^2-r^2}$, d_{xz} or d_{yz} that we cannot distinguish in this experiment).

Fig. 2 shows the polarized Cu L_3-edge XAS of $YBa_2Cu_3O_{6.5}$ (T_c=60K), b) $YBa_2Cu_3O_{6.8}$ (T_c=82.5K) in a larger energy range including the white line and the shoulder at about 1.5 eV above it. The energy separation Δ between the white lines in the two polarizations decreases from 200 meV to 120 meV going from T_c=60K to T_c=82.5 K sample.

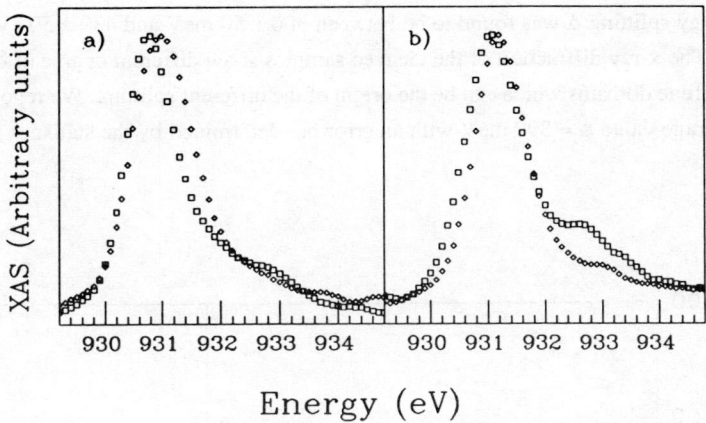

Fig. 2. Polarized L_3-edge XAS spectra measured with electrical field parallel (**E** //**c**) (squares) and perpendicular (**E** ⊥ **c**) (rhombs) to the sample **c** axis, for:
a) $YBa_2Cu_3O_{6.5}$ (T_c=60K), b) $YBa_2Cu_3O_{6.8}$ (T_c=82.5K).

In Fig. 3 the white lines of polarized Cu L_3-edge XAS of a $Bi_2Sr_2Ca_1Cu_2O_8$ single superconducting crystal (T_c=80K) are shown. Similar results have been obtained in several experimental runs and by using crystals grown in different laboratories.

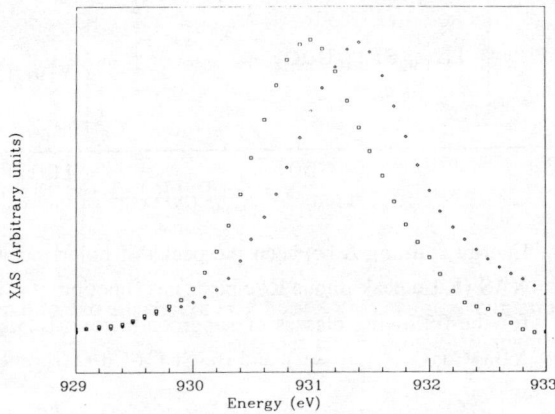

Fig. 3. Polarized L_3 - edge XAS spectra measured with electrical field parallel (**E** //**c**) (squares) and perpendicular (**E** ⊥ **c**) (rhombs) to the sample **c** axis, for for one $Bi_2Sr_2Ca_1Cu_2O_8$ crystal (T_c= 85K).

The energy splitting Δ was found to be between 300 ± 50 meV and 480 ± 50 eV for different crystals. The x-ray diffraction of the cleaved samples show different degree of orientations of the crystalline domains which can be the origin of the different splitting. We report in the figure 4 the average value Δ = 390 meV with an error bar determined by the statistics from different samples.

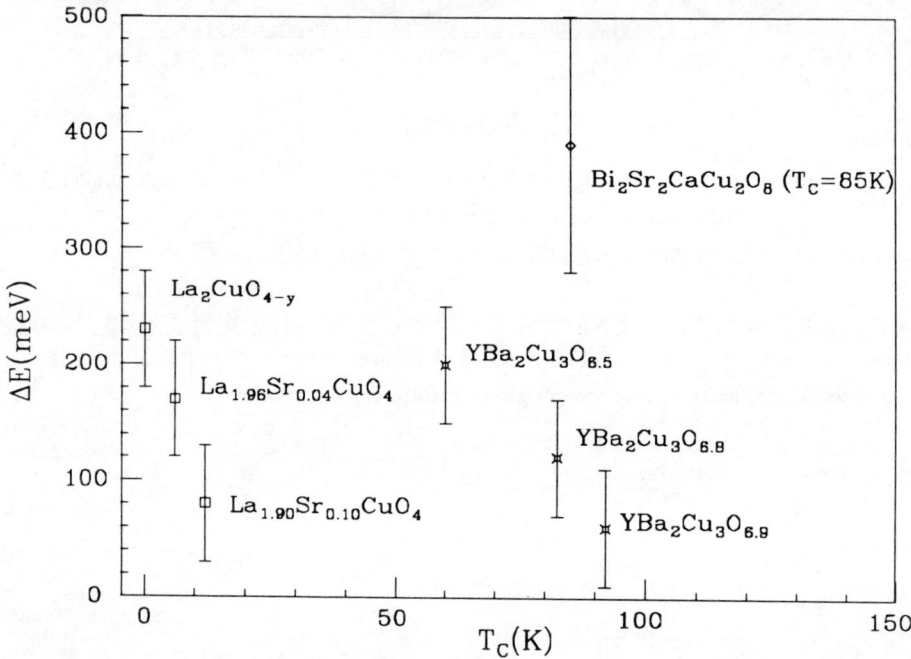

Fig. 4. Energy splitting Δ between the peaks of polarized white lines in L_3-edge XAS ($E \perp c$ peak minus $E//c$ peak) as function of critical temperature (T_c) for the following classes of superconductors: $La_{2-x}Sr_xCu_2O_4$ (squares), $YBa_2Cu_3O_{7-\delta}$ (crosses), and Bi-Sr-Ca-Cu-O (rhombs).

We report in Fig. 4 the energy difference Δ between the maximum of the in-plane (the xy plane of CuO_2 layers) $E \perp z$ polarized white line and the out-of-plane $E//z$ polarized line for the three classes of compounds studied in this work as function of the critical temperature of each sample. We observe that for each class of systems the Δ (T_c) is a decreasing function which is pointing toward zero for the maximum value of T_c. The values of Δ reaches its maximum in the series of

$La_{2-x}Sr_xCuO_4$ for the insulating x=0 system. In conclusion the results given in Fig. 4 show that the doping induces an important change in the d unoccupied states by reducing the energy separation between the unoccupied states with components along **z** and that with components along the **xy** plane.

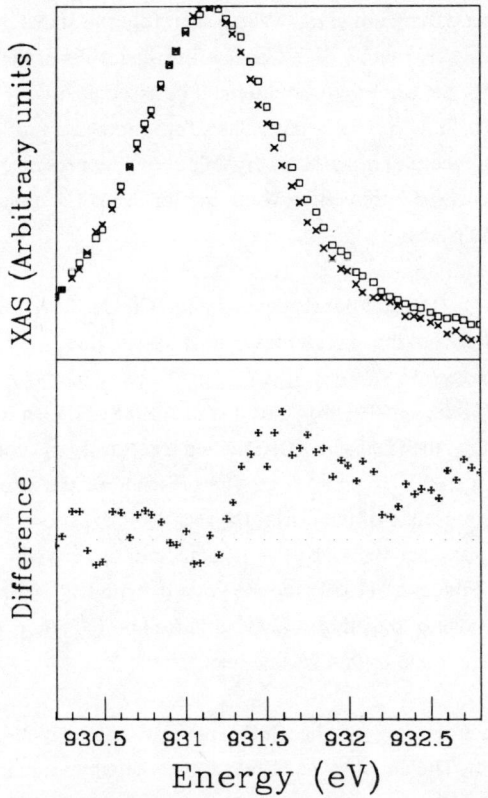

Fig. 5. Comparison between the white lines of L_3-edge XAS spectra measured with an angle of 80^0 between the photon electric field **E** and the **z** axis of $La_{2-x}Sr_xCuO_4$ x=0.04 (crosses) and x=0.1 (squares). In the lower panel the difference between the two spectra is plotted to show the spectrum of strontium induced transitions with a peak at about 931.7 eV. The dotted line indicates the zero of the difference

Finally we want to address the problem of the symmetry of the $3d^9\underline{L}$ states in the doped $La_{2-x}Sr_xCu_2O_4$. In a previous paper (Bianconi *et al.*, 1988a) we have shown that the $3d^9\underline{L}$

states induced by doping give an high energy tail of the white line of unpolarized La_2CuO_4. We have analysed the difference between the $E \perp c$ spectra of $La_{2-x}Sr_xCu_2O_4$ for x=0.04 minus x=0.10 shown in Fig. 5. No detectable variation was found in the difference between the E//c spectra. The spectra were recorded under the same experimental conditions so the instrumental broadening is the same for the two spectra. The spectrum of the difference is similar to that of the unpolarized spectra (Bianconi et al., 1988a) therefore the states induced by doping in the low T_c regime (T_c=12K) are mainly $\underline{d}_x\underline{L}$. Considering the results of the analysis of the shoulder (De Santis et al., 1989) the observed shoulder can be associated with the formation of singlet states $\underline{d}(b_1)\underline{L}(b_1)$ with U_{dL} =1.3 eV. Therefore both $La_{2-x}Sr_xCu_2O_4$ x=0.1 and $Bi_2Sr_2Ca_1Cu_2O_8$ (Bianconi et al., 1988d) crystals show the presence of a resolved shoulder $\underline{2p}3d^{10}\underline{L}$ in the range 1-2 eV above the white line for the $E \perp z$ polarization inducating the presence of $\underline{d}(b_1)\underline{L}(b_1)$ states.

The triplet states $\underline{d}(a_1)\underline{L}(b_1)$ are not detected by the Cu L_3 XAS spectroscopy because the split-off of the shoulder from the energy region of the white line is determined by the Coulomb repulsion $U_{cL} \sim U_{dL}$ which for the $\underline{d}(a_1)\underline{L}(b_1)$ is expected to be much smaller than 1.3 eV for $\underline{d}(b_1)\underline{L}(b_1)$ (Weber, 1988a, and 1988b) (Jarrell et al., 1988) (Shelankov et al., 1988). Because of the small value of U_{cL} the final states determined by $\underline{d}(a_1)\underline{L}(b_1)$ configuration are expected to be unresolved from the E//z white line. The presence of the white line component E//z indicates the possible presence of the $\underline{d}(a_1)\underline{L}(b_1)$ or $\underline{d}(e)\underline{L}(b_1)$ configurations with $U_{cL} \sim U_{dL}$ smaller than 1 eV. In fact the $\underline{L}(b_1)$ was detected by oxygen K-edge absorption of $Bi_2Sr_2Ca_1Cu_2O_8$ by Fink et al. (1989) and the holes $\underline{d}_z$ with Cu 3d orbital components in the z direction ($\underline{d}_z$ indicates one of the 3d orbital: $3d_{z^2-r^2}$ (a_1) or $\underline{d}_{xz}$, $\underline{d}_{yz}$ (e)) are detected by the polarization dependence of the L_3 XAS white line.

The presence of the holes out of the CuO_2 plane is found in the high T_c range of the $YBa_2Cu_3O_{7-\delta}$ system. The shoulder at 1.5 eV above the white line in Fig.2 much weaker in the spectrum of the T_c=60K sample than in the spectrum of the T_c=82.5K sample (Bianconi et al., 1988c). The $E \perp z$ spectrum of the T_c=60K sample shows an asymmetric high energy tail of the white line indicating the presence of a $3d^9\underline{L}$ configuration with $\underline{d}$ unoccupied orbitals in the CuO_2 plane $\underline{d}_x\underline{L}$. Most of the weight of the 1.5 eV shoulder observed in the E//z spectra of the $T_c \sim 80K$ is formed with the transition from the low T_c range (50-60K) to the high T_c range (80-90K). The presence of the hole in the apical O(4) is detected by the x-ray absorption at the Cu(1) ions. These results are in agreement with the theoretical prediction of Fujimori, (1989) that the ligand hole is stabilized in the apical oxygen by decreasing the Cu(2)O(4) distance. Therefore for 50-60K superconductivity both $\underline{d}_x\underline{L}$ and $\underline{d}_z\underline{L}$ configuration have comparable probability while in the 80-90K superconductivity the holes are found to be mostly out of the

CuO_2 plane.

In conclusion we have found evidence in L_3 XAS spectroscopy of a correlation between the splitting Δ of the E//z and E$\perp$z polarized white line and the critical tempearture of the sample iwhich s in agreement with the predictions of the d-d excitation model (Weber, 1988a, and 1988b) (Jarrell et al., 1988). The expected $\underline{d}(b_1)\underline{L}(b_1)$ states have been found but $\underline{d}_z$ holes are also present. In the $YBa_2Cu_3O_7$ system for T_c=80K we have found the presence of the oxygen induced ligand hole out of the CuO_2 planes. The intensity of the $\underline{d}(b_1)\underline{L}(b_1)$ is expected to decrease with decrease of the ratio $\Delta z/U_{cL}$, where Δz is the energy splitting between $d_{x^2-y^2}$ and d_z states, and the configurations $\underline{d}(a_1)\underline{L}(b_1)$ (Weber, 1988a, and 1988b) (Jarrell et al., 1988) or $\underline{d}(b_1)\underline{L}(a_1)$, where the ligand holes are out of the CuO_2 plane (Askenazi, et al. 1988) become more important with increasing T_c.

ACKNOWLEDGEMENTS

We would like to thank M. Matsuda and Y. Hidaka of NTT of Japan and Y.F. Yan and J.H. Wang of Beijing Institute of physics for growing high T_c samples and for their characterization.

REFERENCES

[*] also at Gruppo di Fisica, Dipartimento di Medicina Sperimentale, Università dell'Aquila, Collemaggio, 67100 L'Aquila

Askenazi, J., C.G. Kuper (1988). Band structure, electron correlations, and an interlayer pairing mechanism for high temperature superconductivity. *Physica C* 153-155, 1315-1316.

Askenazi, J., C.G. Kuper (1989). Correlated electrons, a narrow conduction band and interlayer pairing in YBCO. *High T_c Superconductors: Electronic Sructure,* Pergamon Press, London this book. (Proc. International Symposium on the Electronic structures of High T_c Superconductors, Rome 5-7 October 1988, edited by A. Bianconi and A. Marcelli) *and references cited therein.*

Bacci, M. (1986). Jahn-Teller effect in five-coordinated copper(II) complexes. *Chemical Physisc* 104, 191-199.

Bacci, M. (1988). Structural distortion inducted by pseudo Jahn-Teller effect in Y-Ba-Cu-O system. *Japanese Journal of Applied Physics* 27, L1699-L1701.

Bianconi, A., A. Congiu Castellano, M. De Santis, P. Rudolf, A.M. Flank, P. Lagarde, A. Marcelli (1987). $L_{2,3}$ XANES of high T_c superconductor $YBa_2Cu_3O_{\sim 7}$ with variable oxygen content. *Sol. State Commun.* 63, 1009-1013.

Bianconi, A., J. Budnick, A.M. Flank, A. Fontaine, P. Lagarde, A. Marcelli, H.Tolentino, B. Chamberland, C. Michel, B. Raveau, and G. Demazeau (1988a). Evidence of $3d^9$-Ligand hole states in the superconductor $La_{1.85}Sr_{.15}CuO_4$ from L_3 x-ray absorption spectroscopy. *Phys. Lett. A* 127, 285-291.

Bianconi, A., M. De Santis, A.M.Flank, A. Fontaine, P. Lagarde, A. Marcelli, H. Katayama-Yoshida and A. Kotani (1988b). Determination of the symmetry of $3d^9\underline{L}$ states by polarized Cu L_3 XAS spectra of single crystal in $YBa_2Cu_3O_{\sim 6.9}$"*Physica C* 153-155, 1760 -1761.

Bianconi, A., M. De Santis, A. Di Cicco, A.M. Flank, A. Fontaine, P. Lagarde, H. Katayama-Yoshida, A. Kotani, and A. Marcelli (1988c). Symmetry of $3d^9$ ligand hole induced by doping in $YBa_2Cu_3O_{7-\delta}$. *Phys. Rev. B* 38, 7196-7299

Bianconi, A., M. De Santis, A. Di Cicco, A. Fabrizi, A.M. Flank, P. Lagarde, A. Marcelli, C. Politis, H.Katayama-Yoshida, A.Kotani, and Z.X. Zhao (1988d). Simmetry of the hole states in BiCaSrCuO high T_c superconductors. *Modern Physics Letters B,* 2, 1313-1318.

Castellani, C., C. Di Castro and M. Grilli (1988a). Possible occurence of band interplay in high T_c superconductors. *Physica C* 153-155, 1659-1660.

Castellani, C., C. Di Castro and M. Grilli (1988b). Kondo lattice hamiltonian for high Tc superconductors. "Proceedings of the Conf. Towards the theoretical understanding of high T_c superconductors" Trieste, July 1988, edited by S.Lundqvist, E.Tosatti, M. P. Tosi and Yu Lu., *Int. Journal of Modern Physics* 1, 659-665.

De Santis, M., A. Bianconi, A. Clozza, A. Di Cicco, P. Castrucci, M. Desimone, A. Flank, P. Lagarde, J. Budnick, P. Delogu, A. Gargano, R. Giorgi, and T. D. Makris, (1989). Joint analysis of Cu L_3 XAS and XPS spectra of high Tc superconductors for determination of the states inducted by doping, *High T_c Superconductors: Electronic Srueture,* Pergamon Press, London this book. (Proc. International Symposium on the Electronic structures of High T_c Superconductors, Rome 5-7 October 1988, edited by A. Bianconi and A. Marcelli)

Fink, J., N. Nucker, H. Rombeg, and S. Nakai (1989). Electronic structures studies of high T_c superconductors by valence and core excitations. *High T_c Superconductors: Electronic Sructure,* Pergamon Press, London this book. (Proc. International Symposium on the Electronic structures of High T_c Superconductors, Rome 5-7 October 1988, edited by A. Bianconi and A. Marcelli) *and references cited therein.*

Fujimori, A., (1989). *High T_c Superconductors: Electronic Sructure,* Pergamon Press, London this book. (Proc. International Symposium on the Electronic structures of High T_c Superconductors, Rome 5-7 October 1988, edited by A. Bianconi and A. Marcelli).

Geserich, H.P., G. Scheiber, J. Geerk, H.C. Li, G. Linker, W. Assmus, and W. Weber (1988). Optical Spectra of superconductors $YBa_2Cu_3O_{7-\delta}$ films: evidence for Cu^{++} d-d transition. *Europhysics Letters* 6, 277-282.

Jarrell, M., H. R. Krishnamurthy, and D.L. Cox, (1988). Charge-trasfert mechanism for high T_c superconductivity. *Phys. Rev. B* 38, 4584-4587.

Kaindl, G., D. D. Sarma, O. Strebel, C. T. Simmons, U. Neukirch, R. Hoppe, and H. P. Muller (1988). *Physica C* 153-155, 139-140.

Kamimura, H., (1988). Spin-polaron pairing high temperature superconductivity, in "Proceedings of the Conf. Towards the theoretical understanding of high Tc superconductors" Trieste, July 1988, edited by S.Lundqvist, E.Tosatti, M. P. Tosi and Yu Lu., *Int. Journal of Modern Physics* 1, 699-709.

Kurihara, S., (1988). A new pairing mechanism in oxide superconductors. *Physica C* 153-155, 1247-1248.

Nishino, T., M. Kikuchi, and J. Kanamori (1988). A model for superconducting Cu oxides: the oxygen p-band hybridized with $d(3z^2-r^2)$ of Cu. *Solid State Commun.* 68, 455-458.

Taliani, C., R. Zamboni, G. Ruani, A.J. Pal, F.C. Matacotta, Z. Vardeny, and X. Wea (1989). Photoinduced optical excitation in $YBa_2Cu_3O_{7-\delta}$. In *High T_c Superconductors: Electronic Sructure*, Pergamon Press, London this book. (Proc. International Symposium on the Electronic structures of High T_c Superconductors, Rome 5-7 October 1988, edited by A. Bianconi and A. Marcelli)

Weber, W. (1988a). A Cu d-d excitation model for the pairing in the high T_c cuprates. *Z. Phys. B- Condensed Matter* 70, 323-329.

Weber, W., (1988b). On the theory of high T_c superconductors. *Advances in Solid State Physics* 28, 141-162.

Shelankov, A. L., X. Zotos, and W. Weber (1988). A weak coupling limit of the d-d excitation model for high T_c cuprates. *Physica C* 153-155, 1307-1308.

ELECTRONIC STRUCTURE STUDIES OF HIGH-T_c SUPERCONDUCTORS BY VALENCE AND CORE ELECTRON EXCITATIONS

J. Fink, N. Nücker, H. Romberg and S. Nakai

*Kernforschungszentrum Karlsruhe GmbH, Institut für
Nukleare Festkörperphysik, Postfach 3640,
D-7500 Karlsruhe,
Federal Republic of Germany*

ABSTRACT

The electronic structure of high-T_c superconductors has been studied by high-energy electron energy-loss spectroscopy in transmission. From the loss spectra below ~50 eV the dielectric functions could be derived giving information on plasmons, interband transitions and low-lying core levels. From the dispersion of the plasmon related to the charge carriers (holes on O sites) the Fermi velocity could be derived. Orientation dependent measurements on the O 1s and Cu 2p absorption edges gave information on the character and the symmetry of the unoccupied electronic states. In $YBa_2Cu_3O_7$ and $Bi_2Sr_2CaCu_2O_8$ the holes on O sites, formed upon doping, have wave functions lying in the CuO_2 planes (or CuO_3 ribbons). Orbitals perpendicular to these planes can be excluded. The Cu 2p edges reveal for the undoped and doped compounds dominantly in-plane $3d_{x^2-y^2}$ states. There is a 10-15% admixture of lower-lying states, probably with $3d_{3z^2-r^2}$ symmetry.

KEYWORDS

High-T_c superconductivity; electronic structure; electron energy-loss spectroscopy; dielectric functions; plasmons; core-level spectroscopy.

INTRODUCTION

At present the nature of the mechanism for the high-T_c superconductivity in cuprates is by no means clear. Therefore, studies of the electronic structure of these materials are extremely important to increase our knowledge on the parameters which enter into theories on high-T$_c$ superconductivity. Moreover, those studies may help to support or to exclude theoretical models. High energy spectroscopies can yield, although on a rather coarse energy scale, information on the character, the symmetry and the dynamics of electrons at the Fermi level. This work reviews our activities of electronic structure studies of high-T$_c$ superconductors by high-energy electron energy-loss spectroscopy (EELS) in transmission.

High-energy EELS in transmission has long been used to obtain information on the electronic structure of solids (Raether 1980). The differential cross section for high-energy inelastic electron scattering is proportional to

$$\frac{d^2\sigma}{dEd\Omega} \propto \frac{1}{q^2} Im\left| -1/\varepsilon(\mathbf{q},\omega) \right|$$

where $\varepsilon(\mathbf{q},\omega)$ is the dielectric function depending on energy and momentum transfer $\mathbf{q}$. The latter one is composed of a longitudinal (proportional to the energy-loss) and a transverse (proportional to the scattering angle) momentum transfer, i.e., $q^2 = q^2_\| + q^2_\perp$. At low loss energies, $q_\|$ is small and $\mathbf{q} \approx \mathbf{q}_\perp$ can be adjusted by the scattering angle. Thus the loss-function Im[-1/$\varepsilon(\mathbf{q},\omega)$] can be measured for $\mathbf{q}$ perpendicular to the electron beam. By a Kramers-Kronig analysis, Re[-1/$\varepsilon(\mathbf{q},\omega)$] and thus the real and the imaginary part of the dielectric function, ε_1 and ε_2, respectively, can be determined. At energies far above the plasmon energy, the loss function is about ε_2 since ε_1 is about 1. Therefore, for small momentum transfer, EELS measures the same as X-ray absorption spectroscopy (XAS). For a sample tilted by 45° to the electron beam, and chosing |q$_\perp$| (given by the scattering angle) equal to |q$_\|$| (given by the energy-loss) the momentum transfer can be directed parallel and perpendicular to the sample plane by changing the sign of q$_\perp$. So, similar to XAS, orientation dependent measurements of core-electron excitations can be performed probing the unoccupied density of states. Since the scattering angles in EELS are usually extremely small ($\Theta \leq 0.2°$ for primary energy E$_o$ = 170 keV) the observed intensity ratios are not obscured by thickness effects. We emphasize that in XAS measurements using the partial yield mode, it is much more difficult to get exact ratios of core-level absorption edges for two different orientations.

EXPERIMENTAL

The EELS measurements have been performed in transmission with a 170 keV spectrometer described elsewhere (Fink 1989). The full width at half maximum (FWHM) energy and momentum resolution was chosen for the valence electron (core electron) excitations to be 0.15(0.4) eV and 0.04(0.2) Å^{-1}, respectively. Experiments were performed at room temperature or at 30 K.

For EELS measurements in transmission thin films with a thickness of about 1000 Å are needed. Films of $YBa_2Cu_3O_7$ were cut from single crystals by an ultramicrotome using a diamond knife. Subsequently, they were mounted on standard electron microscope grids. Almost single-crystalline films of $YBa_2Cu_3O_7$ were grown epitaxially on freshly cleaved CaO single crystals by sputtering (Xi et al. 1988). The films could be removed from the substrate and were mounted on grids. $YBa_2Cu_3O_6$ films were obtained from $YBa_2Cu_3O_7$ films by annealing them under ultra-high vacuum conditions at temperatures up to 500°C. Thin films of $Bi_2Sr_2CaCu_2O_8$ single crystals were obtained by peeling them from single crystals with a tape.

We emphasize that EELS in transmission is not a surface sensitive method as, e.g., photoelectron spectroscopy or XAS in the partial yield mode. Therefore, the problem of preparing surfaces representing bulk properties is avoided.

RESULTS AND DISCUSSION

<u>Valence Electron Excitations</u>

In Fig. 1 we show typical loss functions of a $YBa_2Cu_3O_7$ single crystal (solid line) and of an epitaxially grown film (dashed line). They were measured at finite momentum transfer $q = 0.1$ Å^{-1} in the **a,b** plane (CuO_2 plane) in order to reduce contributions from surface plasmons. The q value is small in comparison with the extension of the Brillouin zone and, therefore, the data can be directly compared with those derived from optical spectroscopy. The loss functions were obtained by removing contributions from the direct beam (for $E \leq 0.5$ eV) and from double scattering. The absolute value of the loss function was obtained by satisfying $Re[-1/\varepsilon(0,\mathbf{q})] = 0$. As described in the Introduction, the real and the imaginary part

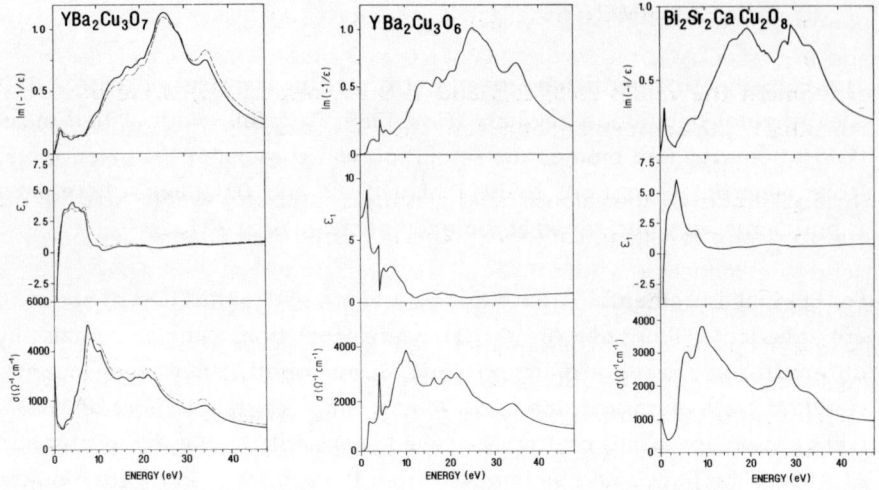

Fig. 1. Loss functions Im(-1/ε(q,ω), real parts of the dielectric functions $\varepsilon_1(q,\omega)$, and optical conductivities $\sigma(q,\omega)$ for single crystals of $YBa_2Cu_3O_7$ (solid line: film cut from a single crystal; dashed line: epitaxially grown film), $YBa_2Cu_3O_6$ and $Bi_2Sr_2CaCu_2O_4$. The direction of the momentum transfer **q** was chosen parallel to the **a,b** plane (CuO_2 planes). The value $q = 0.1$ Å^{-1} is small in comparison to the Brillouin zone. Therefore, the data can be directly compared with optical data.

of the dielectric function were derived by a Kramers-Kronig analysis. For the twinned single crystal $YBa_2Cu_3O_7$ only some not well defined mean value of the components of the dielectric tensor in the **a,b** plane can be obtained. In Fig. 1 we show ε_1 and the optical conductivity σ which is proportional to $\omega\varepsilon_2$. Similar data for $YBa_2Cu_3O_6$ and for $Bi_2Sr_2CaCu_2O_8$ are shown in Fig. 1, as well. All data shown in Fig. 1 are derived from measurements performed at room temperature. EELS data measured in transmission have been already previously published for $YBa_2Cu_3O_7$ by Yuan *et al.* (1988), Chen *et al.* (1988), Tarrio and Schnatterly (1988), Geserich *et al.* (1988b) and Fink (1989). The loss spectra are dominated by broad peaks at energies ranging between 20 and 30 eV. They can be assigned to the collective excitations of all valence electrons, i.e., to highly damped valence band plasmons. A calculation in a free electron model yields for $YBa_2Cu_3O_7$, $YBa_2Cu_3O_6$ and $Bi_2Sr_2CaCu_2O_8$ the values 26.6, 26.0 and 24.5 eV, respectively, while in the experiment the values 25.5, 25.5 and 19.5 eV, respectively, were obtained. For $YBa_2Cu_3O_{7-x}$ the agreement between theory and experiment is rather good.

For $Bi_2Sr_2CaCu_2O_8$ the valence band plasmon is probably shifted to lower energy due to strong low-lying core-level excitations between 23 and 32 eV which can clearly be seen in the loss function and in the optical conductivity. They can partially be assigned to excitations of Bi 5d electrons into unoccupied states having Bi 6p character. Thus, in principle information on the Bi 6p electrons near the Fermi level could be obtained from studies of these excitations. Since there is a 3.1 eV spin-orbit splitting of the $5d_{5/2}$ and $5d_{3/2}$ levels the lower part of these excitations should be related to $5d_{5/2}$-$6p_{3/2}$ transitions (in the atomic limit $5d_{5/2}$-$6p_{1/2}$ transitions are not allowed) while the upper part is predominantly caused by $5d_{3/2}$-$6p_{1/2}$ transitions (in the atomic limit $5d_{3/2}$-$6p_{3/2}$ transitions are five times weaker). Similar spectra on Bi metal and Bi_2O_3 have been published by Benbow and Hurych (1976). One may think that the steep rise in the loss function at 29.1 eV with a total width of only 0.3 eV is related to a Fermi edge of the Bi 6p electrons. However, according to band-structure calculations (Hybertsen and Mattheiss, 1988; Krakauer and Pickett, 1988; Massidda et al. 1988), X-ray absorption spectroscopy (Himpsel, 1988; Kuiper et al. 1988), EELS (Nücker et al. 1988c) and inverse photoelectron spectroscopy (Claessen et al. 1988), there should be a much higher second rise 2 eV above E_F due to flat bands of the BiO planes. As this second rise is not observed, we ascribe the rise at 29.1 eV to transitions from the $5d_{3/2}$ level into these flat BiO bands having $6p_{1/2}$ character. This assignment is supported by the fact that the binding energy of the Bi $5d_{3/2}$ level relative to the Fermi level has been determined to be about 28.2 eV (Meyer et al. 1988). The Fermi edge of the wide bands of the BiO planes is then probably obscurred by $5d_{5/2}$ transitions. The latter excitations, however, show no edge due to flat BiO bands. The reason for this may be that the bottom of these bands has predominantly Bi $6p_{1/2}$ character and, therefore, the $5d_{5/2}$-$6p_{1/2}$ transitions are rather low in intensity and start at 26.0 eV or below. The $5d_{5/2}$-$6p_{3/2}$ transitions are realized at about 28 eV. At present it is not clear whether a small edge at 23.2 eV is caused by the $5d_{5/2}$-$6p_{1/2}$ transition to the Fermi edge of the wide BiO bands or by Ca 3d excitations. In the energy range under consideration, also Sr 4p excitations may appear.

There are various low-lying core-level excitations in the $YBa_2Cu_3O_{7-x}$ spectra which will be discussed elsewhere (Koch et al. 1988) or have already been discussed in the literature. In the superconducting compounds $YBa_2Cu_3O_7$ and $Bi_2Sr_2CaCu_2O_8$ there is a Drude-like decay of the optical conductivity below 1 eV and a plasmon at 1.4 and 1.0 eV, respectively. These plasmons due to the charge carriers, (holes on the O sites, see below) are related to the zero crossing of ε_1. In $YBa_2Cu_3O_7$ there are various interband transitions at 0.5, 1.4 and 3.0 eV which may be partially assigned to d-d excitations enhanced by the coupling to the charge carriers. (Weber 1988, Geserich et al. 1988a, Geserich et al. 1988b) or partially to charge transfer transitions between O 2p and Cu 3d states. In $Bi_2Sr_2CaCu_2O_7$ a broad peak probably due to charge transfer transitions is observed at ~2.5 eV. The sudden rise in the optical conductivity above about 4 eV in all these compounds can probably be ascribed to excitations in BaO, SrO and BiO planes. In the magnetic semiconductor $YBa_2Cu_3O_6$ there is a clear gap of 1.7 eV followed by a very sharp excitation at 4.1 eV, already detected in optical spectroscopy. Since this peak is very pronounced for **q** in the **a,b** plane and almost absent in spectra (not shown) for **q** parallel to the **c** axis we ascribe this excitation to an exciton in the BaO planes and not to the CuO_2 dumbell remaining from the ribbons when removing O from $YBa_2Cu_3O_7$. For CuO_2 excitons a polarization parallel to the **c** axis is expected (Garriga et al. 1988). Finally, we mention that the exciton shows a strong dispersion in momentum transfer which will be described elsewhere (Koch et al. 1988).

In Fig. 2 we show the momentum dependence of the low-energy part of the loss function for the superconductors $YBa_2Cu_3O_7$ and $Bi_2Sr_2CaCu_2O_8$. Since there is a considerable contribution of the direct beam, in particular at higher momentum transfers, the loss function at lower energy ($E \lesssim 1$ eV) is not exactly known and the extrapolation to zero energy is shown by a dashed line. As already mentioned above, the maxima of the loss function in this energy range are caused by the plasmons of the charge carriers which are related to holes on O sites (see below). The direction of the momentum transfer was chosen in the **a,b** plane. For $Bi_2Sr_2CaCu_2O_8$ this means that collective excitations of charge carriers in the CuO_2 and BiO planes with wave vectors in these planes are measured. These plasmons may have two- or three dimensional character depending on the coupling between the planes. In $YBa_2Cu_3O_7$ the situation is more complicated because of the one-dimensional chains. In this case the dielectric functions are also anisotropic in the **a,b** plane and the plasmon energy along the **a** axis should be different from that along the **b** axis. This has been detected by Tanaka et al. (1988)

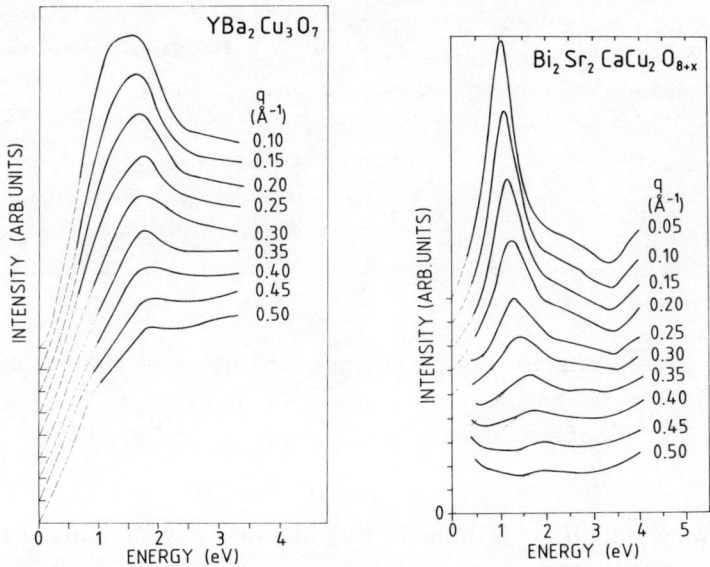

Fig. 2. Low energy loss-function as a function of momentum transfer q for $YBa_2Cu_3O_7$ and $Bi_2Sr_2CaCu_2O_8$ showing the plasmons caused by the charge carriers (holes on O sites). The direction of the momentum transfer **q** is in the a,b plane (CuO_2 plane).

on small untwinned crystals using optical spectroscopy. The present experiments were performed on twinned crystals and no anisotropy of the plasmon was detected up to now, although, contrary to optical spectroscopy on twinned crystals, a difference between the [100] and the [110] direction should be observable in EELS. Two different plasmon energies for **q** along the **a** and the **b** axis may be the reason why in $YBa_2Cu_3O_7$ the plasmon for small momentum transfer has a trapezoidal form with a width of almost 2 eV. This is also in line with the observation of a rather narrow plasmon at small momentum transfer with a width of only 0.7 eV for $Bi_2Sr_2CaCu_2O_8$ where no one-dimensional chains exist. With increasing momentum transfer the plasmon energy shows a dispersion for both systems which is shown in Fig. 3. The dispersion is linear in q^2. For $YBa_2Cu_3O_7$ the plasmon energy is connected with rather large errors because of the large width of the plasmon and because the plasmon intensity strongly decreases at

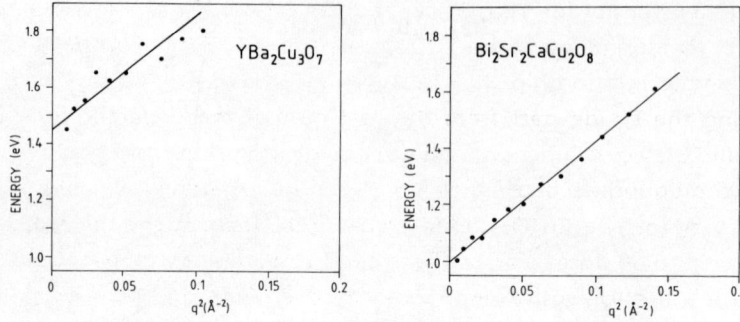

Fig. 3. Dispersion of the energy of charge carrier plasmon as a function of momentum transfer q for the superconductors YBa$_2$Cu$_3$O$_7$ and Bi$_2$Sr$_2$CaCu$_2$O$_8$.

higher momentum transfer probably due to a decay into interband transitions. There is also a large background at higher q which may be explained by incorrect subtraction of the elastic beam or by interband transitions. In Bi$_2$Sr$_2$CaCu$_2$O$_8$ the background is smaller. A strong decay of the plasmon at higher q is observed as well.

As in the present measurements the momentrum transfer q_z perpendicular to the CuO$_2$ layers was rather small ($q_z < 0.01$ Å^{-1} given by the misorientation of the sample) the electrons move in phase on different planes and therefore the measured plasmon dispersion corresponds to three-dimensional behaviour and the dispersion can be evaluated in this case as in a usual three-dimensional sample. The plasmon dispersion is then given in a first approximation by

$$E_p(q) = E_p(0) + \frac{\hbar^2}{m} \cdot \alpha \cdot q^2$$

where the plasmon energy at zero momentum transfer is $E_p(0) = \hbar(ne^2/\varepsilon_\infty \cdot m^*)^{1/2}$ and the dispersion coefficient α is given by $\alpha = 3/10$ $(mv_F^2/E_p(0))$. For a given background dielectric function ε_∞ and a given charge carrier density, the effective mass m^* can be calculated from the experimental value of $E_p(0)$. Furthermore, the experimental value of the dispersion constant α yields the Fermi velocity v_F and with the derived effective mass, the Fermi energy E_F can be calculated. The charge-carrier density for $YBa_2Cu_3O_7$ and $Bi_2Sr_2CaCu_2O_8$ can be obtained from Hall effect measurements to be $5 \cdot 10^{21}$ cm^{-3} and $3 \cdot 10^{21}$ cm^{-3}, respectively, if we apply a simple relationship $R_H = 1/ne$ (Wang et al. 1987; Takagi et al. 1988). Subtracting the Drude part from the real part of the dielectric function the background dielectric function ε_∞ in the energy range 1 to 2 eV could be derived for both compounds to be close to 4.5. Using the experimental values $E_p(0) = 1.4$ eV and 1.0 eV for $YBa_2Cu_3O_7$ and $Bi_2Sr_2CaCu_2O_8$, respectively, the effective mass m^* is derived to be close to one in both cases in this energy range. The dispersion coefficients α are for both compounds near 0.6. Then the Fermi velocity v_F for $YBa_2Cu_3O_7$ and $Bi_2Sr_2CaCu_2O_8$ is derived to be $0.7 \cdot 10^8$ and $0.6 \cdot 10^8$ cm/sec, respectively. Using $m^* = 1$ the Fermi energies for the two compounds are 1.4 eV and 1.0 eV. These values for v_F and E_F are slightly smaller than values for normal metals and also smaller than expected from band-structure calculations ($E_F \sim 2$ eV). The reason for this may be a reduced band width due to electron correlations on the Cu sites.

Because of the discussion on pairing mechanisms due to two-dimensional acoustic plasmons (Kresin and Morawitz, 1988; Ruvalds 1988) it would be highly interesting to perform measurements as a function of the momentum transfer q_z perpendicular to the CuO_2 layers. This would yield information on the existence of acoustic plasmons, i.e., out-of-phase motion of charge carriers on adjacent planes. First measurements with non-zero q_z on $Bi_2Sr_2CaCu_2O_8$ indicate that with increasing q_z the plasmon energy decreases as expected for two-dimensional plasmons. Further systematic studies of this problem are under way.

At the end of the section on valence band excitations we would like to mention that we have also measured the loss function of $Bi_2Sr_2CaCu_2O_8$ at T = 30 K which is well below the superconducting transition temperature $T_c = 83$ K. Neither the loss function nor the plasmon dispersion shows a significant difference between room temperature and 30 K. This indicates that changes of the electronic structure as a function of temperature are well below our energy resolution of 0.15 eV.

Core Electron Excitations

In this section we describe core-level absorption edges of the O 1s and Cu $2p_{3/2}$ levels of the high-T_c superconductors $YBa_2Cu_3O_7$ and $Bi_2Sr_2CaCu_2O_8$ and of the antiferromagnetic semiconductor $YBa_2Cu_3O_6$. These absorption edges are related to transitions from the narrow core level into unoccupied states. Since the transitions start from a rather localized core level, the transition matrix element is non-zero only for final states overlapping with the core level. Therefore, the *local* unoccupied density of states is probed. Furthermore, because we measure at momentum transfers small compared to the inverse mean value of the core-orbital radii, dipole selection rules apply and in the O 1s and Cu 2p edges, unoccupied O 2p and Cu 3d states, respectively, are probed. Both states are extremely important in the electronic structure of the high-T_c superconductors because, according to LDA band structure calculations, they are supposed to form the density of states at the Fermi level by a $3d_{x^2-y^2}$-O $2p_{x,y}$ band from the CuO_2 planes. Deviations from the single-particle picture are expected due to the on-site Coulomb energy (U) of the 3d electrons. In the case of a large U and a smaller charge transfer energy Δ (for transitions from O to Cu atoms) Zaanen *et al.* (1985) have predicted a picture of a charge transfer insulator where a band gap is formed between an occupied O 2p band and an unoccupied upper Cu 3d Hubbard band. The lower occupied Cu 3d Hubbard band should be below the O 2p band. Upon doping, holes should be formed in the O 2p band, i.e., the charge carriers should have predominant O 2p character. Experimentally, this has been derived indirectly from the La absorption edges (Tranquada *et al.* 1987), from photoemission spectroscopy (Fujimori *et al.* 1987), and from Cu $2p_{3/2}$ absorption edges (Bianconi *et al.* 1987). The most direct evidence for holes on O sites comes from the O 1s edges as a function of dopant concentration measured by EELS (Nücker *et al.* 1987, Nücker *et al.* 1988a). Typical spectra are shown in Fig. 4. The Fermi level is indicated by the dashed line. It has been determined by measuring the binding energies E_B of the O 1s level relative to the Fermi level by X-ray induced photoelectron spectroscopy (XPS). In both systems E_B is close to 528 eV independent of the dopant concentration.

In the undoped antiferromagnetic semiconductors ($x = 0$ and $y \geq 0.6$) there is no spectral weight at the Fermi level. This indicates a gap in the order of several eV in agreement with optical data and low-energy EELS data described in the previous section but in disagreement with band-structure calculations. The steep rise of intensity about 3 eV above E_F is due to La 5d and 4f or Ba 5d and 4f and Y 4d states hybridized with O 2p states. With increasing dopant concentration there is more and more spectral weight at the Fermi level which was ascribed to transitions from the O 1s level into the holes formed upon doping in the upper part of the O 2p band. Thus, the density of states of the holes on O sites can be probed directly. Since up to now, no changes of the density of states of the Cu 3d states upon doping (measured by XPS) have been observed (Nücker *et al.* 1987) this leads to

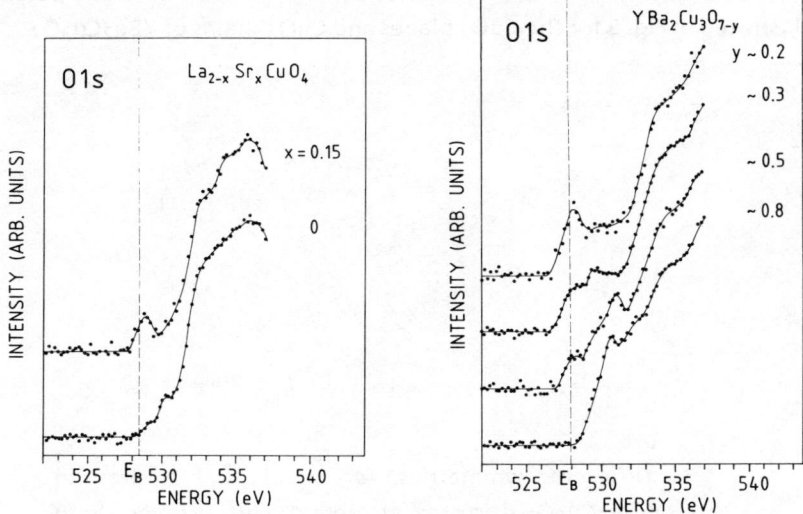

Fig. 4. O 1s absorption edges for polycrystalline cuprates measured by electron energy-loss spectroscopy. Left: La_2CuO_4 and $La_{1.85}Sr_{0.15}CuO_4$; right: $YBa_2Cu_3O_{7-y}$ for $y = \sim 0.2$–0.8.

the picture for the electronic structure of the high-T_c superconductors described at the beginning of this section. The main point of this picture is that the holes formed upon doping have predominant O 2p character.

As outlined in the Introduction, the symmetry of the unoccupied density of states can be probed by orientation dependent measurements of core-electron excitations on single crystals. These measurements are of particular interest because at present there is a considerable discussion on the symmetry of holes on the O sites. The LDA band-structure calculations which neglect short range correlations and the more delocalized theories predict holes in the CuO_2 planes (or CuO_3 ribbons) which have $p_{x,y}$ character on the O sites along the Cu-O-Cu axis (σ-holes). In a rather localized picture derived by calculations on finite CuO clusters (Guo et al. 1988) the holes on O are again in the CuO_2 planes (or CuO_3 ribbons) but now they are perpendicular to the Cu-O-Cu axis (in-plane π-holes). Finally, Johnson et al. (1988) postulate the formation of holes on O sites in p_z orbitals perpendicular to the CuO$_2$ planes (out-of-plane π-holes). The three possibilities are illustrated in Fig. 5 for the CuO$_2$ planes and CuO$_3$ chains of YBa$_2$Cu$_3$O$_7$.

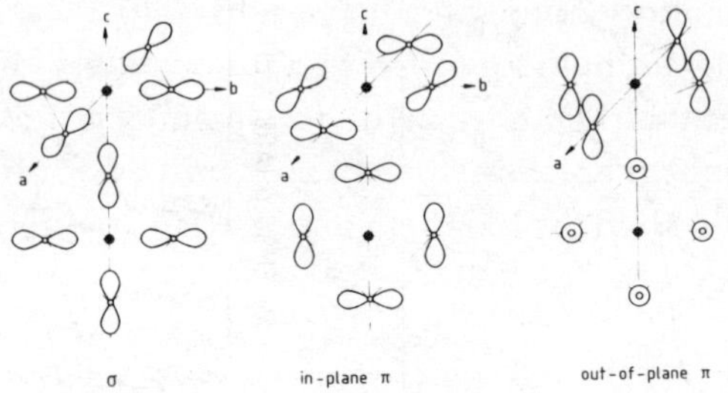

Fig. 5. Different symmetries for holes on O sites in YBa$_2$Cu$_3$O$_7$ discussed at present. Left: σ holes along the CuO bond; middle: in-plane π holes perpendicular to the Cu-O bond; right: out-of-plane π holes perpendicular to the Cu-O bond. For each type of holes, a square of CuO$_2$ planes and a square of the CuO$_3$ ribbons are shown. Closed circles: Cu atoms; open circles: O atoms.

Assuming σ holes, only O(4) atoms (in the ribbons on the **c** axis) contribute to the O 1s spectrum for **q**∥**c** while the O(2) and O(3) sites (in the CuO$_2$ planes) and the O(1) site (in the ribbons on the **b** axis) contribute to the **q**∥**a,b** spectrum. For an equal distribution of holes among the four O sites, the intensity ratio ($I_{x,y}/I_z$) between hole states reached for **q**∥**a,b** to those reached for **q**∥**c** should be 2. Assuming in-plane π holes this intensity ratio should be 1.25. Finally, for out-of-plane π holes $I_{x,y}/I_z$ should be 3/8. Recently, we have measured orientation dependent O 1s absorption edges on single crystals (Nücker *et al*. 1988b, Nücker *et al*. 1988c). The direction of the momentum transfer **q** could be changed to be parallel or perpendicular to the CuO$_2$ planes. Thus, information on the symmetry of the holes on O sites could be obtained. In Fig. 6 we show typical spectra for YBa$_2$Cu$_3$O$_7$ and YBa$_2$Cu$_3$O$_6$. The spectra were corrected for finite momentum transfer resolution. While the states 3 eV above E_F show only a small anisotropy, considerable differences are realized for the states just above the Fermi level (density of states of holes on O sites) for YBa$_2$Cu$_3$O$_7$. The threshold energy is slightly lower for **q∥c** than for **q∥a,b**. This may result from different O 1s binding energies for the 4 different O sites or from different energy positions of the final states relative to E_F. Again there is no density of states in this energy range for $YBa_2Cu_3O_6$ indicating a gap of several eV. In $YBa_2Cu_3O_7$ there are holes with O $2p_{x,y}$ and $2p_z$ symmetry. Comparing the results with those of $YBa_2Cu_3O_6$ the intensity ratio ($I_{x,y}/I_z$) for hole states reached for **q∥a,b** to those reached for **q∥c** is about 2. For σ holes this ratio is too large for an equal distribution of holes and, therefore, the number of holes on the O(4) sites must be decreased relative to that of the other 3 O sites. Then the hole distribution comes out close to that derived in

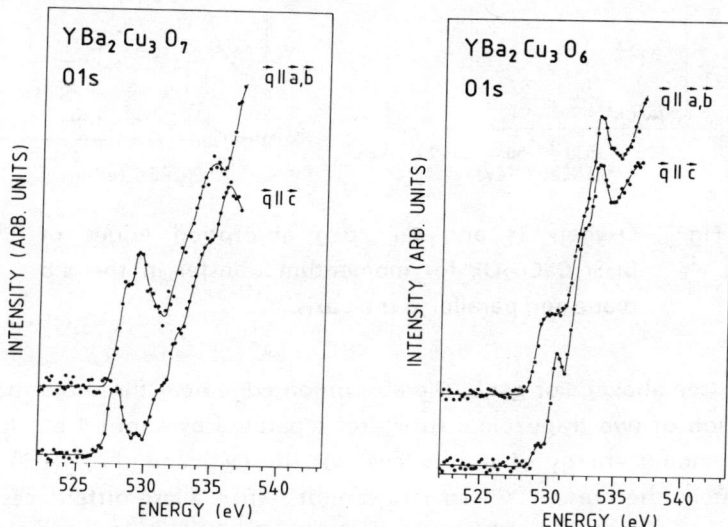

Fig. 6. Oxygen 1s absorption edges of $YBa_2Cu_3O_7$ and $YBa_2Cu_3O_6$ for momentrum transfer in the **a,b** plane an parallel to the **c** axis.

band-structure calculations. It is interesting to note that LDA band-structure calculations predict for the absorption edges for the four differnt O sites almost the same trapezoidal form (typical of two-dimensional bands) with a width of about 2 eV. This trapezoidal form is observed for **q∥c**, indeed. The width of this structure is, however, only about 1.3 eV which may be caused by a slightly higher Fermi level or by strong correlation effects which reduce hopping between Cu and O sites thus decreasing the width of the dpσ band. It is remarkable that the observed width is almost the same as the Fermi energy derived from the plasmon

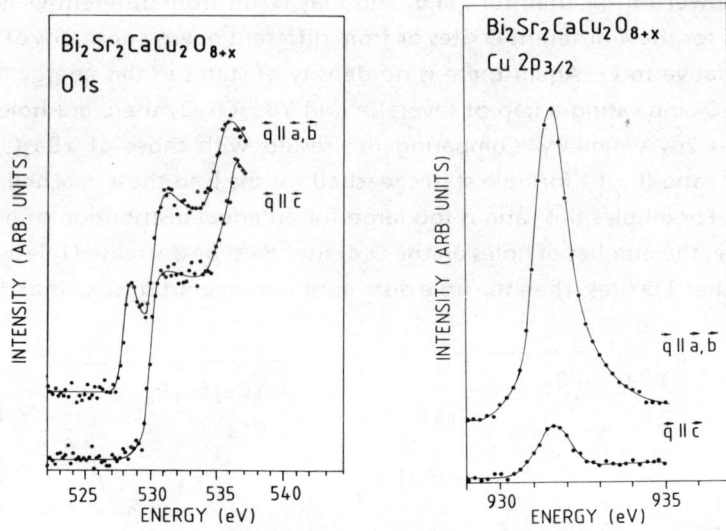

Fig. 7. Oxygen 1s and Cu $2p_{3/2}$ absorption edges of $Bi_2Sr_2CaCu_2O_8$ for momentum transfer in the **a,b** plane and parallel to the **c** axis.

dispersion (see above). For **q**∥**a,b**, the absorption edge near threshold may be a superposition of two trapezoidal structures separated by about 1 eV due to a different binding energy of the 1s level for the O(1) and the O(2,3) atoms. Unfortunately, the present XPS spectra cannot resolve these differences in the binding energies, but recent XPS work on $YBa_2Cu_3O_7$ single crystals (Weaver *et al.* 1988) indicates different binding energies of the four O sites in the order of 1 eV. For in-plane π holes the measured intensity ratio $I_{x,y}/I_z$ is too small for an equal distribution of holes and, therefore, the holes on the O(1) sites should be increased by about 50% to the average number of holes on the other 3 sites. Finally, for out-of-plane π holes the measured $I_{x,y}/I_z$ is by far too large for an equal distribution of holes. In this case, there should be almost no holes in the planes and almost all holes in the ribbons, which is very unlikely.

Orientation dependent O 1s absorption edges of $Bi_2Sr_2CaCu_2O_8$ are shown in Fig. 7. No O $2p_z$ states are observed near threshold indicating that there are no holes on O orbitals perpendicular to the CuO_2 and BiO planes. This clearly rules out all models for high-T_c superconductivity based on out-of-plane π-holes in the CuO_2 planes or holes in p_z orbitals of the apex oxygens. . Our measurements, however, cannot differentiate between O σ-holes or in-plane π-holes. At energies 2 eV above threshold, a strong rise in intensity is observed due to the flat Bi 6p-O 2p bands already discussed in the section on valence band excitations. As these bands are caused by O 2p states of the O(2) sites (in the BiO planes) and O(3) sites (in the SrO planes) hybridized with Bi 6p states, the anisotropy should be small in agreement with the experimental data. The peak 8 eV above E_F is probably caused by Ca 3d states hybridized with O 2p states. To summarize our measurements of the O 1s edges in the high-T_c superconductors we favour at present the existence of holes on O with σ symmetry. On the other hand, we cannot exclude from our experiments holes with in-plane π symmetry. Holes with out-of-plane π symmetry are very unlikely for $YBa_2Cu_3O_7$ or can be excluded for $Bi_2Sr_2CaCu_2O_7$.

In Fig. 7 we show the Cu $2p_{3/2}$ edges of single crystalline $Bi_2Sr_2CaCu_2O_7$ with **q∥a,b** plane and **q∥c** probing unoccupied Cu 3d states. According to band-structure calculations, the Cu holes have predominantly in-plane $3d_{x^2-y^2}$ character. This is in agreement with the experimental results that strong spectral weight is observed at 931.3 eV for **q∥a,b**. For **q∥c**, the intensity is only 10-15% of that for **q∥a,b**. This spectral weight is probably due to $3d_{3z^2-r^2}$ states admixed to the $3d_{x^2-y^2}$ band or due to a partially unoccupied $3d_{3z^2-r^2}$ band. The latter explanation has been given recently by Bianconi et al. (1988a) because in their spectra the line for **q∥c** appears at 400 meV lower energy compared to that for **q∥a,b**. In our spectra we have never observed any energy shift within error bars ($\Delta E = 50$ meV) and, therefore, we explain the line for **q∥c** by an admixture of $3d_{3z^2-r^2}$ to the $3d_{x^2-y^2}$ band. The observed asymmetry of the main line for **q∥a,b** may be explained by the influence of holes on O sites onto the Cu 2p-3d transitions (Bianconi et al. 1987). We emphasize, that this asymmetry, probably related to a second line at 932.8 eV, appears only for **q∥a,b**. This is in line with the result from the O 1s edges, where only holes having x,y symmetry in the a,b planes have been observed.

In the spectra for $YBa_2Cu_3O_7$, shown in Fig. 8, the main line shows strong orientation dependence in the intensity and in the spectral form. For **q∥a,b**, where dominantly $3d_{x^2-y^2}$ states from the planes and to a lesser extent $3d_{z^2-y^2}$ states from the ribbons are reached, the line at 931.3 eV is slightly asymmetric with a tail extending to higher energies probably due to an additional line in this energy range. For **q∥c**, where dominantly $3d_{z^2-y^2}$ states from the ribbons are reached, the total intensity is reduced by a factor of two and a more pronounced shoulder at

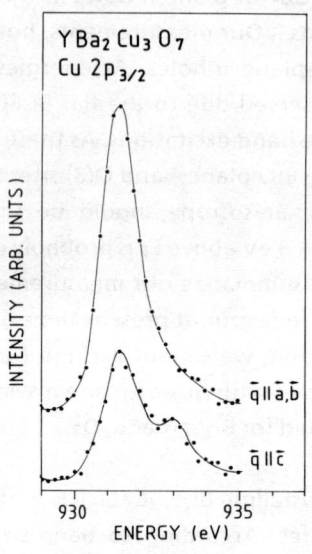
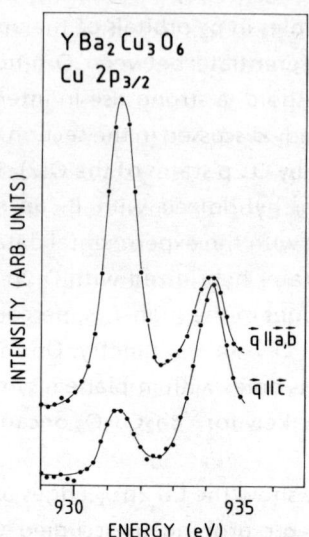

Fig. 8. Copper $2p_{3/2}$ absorption edges of $YBa_2Cu_3O_7$ and $YBa_2Cu_3O_6$ for momentum transfer parallel to the **c** axis and in the **a,b** plane.

932.8 eV is realized. The reduction in intensity by a factor of 2 is not far from the expected value of 2.5 assuming Cu holes only in the Cu b_{1g} states ($3d_{x^2-y^2}$ states in the planes and $3d_{z^2-y^2}$ states in the ribbons). The shoulder at 932.8 eV appears mainly for **q**∥**c**. This result is in agreement with recent results from Bianconi et al. (1988b). From this the authors concluded that the holes on O are mainly formed in $2p_z$ orbitals on the O(4) sites in the BaO layers. As we know, however, from the O 1s spectra, where the density of states of the holes can be measured *directly*, that the number of O $2p_{xy}$ holes is twice that of O $2p_z$ holes, we cannot agree with the interpretation of the orientation dependence of the shoulder at 932.8 eV given by Bianconi et al. (1988b). A possible explanation of the orientation dependence of the shoulder may be that due to the shielding of the Coulomb repulsion, the shift of the shoulder depends strongly on the distance between a hole on O and the Cu atom. Then the holes on O(4) sites should cause the strongest energy shift since the Cu(1)-O(4) distance is only 1.84 Å while the other Cu(1)-O(1) and Cu(2)-O(2,3) distances are about 1.94 Å. A further explanation may be that holes localized in the squares of the ribbons cause a larger energy shift than the more delocalized holes in the planes. Finally, we cannot exclude that the shoulder may be related to holes in the dpπ band which according to LDA band-structure calculations crosses the Fermi level only for the electrons in the ribbons.

In the non-superconducting $YBa_2Cu_3O_6$, the shoulder and the asymmetry of the main lines in the Cu $2p_{3/2}$ spectra both have disappeared confirming that they are caused by the formation of holes due to self-doping with O. For **q∥c** there remains a line which has about 20 % of the intensity for **q∥a,b**. Again we have never observed any energy shift within error bars ($\Delta E = 50$ meV) between the main lines for **q∥a,b** and for **q∥c**. It is generally believed that this line should completely disappear because the Cu(1) atoms in the ribbons should have transformed from divalent to monovalent Cu. This line may again be interpreted by an admixture of $3d_{3z^2-r^2}$ states to the $3d_{x^2-y^2}$ band. Another explanation may be that there is still some divalent Cu in a not perfectly reduced O_6 sample. The strong spectral weight at 934 eV implies the transformation of Cu(1) atoms from divalent ($3d^9$) to monovalent ($3d^{10}$) because in monovalent Cu compounds, always a line near 934 eV is observed. The resonance like structure indicates that also for monovalent Cu there is a considerable amount of unoccupied 3d states. The orientation dependence of this line can be explained by the existence of CuO_2 molecules in $YBa_2Cu_3O_6$ remaining from the ribbons in $YBa_2Cu_3O_7$ upon removing the O(1) atoms. Then the unoccupied states should have predominantly $3d_{3z^2-r^2}$ character in agreement with the experimental results.

ACKNOWLEDGEMENTS

We thank D. Ewert, T. Wolf, X.X. Xi, H.C. Li, B. Scheerer, B. Gegenheimer, and Z.X. Zhao for sample preparation. We gratefully acknowledge stimulating discussions with W. Weber, H. Rietschel, G. Roth, J. Zaanen, G.A. Sawatzky, and J. Fuggle.

REFERENCES

Benbow, R.L. and Z. Hurych (1976). Ultraviolet photoemission studies of the oxidation of thin Bi films using synchrotron radiation. *Phys. Rev. B* **14**, 4295.

Bianconi, A., A. Congin Casatellano, M. De Santis, P. Rudolf, P. Lagarde, A.M. Flank and A. Marcelli (1987). $L_{2,3}$ XANES of the High-T_c Superconductor $YBa_2Cu_3O_{\sim 7}$ with Variable Oxygen Content. *Solid State Commun.* **63**, 1009.

Bianconi, A., P. Castrucci, M. De Santis, A. Di Cicco, A. Fabrizi, A.M. Flank, P. Lagarde, H. Katayama-Yoshida, A. Kotani, A. Marcelli, Zhao Zhongxian and C. Politis (1988a). Symmetry of the Hole States in BiCaSrCuO High-T_c Superconductors. *Mod. Phys. Lett.* B2, *1313.*

Bianconi, A., M. DeSantis, A. Di Cicco, A.M. Flank, A. Fontaine, P. Lagarde, H.Katayama-Yoshida, A. Kotani, A. Marcelli (1988b). Symmetry of the 3d9 ligand hole induced by doping in YBa$_2$Cu$_3$O$_{7-\delta}$. *Phys. Rev. B* 38, 7196.

Chen, C.H., L.F. Schneemeyer, S.H. Liou, M. Hong, J. Kwo, H.S. Chen and J.V. Waszczak (1988) Electronic Excitations of YBa$_2$Cu$_3$O$_{7-x}$ Superconductor: A Study by Transmission Electron Energy-Loss Spectroscopy with an Electron Microprobe. *Phys. Rev. B* 37, 9780..

Claessen, R., R. Manzke, H. Carstensen, B. Burandt, T. Buslaps, M. Skibowski, J. Fink (1988). A Surface Study of the 83 K Superconductor Bi$_2$Sr$_2$CaCu$_2$O$_8$ by LEED and Angle-Resolved Inverse Photoemission Spectroscopy. Preprint.

Fink, J. (1989). Recent Development in Energy-Loss Spectroscopy. *Adv. Electron. Electron Phys.* 75, 121.

Fujimori, A., E. Takayama-Muromachi, Y. Uchida and B. Okai (1987). Spectroscopic Evidence forStrongly Correlated Electronic States in La-Sr-Cu and Y-Ba-Cu Oxides. *Phys. Rev. B* 35, 8814.

Garriga, M., J. Humliček, M. Cardona, E. Schönherr (1988). Effects of Oxygen Deficiency on the Optical Spectra of YBa$_2$Cu$_3$O$_{7-x}$. *Solid State Commun.* 66, 1231.

Geserich, H.P., G. Scheiber, J. Geerk, H.C. Li, G. Linker, W. Assmus and W. Weber (1988a). Optical Spectra of Superconducting YBa$_2$Cu$_3$O$_{7-\delta}$ Films: Evidence for Cu^{++} d-d Transitions *Europhys. Lett.* 6, 277.

Geserich, H.P., G. Scheiber, J. Geerk, H.C. Li, W. Weber, H. Romberg, N. Nücker, J. Fink and B. Gegenheimer (1988b). Cu d-d Orbital Transitions and Charge-Transfer Excitations in High-T_c Superconductors. In: *High-T_c Superconductors*, Plenum, New York, in print.

Guo, Y., J.-M. Langlois and W.A. Goddard III. (1988). Electronic Structure and Valence-Bond Band Structure of Cuprate Superconducting Materials. *Science* 239, 896.

Himpsel, F.J., G.V. Chandrashekhar, A.B. McLean and Shafer (1988). Orientation of the O 2p Holes in Bi$_2$Sr$_2$Ca$_1$Cu$_2$O$_8$. *Phys. Rev. B* 38, 11946.

Hybertsen, M.S. and L.F. Mattheiss (1988). Electronic Band Structure of CaBi$_2$Sr$_2$Cu$_2$O$_8$. *Phys. Rev. Lett.* 60, 1661.

Johnson, K.H., M.E. McHenry, C. Counterman, A. Collins, M.M. Donovan, R.C.

O'Handley and G. Kalonji (1988). Quantum Chemistry and High-T_c Superconductivity. *Physica C* 153-155, 1165.

Koch, B., H.P. Geserich, M. Dürrler, H. Romberg, N. Nücker, J. Fink, X.X. Xi, T. Wolf, W. Assmus and B. Gegenheimer (1988). to be published.

Krakauer, H. and W.E. Pickett (1988). Effect of Bismuth on High-T_c Cuprate Superconductors: Electronic Structure of $Bi_2Sr_2CaCu_2O_8$. *Phys. Rev. Lett.* 60, 1665.

Kresin, V.Z. and Morawitz (1988). Layer Plasmons and High-T_c Superconductivity. *Phys. Rev. B* 37, 7854.

Kuiper, P. and G.A. Sawatzky (1988). Private communication.

Meyer III, H.M., D.M. Hill, J.H. Weaver, D.L. Nelson and C.F. Gallo (1988). Occupied electronic states of single-crystal $Bi_2Ca_{1+x}Sr_{2-x}Cu_2O_{8+y}$. *Phys. Rev. B* 38, 7144.

Massidda, S., J. Yu and A.J. Freeman (1988). Electronic Structure and Properties of $Bi_2Sr_2CaCu_2O_8$, the Third High-T_c Superconductor. *Physica C* 152, 251.

Nücker, N., J. Fink, B. Renker, D. Ewert, C. Politis, P.J.W. Weijs and J.C. Fuggle (1987). Experimental Electronic Structure Studies of $La_{2-x}Sr_xCuO_4$. *Z. Phys. B* 67, 9.

Nücker, N., J. Fink, J.C. Fuggle, P.J. Durham and W.M. Temmerman (1988a). Evidence for Holes on Oxygen Sites in the High-T_c Superconductors, $La_{2-x}Sr_xCuO_4$ and $YBa_2Cu_3O_{7-y}$. *Phys. Rev. B* 37, 5158.

Nücker, N., J. Fink, J.C. Fuggle, P.J. Durham and W.M. Temmerman (1988b). Electronic Structure of $La_{2-x}Sr_xCuO_4$ and $YBa_2Cu_3O_{7-y}$. *Physica C* 153-155, 119.

Nücker, N., H. Romberg, X.X. Xi, J. Fink, B. Gegenheimer and Z.X. Zhao (1988c). On the Symmetry of Holes in High-T_c Superconductors. *Phys. Rev. B*, submitted.

Nücker, N., H. Romberg, S. Nakai, B. Scheerer, J. Fink and Z.X. Zhao (1988d). to be published.

Raether, H. (1980). Excitation of Plasmons and Interband Transitions by Electrons. In: *Springer Tracts in Modern Physics*, Springer Verlag, Berlin, Vol. 88.

Ruvalds, J. (1987). Plasmons and High-Temperature Superconductivity in Alloys of Copper Oxides. *Phys. Rev. B* 35, 8869.

Takagi, T., E. Eisaki, S. Uchida, A. Maeda, S. Tajima, K. Uchinokura and S. Tanaka (1988). Transport and Optical Studies of Single Crystalline Bi-Sr-Ca-Cu-O. *Nature* 332, 236.

Tanaka, J., K. Kamiya, M. Shimizu, M. Simada, C. Tanaka, H. Ozeki, K. Adachi, K. Iwahashi, F. Sato, A. Sawada, S. Iwata, H. Sakuma and S. Uchiyama (1988). Optical Spectra and Electronic Structures of High-T_c Oxide Superconductors. Preprint.

Tarrio, C. and S.E. Schnatterly (1988). Inelastic Electron Scattering in the High-T_c

Compound YBa$_2$Cu$_3$O$_{7-x}$. *Phys. Rev. B* <u>38</u>, 921.

Tranquada, J.M., S.M. Heald, A. Moodenbaugh and M. Suenaga (1987). X-Ray Absorption Studies of La$_{2-x}$(Ba,Sr)$_x$CuO$_4$ Superconductors. *Phys. Rev. B* <u>35</u>, 7187.

Wang, Z.Z., J. Clayhold, N.P. Ong, J.M. Tarascon, L.H. Greene, W.R. McKinnon and G.W. Hull (1987), *Phys. Rev. B* <u>36</u>, 7222.

Weaver, J.H., H.M. Meyer III, T.J. Wagener, D.M. Hill, Y. Gao, D. Peterson, Z. Fisk and A.J. Arko (1988). Valence Bands, Oxygen in Planes and Chains, and Surface Changes for Single Crystals of M$_2$CuO$_4$ and MBa$_2$Cu$_3$O$_x$ (M = Pr,Nd,Eu,Gd). *Phys. Rev. B* <u>38</u>, 4668.

Weber, W. (1988). A Cu d-d Excitation Model for the Pairing in the High-T$_c$ Cuprates. *Z. Phys. B* <u>70</u>, 323.

Xi, X.X., H.C. Li, J. Geerk, G. Linker, O. Meyer, B. Obst, F. Ratzel, R. Smithey and F. Weschenfelder (1988). Growth and Properties of YBaCu-Oxide Superconducting Thin Films Prepared by Magnetron Sputtering. *Physica C* <u>153-155</u>, 794.

Yuan, J., L.M. Brown and W.Y. Liang (1988). Electron Energy-Loss Spectroscopy of the High-Temperature Superconductor Ba$_2$YCu$_3$O$_{7-x}$. *J. Phys. C: Solid State Phys.* <u>21</u>, 517.

Zaanen, J., G.A. Sawatzky and J.W. Allen (1985). Band Gaps and Electronic Structure of Transition-Metal Compounds. *Phys. Rev. Lett.* <u>55</u>, 418.

JOINT ANALYSIS OF Cu L_3 XAS AND XPS SPECTRA OF HIGH T_C SUPERCONDUCTORS FOR DETERMINATION OF THE STATES INDUCED BY DOPING

M. De Santis, A. Bianconi, A. Clozza, P. Castrucci,
A. Di Cicco, M. De Simone

Dipartimento di Fisica, Università degli Studi di Roma "La Sapienza" I-00185 Roma, Italy

A.M. Flank, P. Lagarde

LURE CNRS-CEA-MEN, Batiment 209D, Université Paris sud, F-91405 Orsay, France

J. Budnick

Department of Physics and Institute of Material Science, University of Connecticut, Storrs, CT 06268 Connecticut, USA

P. Delogu, A. Gargano, R. Giorgi, T. D. Makris

ENEA, Centro Ricerche Casaccia, Laboratorio Superfici e Metallurgia Fisica, Via Anguillarese 301, I-00060 Roma, Italy

ABSTRACT

The joint analysis of the Cu $2p_{3/2}$ core level x-ray absorption and photoemission spectra is applied to the divalent compound La_2CuO_4 to determine the energies to create a d hole $\varepsilon_d(T)$ and an oxygen hole $\varepsilon_L(T)$, where T is the hybridization energy, in the ground state.
The same approach is applied to the trivalent Cu compound $NaCuO_2$ which is described by mixing of the two-hole configurations $3d^9\underline{L}$, $3d^{10}\underline{L}^2$ and $3d^8$ in the ground state.
The two hole configurations of the states induced by Sr doping in the superconducting material $La_{2-x}Sr_xCuO_4$ are obtained with the same approach. The $3d^8$ configuration contributes with only 1% to the ground state configuration, with hybridization energy T=2.2 eV, because it is found at high energy.
In this work we discuss the important role of the Coulomb interaction U_{cL} between the core hole $\underline{c}$ and the ligand hole $\underline{L}$ in the final state and the interaction U_{dL} between the $\underline{d}$ hole and the ligand hole $\underline{L}$ in the ground state.

INTRODUCTION

The electronic structure of insulating correlated oxides such as NiO, CeO_2, CuO, and PrO_2 has been object of large interest in these last years (Zaanen *et al.* 1985, Bianconi *et al.*, 1988a; and 1989; and references cited therein). In these compounds a strong hybridization between a localized metal orbital and the delocalized 2p oxygen states coexists with a large electronic d-d or f-f Hubbard repulsion. The cuprate high-T_c superconductors belong to this

The $|3d^{10}\underline{L}\rangle$ is the first excited state of the ionic configuration obtained by a charge transfer excitation giving one hole in a 2p orbital. The XAS white line is due to the $\underline{c}3d^{10}$ final state. The two XPS peaks are due to the hybridization of the two configurations $\underline{c}3d^9$ and $\underline{c}3d^{10}\underline{L}$.

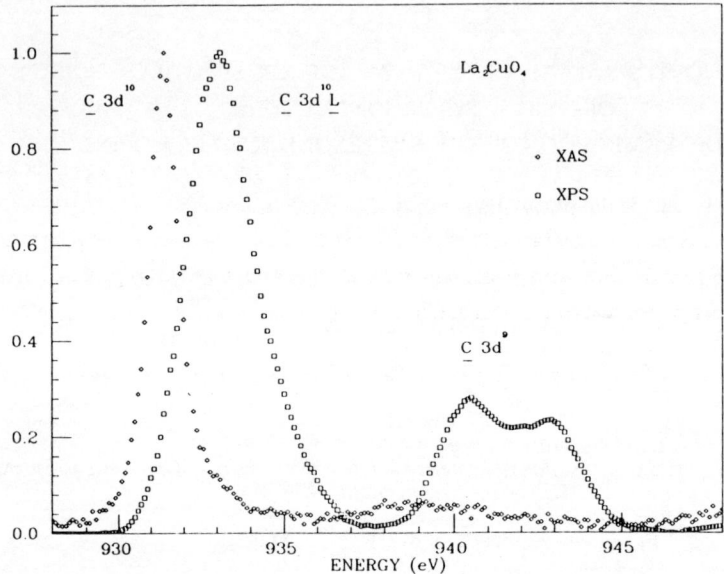

Fig. 1. La_2CuO_4 L_3-edge x-ray absorption and x-ray photoemission spectra.

In the frame of this simple molecular orbital model we can calculate the final state and ground state energies as a function of the parameters Δ, ε_L, ε_d, Q, U_{cL} and T, where $\Delta = \varepsilon_L - \varepsilon_d$ is the $\underline{d}\rightarrow\underline{L}$ transfer energy, ε_L and ε_d are the energies for creating one hole in the oxygen 2p orbital and in the Cu 3d orbital respectively; Q is the $\underline{d}$ - $\underline{c}$ hole interaction; T is the hybridization integral between the two configurations (T=$\langle 3d^9|H|3d^{10}\underline{L}\rangle$); U_{cL} is the pair hole interaction between $\underline{c}$ and $\underline{L}$ orbital.

The zero of the energy scale is taken to be the energy of the pure copper $3d^{10}$ (Cu^+) oxygen $2p^6$ (O^{2-}) ionic configurations, which is often considered as the vacuum state $|o\rangle$ (i.e. the space without any electronic hole), therefore the energies ε_d, ε_L are the energies to add one non interacting d or L hole particle in this space.

We have measured the intensity ratio I_2/I_1=0.43 and the energy separation $E_2^{XPS}-E_1^{XPS}$=8.5 eV between the two x-ray photoemission peaks: E_2^{XPS} is the energy of the $\underline{2p}3d^{10}\underline{L}$ main line in class of materials. Since the first experimental investigation by x-ray absorption and x-ray photoemission spectroscopy (Bianconi *et al.* 1987; Fujimori *et al.* 1987) the d-d Coulomb repulsion and the Cu-O hybridization energies were found to be about 6-8 eV and 2-3 eV respectively.

In this work we have extracted from the joint analysis of Cu L_3 x-ray absorption and photoemission spectra the relevant energy levels of the many body configurations of the divalent antiferromagnetic La_2CuO_4 and of the trivalent diamagnetic $NaCuO_2$ compound. This approach is finally used to obtain the ground state energies for the two hole configurations of the strontium induced states in the superconducting $La_{2-x}Sr_xCu_2O_4$. The results of this work shows that the final states determined by the $3d^8$ contribution to the initial state are so weak that cannot be detected experimentally and the two hole d-L and c-L Coulomb interactions play an important role both in the energy position of the final and initial states.

EXPERIMENTAL

The Cu L_3 x-ray absorption measurements have been performed at the ACO storage ring of the synchrotron radiation facility LURE (Orsay) using a double crystal $1\,0\,\overline{1}\,0$ Beryl monochromator, which gives an energy resolution of about 0.35 eV at 900 eV. The absorption coefficient has been measured by detecting the emitted electrons with a channeltron (total yield mode). The x-ray photoemission measurements have been carried out at the ENEA-Casaccia laboratories using a Vacuum Generator ESCA LAB MK II system and the Mg K_α line as x-ray source.

RESULTS AND DISCUSSION

In Fig.1 the La_2CuO_4 L_3-edge x-ray absorption spectrum (XAS), and the Cu $2p_{3/2}$ x-ray photoemission spectrum (XPS) are shown.

This compound shows hopping-conductivity behavior and formal divalent copper valence. Strontium doping in $La_{2-x}Sr_xCuO_4$ induces both the increases of the Cu formal valence and superconductivity.

We can represent the one hole ground state wave function for the divalent copper site as a mixing of the two pure configurations: $|3d^9\rangle$ with one hole on the copper site and $|3d^{10}\underline{L}\rangle$ with one hole on the oxygen site

$$\psi_g = a_g\,|3d^9\rangle + b_g\,|3d^{10}\underline{L}\rangle$$

XPS and E_1^{XPS} is the energy of the 2p 3d^9 satellite line in XPS. Using these values we have determined Q-U_{cL} and Δ as a function of T. In this first part U_{cL} determines simply a Q renormalization. Then we have used the equation which gives the energy separation between the energy E^{XAS} of the XAS white line (2p3d^{10} final state) and the XPS main line (2p3d^{10}L final state), which is strongly U_{cL} dependent, to calculate ε_L. The experimental value is E_1^{XPS}-E^{XAS}=1.8 eV

$$E_1^{XPS}\text{-}E^{XAS}= \varepsilon_L+U_{cL}+[(Q-U_{cL})-\Delta]/2-\{[(Q-U_{cL})-\Delta]^2+4T^2\}^{1/2}/2$$

The introduction in our calculation of the parameter U_{cL} reduces the ε_L value, thus the measured energy separation becomes mainly a final state effect. The ε_L, ε_d and Q values have been calculated for several values of U_{cL}. The energy of the ground state with respect to the vacuum level Io> is found to decreases by increasing the c-L Coulomb repulsion in the final state. A characteristic set of data obtained by this approach ε_L=1.3 eV, ε_d=0.3 eV and Q=9.5 eV for U_{cL}=1.3 eV and T=2.2 eV.

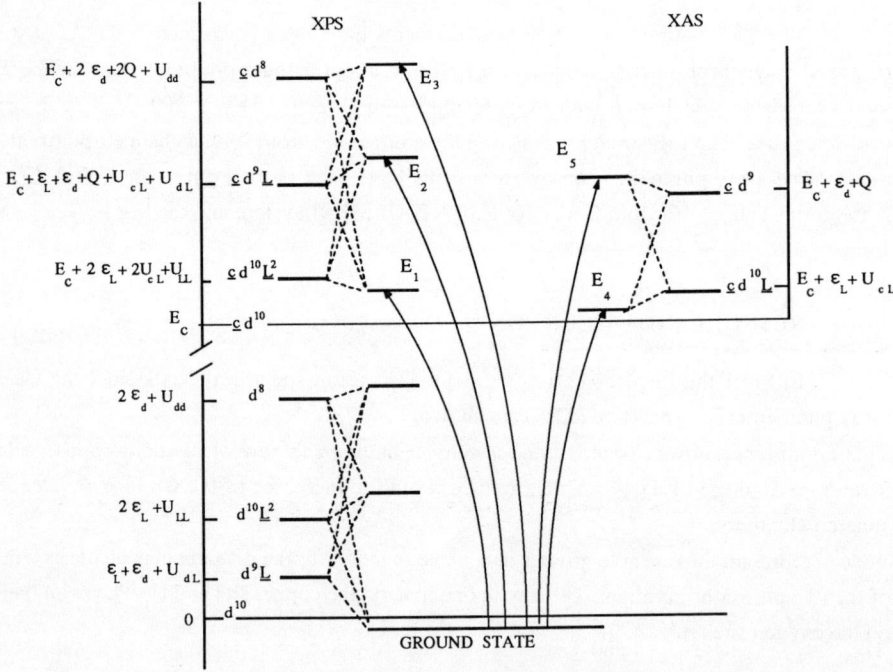

Fig. 2.   Schematic view of the energy levels of the ground state configurations (lower part) and of the excited states for XPS and XAS of the trivalent copper compound NaCuO$_2$.

The formally trivalent copper NaCuO$_2$ compound has been studied by extending this approach two the case of two holes configurations in the ground state and XAS final state and to the three hole XPS final state.

The cluster appropriate to schematize NaCuO$_2$ is a square plane CuO$_4$, where the diamagnetic ground state is considered to be a singlet pair of two holes given by the linear combination of the $3d^9\underline{L}$, $3d^{10}\underline{L}^2$ and $3d^8$ configurations. The final states are formed by the hybridization of $\underline{c}3d^9\underline{L}$, $\underline{c}3d^{10}\underline{L}^2$ and $\underline{c}3d^8$ configurations for the x-ray photoelectron spectroscopy and of $\underline{c}3d^9$, $\underline{c}3d^{10}\underline{L}$ for the absorption spectroscopy.

A diagram of the energy levels probed by X-ray photoemission and absorption is shown in Fig.2 for the trivalent case. The experimental XPS (Steiner *et al.*, 1989) and XAS (Bianconi *et al.*, 1988c) spectra of NaCuO$_2$ are shown in Fig.3. The white line in the absorption spectrum is due mainly to the $\underline{c}3d^{10}\underline{L}$ and the two XPS peaks come principally from $\underline{c}3d^{10}\underline{L}^2$ and $\underline{c}3d^9\underline{L}$ final states.

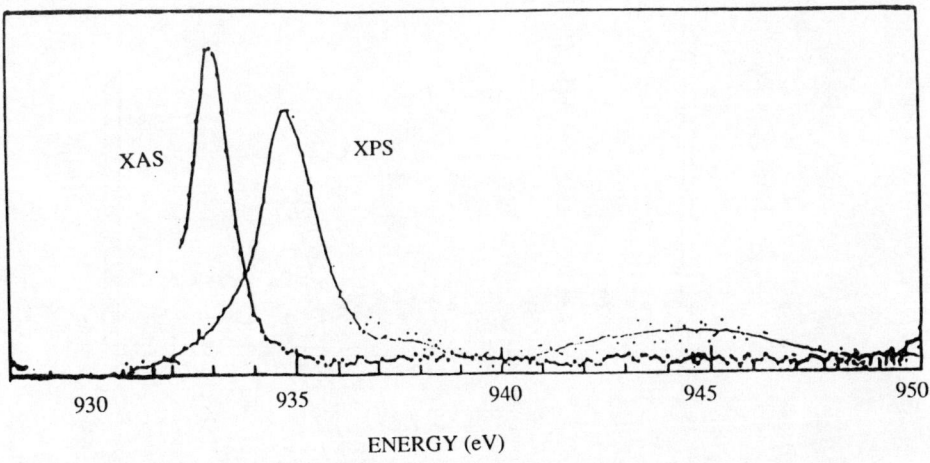

Fig. 3. NaCuO$_2$ L$_3$-edge x-ray absorption (thick line) and x-ray photoemission (thin line) spectra

In order to extract the energy values of ε_L, ε_d and Q as a function of the pair-hole interactions (U_{dd}, U_{cL}, U_{dL} and U_{LL}) and T we have developed a fortran code for simulating the energy separation and the relative peak intensities. This calculation requires as input the experimental energy separation $E_2^{XPS}-E_1^{XPS}=9.2$ eV of the photoemission peaks, their intensity ratio $I_2/I_1=0.24$ and the energy difference between XPS and XAS main lines $E_1^{XPS}-E_4^{XAS}=1.8$ eV (see Fig.3), where E_1^{XPS} E_2^{XPS} and E_4^{XAS} are the energy of the transitions defined in Fig.2. The computer code requires also as input the effective parameters U_{dd}, U_{cL}, U_{dL} and U_{LL}.

We have chosen U_{dd}=8 eV, in agreement with the experimental valence band photoemission spectra, and we have performed several calculations changing the U_{cL}, U_{dL} and U_{LL} values. As in the divalent case, the introduction of hole interactions involving ligand holes reduces the single-hole energy values and the ground state energy. In Fig.4 we show the values Δ, ε_L, ε_d, Q as function of T obtained with U_{cL}=U_{dL}=1.5 eV and U_{LL}=0.5 eV. We remark that the energy parameters ε_L and ε_d are lowered in this case respect to the calculation without pair hole interactions U_{cL} U_{dL} U_{LL}, giving the values (at T=2.2 eV) ε_L=0.3 eV, ε_d=0.2 eV Δ=0.1 eV. Thus, the cost for creating one hole on the ligand orbital is reduced in the trivalent compound as well as the charge transfer energy Δ.

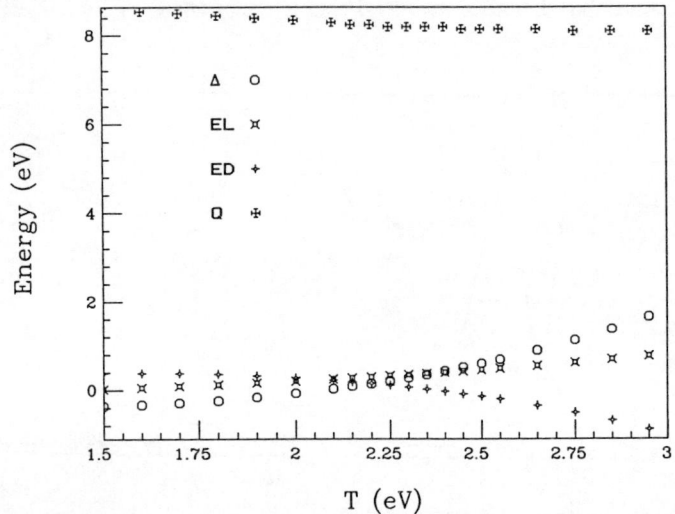

Fig. 4. Variation of the energy parameters Q, D, ε_L (EL), ε_d (ED) as a function of the hybridization energy T in the case of the formally trivalent compound $NaCuO_2$.

The XAS and XPS transitions E_3^{XPS} and E_5^{XAS} at higher energy are due mainly to the $3d^8$ contribution are not detected experimentally.
Using this parameters we can calculate the predicted to be of very low intensity (about 2% for the XAS, and again less for the third XPS peak).

Using these parameters we have calculated the energies of the ground state and of the two excited levels obtained from the hybridization of the ionic configurations $d^{10}\underline{L}^2$ and d^8 (as indicated in Fig.2) and the results are plotted in Fig. 5 as function of the hybridization energy. Fig. 6 shows the weight of the ionic $3d^9\underline{L}$ (A), $3d^{10}\underline{L}^2$ (B) and $3d^8$ (C) configurations in the ground state. The contribution of the $3d^9\underline{L}$, $3d^{10}\underline{L}^2$ is nearly the same while the weight of the $3d^8$ is about 4%.

For doped La-Sr-Cu-O superconductor with formally mixed copper valence we can describe the ground state as a superposition of the divalent $3d^9$, $3d^{10}\underline{L}$, and trivalent $3d^8$, $3d^9\underline{L}$, and $3d^{10}\underline{L}^2$ states. The ground state should be constructed by hybridization of the two-hole and one-hole configurations. However in a first approximation we can omit the mixing between divalent and trivalent configurations. By subtracting the La_2CuO_4 spectra to the superconductor ones we have obtained a simulation of the spectra of a trivalent compound. In Fig.7 the difference between the L_3 core level XAS and XPS spectra of $La_{1.85}Sr_{0.15}CuO_4$ and La_2CuO_4 are plotted. Similar experimental results are obtained by Steiner *et al.* 1989 for the differences of XPS spectra of Y-Ba-Cu-O system.

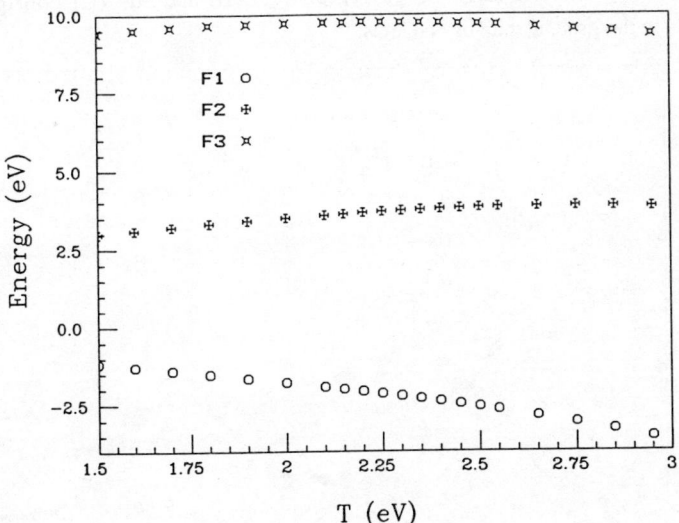

Fig. 5. Energies of the ground state (F1) and of the two excited levels F2 and F3 obtained from the hybridization of the ionic configuration $d^{10}\underline{L}^2$ and d^8 as a function of the hybridization energy T for the formally trivalent compound $NaCuO_2$.

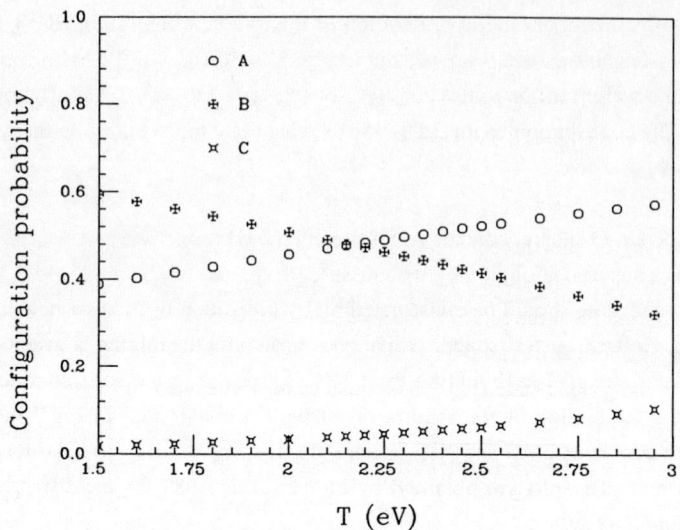

Fig. 6. Weight of the ionic $3d^9\underline{L}$ (A), $3d^{10}\underline{L}^2$ (B) and $3d^8$ (C) configurations in the ground state of $NaCuO_2$.

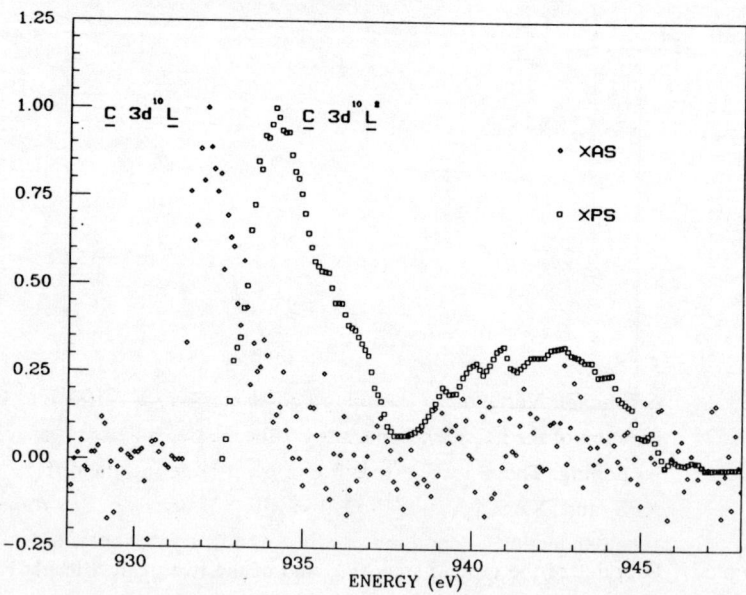

Fig. 7. Difference between the L_3 core level XAS and XPS spectra of $La_{1.85}Sr_{0.15}CuO_4$ and La_2CuO_4.

Starting from these difference spectra, we have performed the same kind of analysis carried out for the NaCuO$_2$ case. The experimental values were found to be E_2^{XPS}-E_1^{XPS} =8.1 eV, I_2/I_1=0.3 and E_1^{XPS}-E_4^{XAS}=2.1 eV. In Fig.8 we show the ε_L, ε_d, Δ and Q curves obtained for U_{cL}=U_{dL}=1.3 eV and U_{LL}=0.8 eV (U_{dd}=8 eV). By choosing T=2.2eV we have ε_L=0.5 eV and ε_d=0 eV, Δ=0.5 eV.

The energy values of the three hybridized states for the two hole states obtained by our experimental approach are shown in Fig. 9. The configuration F1 is the ground state. The excited configuration F3 at the highest energy is mainly due to the $3d^8$ configuration. Its energy is found to be at about 9 eV above the vacuum level. The calculated contribution the $3d^8$ configuration to the ground state (Fig.10) is found to be of the order of 6%.

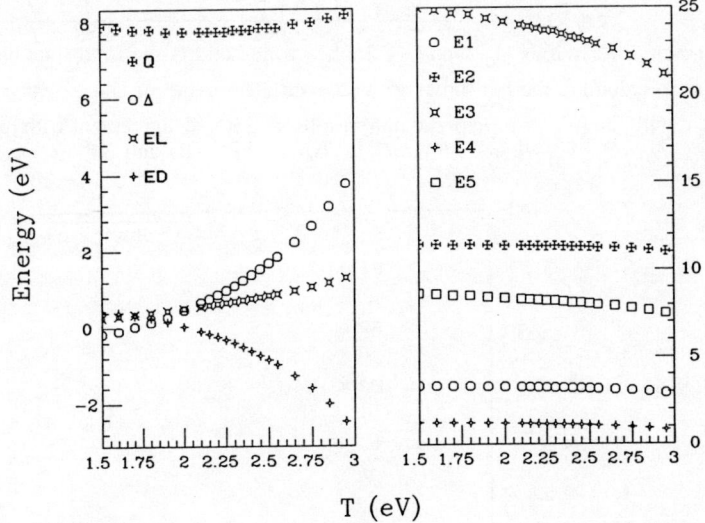

Fig. 8. *Left panel:* Variation of the energy parameters Q, Δ, ε_L (EL), ε_d (ED) as a function of the hybridization energy T for the two hole states states induced by doping. These curves were found by considering the difference of the XPS and XAS spectra shown in Fig. 7 between the mixed-valent superconducting compound La$_{1.85}$Sr$_{0.15}$CuO$_4$ and the divalent one La$_2$CuO$_4$. *Right panel:* XAS and XPS final state energies where E1, E2, E4 are the experimental input data and E3 and E5 are the predicted values for the transitions not observed in the experiment as obtained by values in left side panel.

CONCLUSIONS

We have presented a simple experimental approach to obtain the energies of two hole configurations in a formally trivalent compound and for the states induced by doping in $La_{1.85}Sr_{0.15}CuO_4$ from the joint analysis of XPS and XAS spectra of the same core level. The results have to be valuated considering the limitations of this approach which does not consider multiplet splitting and bandwidth.

We have shown that in the copper-oxygen compounds the pair-hole interaction in the final state U_{cL} affects the energy difference between the x-ray absorption and x-ray photoemission main lines. The value of the charge transfer energy $\Delta \sim 1$ eV found for La_2CuO_4 is in agreement with Annett et al. 1989, calculations.

The pair-hole interactions U_{dL} and U_{LL} in the ground and in the final state have found to be important to calculate the two holes ground state. The value of $U_{dL} \sim 1$ eV is found to be in agreement with the full set of experimental results, and it is in agreement with recent theoretical prediction. (Annett et al. 1989).

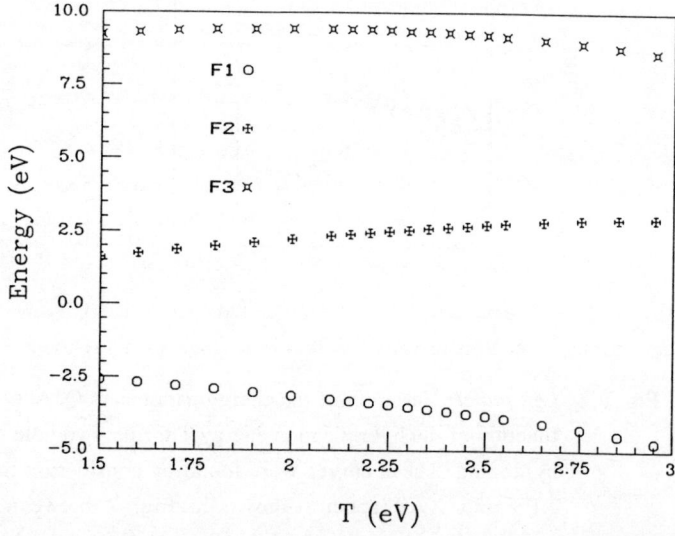

Fig. 9. Energies of the ground state (F1) and of the two excited levels (F2 and F3) obtained by the hybridization of the ionic configuration $d^{10}\underline{L}^2$ and d^8 as a function of the hybridization energy T for the two hole states states induced by doping in the superconducting compound $La_{1.85}Sr_{0.15}CuO_4$.

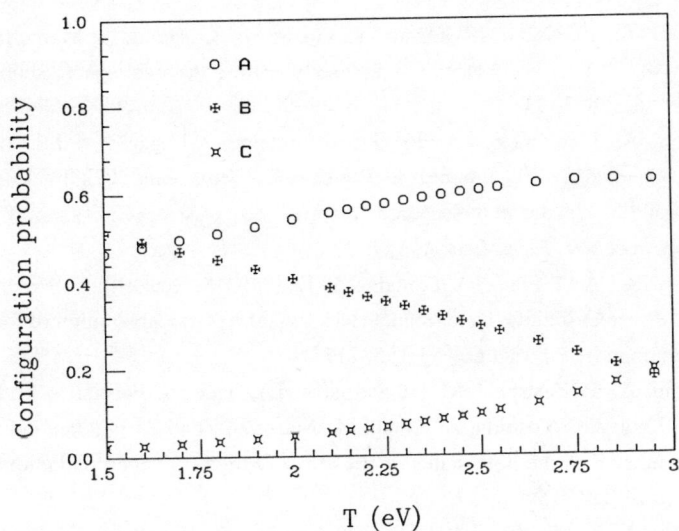

Fig. 10. Weight of the ionic $3d^9\underline{L}$ (A), $3d^{10}\underline{L}^2$ (B) and $3d^8$ (C) configurations in the ground state of $La_{1.85}Sr_{0.15}CuO_4$.

We have considered only one type of orbital symmetry for the d and L holes. The inclusion of configuration where d holes and L holes have different symmetries can be of relevance in some superconducting compounds. Future efforts will be devoted to this subject of study.

REFERENCES

Annett, J.F., R.M. Martin, A.K. McMahan and S. Satpathy (1989). The electronic Hamiltonian and antiferromagnetic interactions in La_2CuO_4, *Phys. Rev. B*, to be published.

Bianconi, A., A. Congiu Castellano, M. De Santis, P. Lagarde, A.M. Flank, A. Marcelli (1987a). $L_{2,3}$ XANES of the high T_c superconductor $YBa_2Cu_3O_7$ with variable oxygen content, *Solid State Commun.*, 63, 1009.

Bianconi, A., A. Clozza, A. Congiu Castellano, S. Della Longa, M. De Santis, A. Di Cicco, K. Garg, P. Delogu, A. Gargano, R. Giorgi, P. Lagarde, A. M. Flank, and A. Marcelli (1987b). Experimental evidence of itinerant Cu $3d^9$ oxygen hole many body configuaration in the high T_c superconductor $YBa_2Cu_3O_7$, *International Journal of Modern Physics*, 1,

205.

Bianconi, A., A. Kotani, K. Okada, R. Giorgi, A. Gargano, A. Marcelli and T. Miyahara (1988a). Many body effects in praesodymium core-level spectroscopies of PrO_2, *Phys. Rev. B*, <u>38</u>, 3433.

Bianconi, A., J. Budnick, A.M. Flank, A. Fontaine, P. Lagarde, A. Marcelli, H. Tolentino, B. Chanberland, C. Michel, B. Raveau, G. Demazeau. (1988b). Evidence of the $3d^9$-Ligand hole states in the superconducting $La_{1.85}Sr_{0.15}CuO_4$ from L_3 x-ray absorption spectroscopy, *Phys. Lett. A*, <u>127</u>, 285.

Bianconi, A., A.M. Flank, A. Fontaine, P. Lagarde, A. Marcelli, M. Verdaguer, J. Jegoudez, A. Revcolevschi, G. Demazeau. (1988c). Cu L_3 X-ray absorption of formally trivalent Cu compounds, *Physica C*, <u>153-155</u>, 117.

Bianconi, A., P. Castrucci, M. De Santis, A. Di Cicco, A. Fabrizi, A.M. Flank, P. Lagarde, H. Katayama Yoshida, A. Kotani, A. Marcelli, Zhao Zhongxian and C. Politis (1988d). Symmetry of the hole states in BiCaSrCuO high T_c superconductors, *Modern Physics Lett. B*, <u>2</u>, 1313.

Bianconi, A., T. Miyahara, A. Kotani, Y. Kitajima, T. Yokoyama, H. Kuroda, M. Funabashi, H. Arai, and T. Ohta (1989). Correlation satellites in deep metal 3p core x-ray photoemission of tetravalent oxides MO_2 (M=Ce, Pr, Tb, Hf) and of LaF_3, *Phys. Rev. B*, <u>39</u>, 3380.

Fujimori A., E. Takayama-Muromachi, Y. Uchida, and B. Okai (1987). Spectroscopic evidence for strongly correlated electronic states in La-Sr-Cu and Y-Ba-Cu oxides, *Phys. Rev. B*, <u>35</u>, 8814.

Nücker N., J. Fink, J.C. Fuggle, P.J. Durham, W.M. Temmermann, Evidence on oxygen sites in High T_c superconductors, (1988). *Phys. Rev. B*, <u>37</u>, 5158.

Steiner P., S. Hufner, A. Jungmann, V. Kinsinger, and I. Sander, (1989). Photoemission on high T_c superconductors, *in this book and references cited therein*.

Zaanen, J., G.A. Sawatzky and J.W. Allen (1985) Band gaps and electronic structure of transition metal compounds *Phys. Rev. Lett.*, <u>55</u>, 418.

SOFT X-RAY EMISSION SPECTRA OF HIGH T_c SUPERCONDUCTORS

C.F. Hague, V. Barnole, J.-M. Mariot

*Laboratoire de Chimie Physique (unité associée au CNRS), Université Pierre et Marie Curie
11 rue Pierre et Marie Curie, F-75231 Paris Cedex 05*

C. Michel and B. Raveau

°*Laboratoire de Cristallographie et Sciences des Matériaux (unité associée au CNRS), ISMRa
Bld du Maréchal Juin, F-14032 Caen Cedex*

ABSTRACT

Information on Cu 3d and O 2p states obtained from the study of the Cu Lα and O K$_\alpha$ x-ray emission bands are presented for several high Tc superconductors. The spectra for $YBa_2Cu_3O_7$ are discussed in some detail in relation to other high energy spectroscopies. It is shown that the broad O 2p bands in La_2CuO_4, $La_{1.85}Sr_{0.85}CuO_4$, $YBa_2Cu_3O_7$ and $Bi_2Sr_2CaCu_2O_8$ are well described by local density functional theory but, as expected, Cu 3d states are highly correlated. Non-superconducting phases are observed to have spectral features typical of the Cu^I valence.

KEYWORDS

High-T_c superconductors; electronic structure; x-ray emission spectroscopy.

INTRODUCTION

X-ray emission spectroscopy (XES) in the soft (low) energy region has long been used to obtain information on the electronic structure of materials (see Sellmyer, 1978). The intensity of the x-ray emission at a frequency ω which results from the transition from an initial core hole |i⟩ (energy ε_i) to a final valence hole strate |f⟩ (energy ε_f) can be expressed as

$$I_i(\omega) = \Sigma_f |\langle f | H_{int} | i \rangle|^2 \delta(\varepsilon_i + \omega - \varepsilon_f) 2\pi \rho(\omega) \tag{1}$$

where $\rho(\omega)$ is the photon density of states. The transition matrix element varies quite monotonously with photon energy over the width of a typical valence band but it has important properties. At least as far as concerns us here, the core level wavefunction is highly localized within the potential well of the emitting atom so that, in a multicomponent system, the averaged densities of states (DOS) of each constituent can be probed separately (Durham et al., 1979; Bianconi, 1980). Also dipole selection rules apply i.e., the transition must involve a change in l such that $\Delta l = \pm 1$. This implies that the symmetry of the valence state probed is known unambiguously.

Empirically it is found that x-ray emission bands are, in general, well described by the partial DOS calculated within the one-electron approximation. Hence the formulation of the final state rule which says that the transition matrix element should be calculated from wavefunctions in the final state of the x-ray process. This means that, for x-ray emission, the perturbation due to the core hole in the initial state can be neglected (von Barth and Grossman, 1979). The effect of the presence of an initial core hole is then simply to introduce an intrinsic limit to the resolving power of the experiment because of its finite lifetime. In passing, it is also useful to keep in mind that the thickness of the sample probed by such an experiment is related to the absorption of the outgoing photon in the target material (the penetration of the incoming photons or electron beam used to create the initial-state core-hole can be made large). Generally speaking the effective thickness probed is at least one or two orders of magnitude larger than in electron spectroscopies where the limiting factor is the mean free path of the outgoing electron.

Except in the case where the DOS fall off sharply from a maximum value at the Fermi energy ε_F (e.g, in a simple metal), no direct information is provided concerning the valence band binding energy ε_V because ω is given by the difference in energy between the initial and final states. This is easily solved, however, if the core level binding energy ε_c is known from x-ray photoelectron spectroscopy (XPS) by means of the expression

$$\varepsilon_v = \varepsilon_c - \omega \tag{2}$$

The validity of the procedure is shown in Fig. 1 for Zn 3d valence states. We have used the Zn $2p_{3/2}$ core level binding energy measured by XPS to place the Zn L_α ($3d \rightarrow 2p_{3/2}$) emission on a binding energy scale.

Comparison with valence band XPS data which give the 3d binding energy directly (Ley et al., 1973) reveals excellent agreement. We have chosen this example to indicate that such x-ray emission may also show up complex satellite structure as a result of multiple-hole initial and final states. It also illustrates that comparison with the one-particle ground-state calculation is not necessarily satisfactory. Here the calculated d DOS (Moruzzi et al., 1978) lie $\approx$ 2 eV closer to ε_F than in the experiments. The discrepancy can be seen as a failure of the theory to take into account the on-site atomic-like correlations resulting from the localized nature of the Zn 3d states. Information from high energy spectroscopies have already demonstrated that likewise the obervable properties of high T_c superconductors cannot be systematically related to the calculated ground states (for a review see Fuggle et al., 1988).

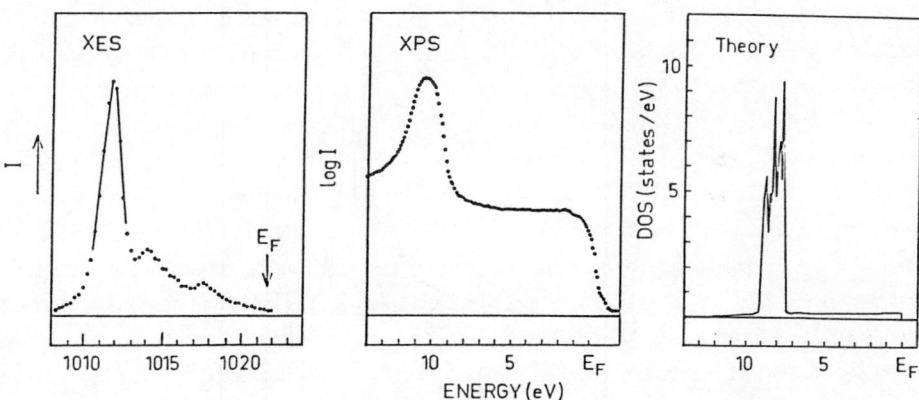

Fig. 1. Zn L_α set relative to ε_F by means of the $2p_{3/2}$ XPS binding energy compared to XPS valence band (Ley et al., 1973) and calculated DOS (Moruzzi et al., 1978).

In this summary of our x-ray emission experiments we present spectra relating to the Cu 3d and O 2p valence states of several high T_c compounds. They are compared to similar data on the simple cuprons and cuprous oxides and on two non-superconducting phases. We will conclude the discussion by a comparison with calculated partial DOS.

EXPERIMENTAL

Sintered slabs (10 x 5 x 1 mm^3) of La_2CuO_4, $La_{1.85}Sr_{0.15}CuO_4$, $YBa_2Cu_3O_{6+\delta}$ ($\delta<0.5$) were prepared by the usual techniques (Michel et al., 1987; Beille et al., 1987). A superconducting phase of La_2CuO_4 was obtained by annealing at 400°C for 24 h under O_2 at atmospheric pressure; the non-superconducting phase was obtained by quenching from a high temperature. Cu_2O, CuO and $Bi_2Sr_2CaCu_2O_8$ targets were prepared in the form of pressed-powder pellets.

Fresh surfaces were prepared by light scraping with a diamond file just prior to measurement.

Experiments were carried out with a Johann-type bent-crystal spectrometer (Hague and Laporte, 1980) fitted with a beryl ($10\overline{1}0$) or RbAP crystal for the analysis of the Cu L_α ($3d \rightarrow 2p_{3/2}$) and O K_α ($2p \rightarrow 1s$) emission bands respectively. Instrumental resolution was ≈ 0.3 eV. The core level ionizations were obtained with an electron gun usually operating at 4 kV and 0.3 mA. During acquisition the pressure was better than 10^{-7} mbar. The curves presented are the sum of several scans taken in a step-by-step mode.

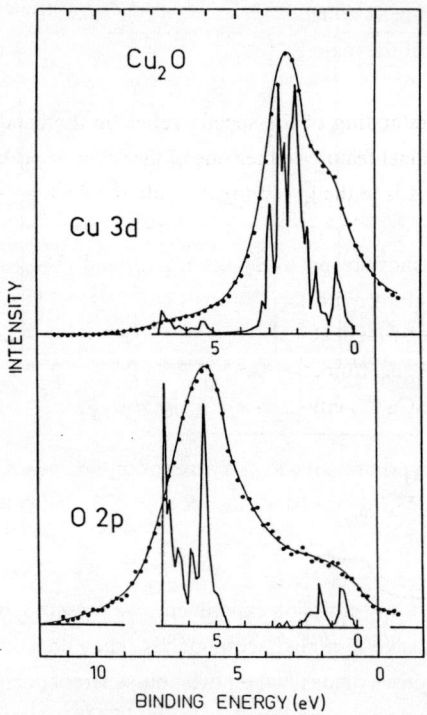

Fig.2. Cu L_α and O K_α for Cu_2O compared to the calculated partial DOS (Marksteiner et al., 1986). The theoretical spectra have been shifted by ≈ 0.8 eV to give best visual fit.

RESULTS AND DISCUSSION

In Fig. 2 we present our Cu L_α and O K_α data for Cu_2O (Mariot et al., 1989) compared to the calculated partial DOS of Marksteiner et al., (1986). The XES spectra are set on a binding scale energy using the Cu $2p_{3/2}$ and O 1s XPS binding energies. The theoretical curves have been placed to give the best visual fit with experiment. As Cu_2O is a wide gap semiconductor, we may consider that the 0.8 eV shift in the energy scale is justified by the difference in defining the origin (zero energy is set at the top of the valence band in the calculation).

Subject to this condition we see that the calculated DOS are in very satisfactory agreement with our observations. Experiment and theory reveal that this compound is not completely ionic since some hydridization occurs at the top and bottom of the band. In the Cu L_α spectrum this shows up as a shoulder to the low binding energy side of the main peak and as a very weak structure at ≈ 6 eV. The main O 2p peak coincides with the 6 eV structure in the 3d states and its intensity drops off in the region of the main 3d peak.

A more detailed understanding of the spectra relies on the band structure calculation which indicates that the additional features either side of the main 3d peak result from $d(z^2,xz,yz) - p(z)$ hybridization (the z axis is in the [111] direction along which a Cu atom is coordinated by two oxygen atoms). The Cu $d(x^2 - y^2,xy)$ states are non-bonding.

Figures 3(a) and 5(a) show respectively the Cu L_α and O K_α spectra for $YBa_2Cu_3O_7$ on a binding energy scale along with the Cu $2p_{3/2}$ and O XPS 1s core level spectra taken from Steiner *et al.* (1987). We first deal with the Cu-site data. Other than the strong satellite structure at ≈ 942 eV in the XPS spectrum, the most striking feature is that the Cu $2p_{3/2}$ XPS core line is in fact broader than the Cu L_α emission.

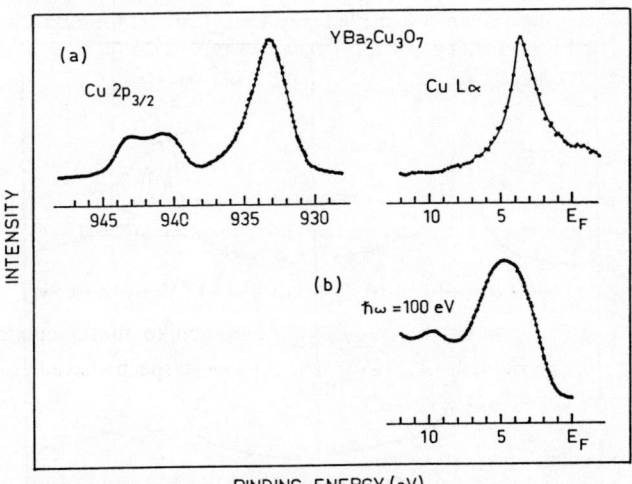

Fig. 3. (a) Cu $L\alpha$ and Cu $2p_{3/2}$ XPS spectra for $YBa_2Cu_3O_7$ both on a binding energy scale; (b) Photoemission spectroscopy (PES) valence band recorded with ω=100 eV (Stoffel *et al.*, 1987).

It is widely accepted that the XPS spectrum can be explained in terms of configuration interaction (van der Laan et al., 1981, Fujimori et al., 1987) with a $2p^5 d^{10}\underline{L}$ configuration for the main XPS peak as a result of charge transfer ($\underline{L}$ indicates a hole in the oxygen-derived states). The XPS satellite can then be assumed to reflect multiplet structure arising from hole-hole interaction in the $2p^5 3d^9$ configuration.

Clearly the $2p_{3/2}$ XPS core line cannot represent the core-level lifetime broadening of the x-ray emission process. This may reasonably be taken to argue in favour of assuming the validity of the final state rule.

Wassdahl et al., (1987) have pointed out that the ligand-hole state would add about the same energy to the initial and final states of the x-ray transition i.e., the x-ray transition energy should not be measurably different in the presence or absence of the ligand hole. The implication is that the energy of the Cu L_α emission cannot provide information on the formal valence of copper since the energy would be the same for both $2p^5 3d^n \rightarrow 3d^{n-1}$ and $2p^5 3d^{n+1}\underline{L} \rightarrow 3d^n \underline{L}$ transitions (Barnole et al., 1988). Nevertheless we can use XES to interpret the valence band photoelectron spectroscopy data, since if we accept that the main XPS $2p_{3/2}$ line has a $2p^5 3d^{10} \underline{L}$ configuration then we can expect a $3d^9\underline{L}$ peak at $\varepsilon_c - \omega \approx 4$ eV in the valence band spectrum. This indeed identifies the main peak in spectra taken with high photon energies where the photoinization cross section is much lager for Cu 3d than for O 2p [e.g., Fig. 3 (b)].

Figure 4 (a) illustrates the change in shape between Cu L_α spectra for $YBa_2Cu_3O_7$ and CuO.

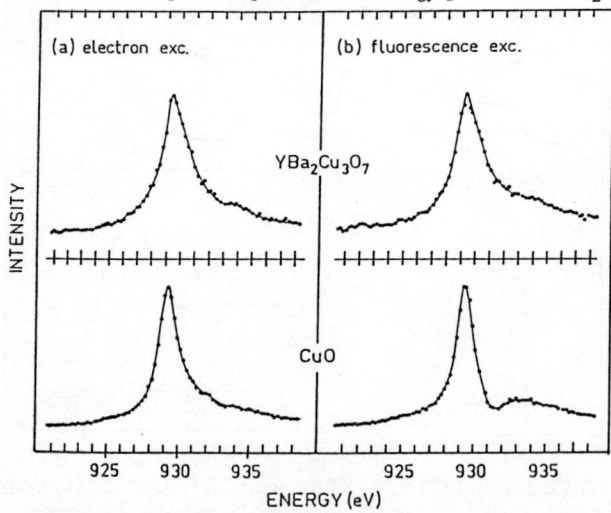

Fig. 4. Cu L_α for $YBa_2Cu_3O_7$ and CuO produced by a) electrons b) Al K_α photons.

At this point an experimental difficulty arising from the effect of self-absorption (Bonnelle, 1966) has to be taken into consideration when analysing the data. The presence of high Z elements in the high T_c compounds reduces the effective target thickness contributing to the x-ray emission signal because of the higher overall absorption of the outgoing photons. This in turn reduces the effect of the sharp change in photoabsorption close to ε_F. An extreme example is given in Fig. 4 (b) taken from our earlier data (Mariot et al., 1987a, 1987b) where the fluorescence mode was used (the x-ray emission was produced by Al K_α photons). For the data in Fig. 3 (a) a lower electron-beam excitation energy was used for CuO to reduce the probed thickness to the same value as in $YBa_2Cu_3O_7$. Observation conditions are not so critical for Cu_2O because there are no empty 3d states close to ε_F.

The Cu L_α emission seems to tail-off more gradually towards low x-ray energies for $YBa_2Cu_3O_7$ than for CuO or Cu_2O but this can be accounted for by increased band width. We find that the high binding energy side of the emissions can be accurately fitted by a Lorentzian profile with 1.6 eV full-width at half-maximum (FWHM) for CuO, 1.8 eV for Cu_2O and 2.2 eV for $YBa_2Cu_3O_7$. CuO has weak structure 3-4 eV below the peak due to hybridization with oxygen states as in Cu_2O (Mariot et al., 1989). No such structure is observed for $YBa_2Cu_3O_7$ but its intensity in CuO is < 5% of the main peak so its presence could be masked by the increased width of the emission.

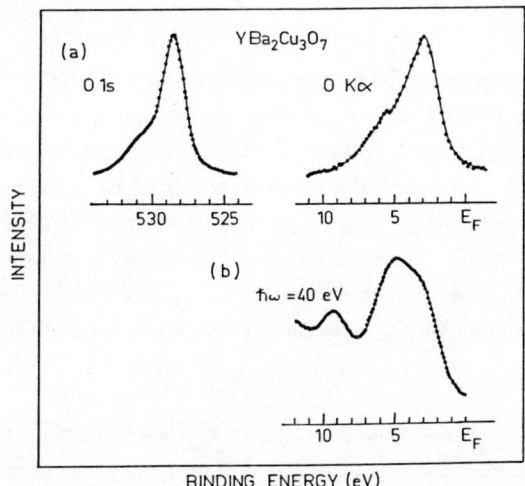

Fig. 5. (a) O K_α and O 1s XPS spectra for $YBa_2Cu_3O_7$ both on a binding energy scale; (b) PES valence band recorded with $\omega=40$ eV (Stoffel et al., 1987).

The spectra are asymmetrical and broadened out to the high x-ray energy side due to satellite structure. Most of the structure, as confirmed by Wassdahl et al. (1987) arises from Coster-Kronig transitions (Guennou et al., 1981). The extra intensity above Lorentzian fits to the main line is the same for CuO and $YBa_2Cu_3O_7$. Going on to the O K_α data for $YBa_2Cu_3O_7$ we see [Fig. 5 (a)] that the main O 2p peak has shifted to lower binding energies compared to Cu_2O reflecting the change in O - O and O - Cu coordination.

The total width of the O $K\alpha$ emission is about the same in both cases. Here the main O 1s peak of the XPS data relates clearly to the initial state of the x-ray emission. The structure at ≈ 531 eV has diversely been attributed to contaminations (Steiner et al., 1988) or to oxygen dimerization (Dauth et al., 1987, Sarma and Rao, 1987). We can expect little contribution from contaminants in our experiments because of reduced surface sensitivity compared to photoemission experiments. Also the signal from possible dimerized states would be superimposed on the main O K_α emission so its detection seems improbable. The only sign of correlation effects is the presence of a satellite some 6 eV above the main O K_α peak position which has been attributed to the $1s^{-1} 2p^{-2} \rightarrow 2p^{-3}$ transition (Valjakka et al., 1985). We found that its intensity was the same for all the compounds.

Photoelectron spectra taken with low photon energies have large photoionization cross sections for O 2p states and indeed both experiments show a peak 2.5 eV below ε_F [Fig. 5 (b)]. Towards higher binding energies the difference in shape is explained by the contribution from the Cu 3d states in the photoemission data.

Our XES results for four superconducting compounds are presented Fig. 6 on an x-ray energy scale (Barnole et al., 1989). The Cu L_α emission is practically identical in each case. The slight broadening for $Bi_2Sr_2CaCu_2O_8$ compared to the other compounds might be due to slightly reduced self-absorption because the stopping power of Bi is higher than for La and Ba (photoabsorption in this energy range is practically identical for the three compounds). The shape of the O K_α emission from the La-based compounds is, however, somewhat different from the other two cuprates. Figure 7 shows the Cu L_α emission for the non-superconducting phases La_2CuO_4 and $YBa_2Cu_3O_{6+\delta}$. Here a shoulder, reminiscent of Cu_2O, is apparent towards high x-ray energies.

The large difference observed between the O K_α spectrum from Cu_2O and the high T_c compounds can be understood as a consequence of the tetrahedral coordination of the Cu atoms around an oxygen in Cu_2O leading to Cu $d(z^2,xz,yz)$ - O $p(z)$ hybridization while the coplanar arrangements in the other oxides lead to Cu $d(x^2 - y^2)$ - O $p(x,y)$ hybridization. [The spectrum for CuO which also has oxygen atoms tetrahedrally coordimated to Cu is intermediate between

Cu$_2$O and the other cuprates (Mariot et al., 1989)]. The smaller differences which are observed between signals from the high T$_c$ compounds can reasonably be equated with the varying environment of the other oxygen sites.

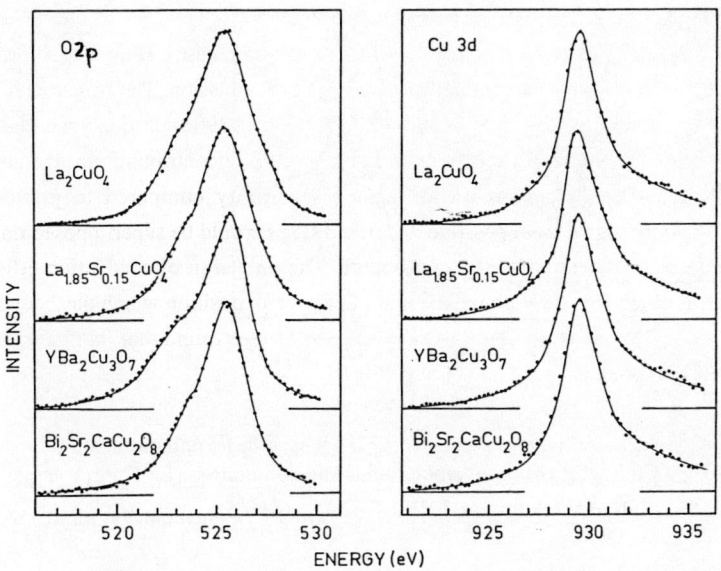

Fig.6. Cu L$_\alpha$ and O K$_\alpha$ data for four high T$_c$ superconductors.

Qualitatively the overall broadening of the Cu Lα emission in the high T$_c$ compounds compared to CuO, despite their similar Cu - O coplanar arrangements, might explained by the increased nominal valence of the Cu atoms. This should be understood in terms of the increased O - Cu charge transfer and therefore to the presence of less localized Cu 3d states. Raveau et al. (1988) have argued, on the basis of the possible coordination for Cu^I, Cu^{II} or Cu^{III} cations that a continuous transitions from YBa$_2$Cu$_3$O$_7$ to YBa$_2$Cu$_3$O$_6$ must necessarily involve the intergrowth of "O$_7$" and "O$_6$" structures. This can best be formulated as YBa$_2$[Cu$_2^{II}$Cu^{III}O$_7$]$_\delta$[Cu$_2^{II}$Cu^{I}O$_6$]$_{1-\delta}$. Remembering that a nominally Cu$_2$O, CuO or (CuO)$^+$ configuration would lead to about the same Cu Lα emission energy, the x-ray signal would then contain varying proportions of superimposed information from Cu$_2$O-like or (CuO)$^+$-like states. Such a model is perfectly compatible with our findings and explains the appearance of the

Cu_2O-like shoulder for $\delta<0.5$. A more detailed study as a function of δ is required however to establish the validity of the model more fully.

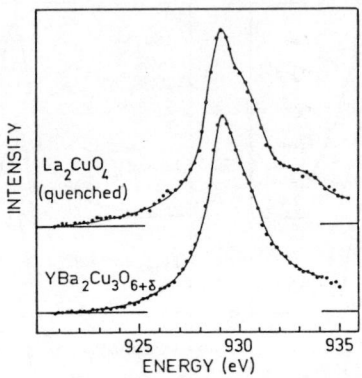

Fig. 7. Cu L_α for two non-superconducting compounds: La_2CuO_4 (quenched from a high temperature) and $YBa_2Cu_3O_{6+\delta}$ ($\delta<0.5$).

Similarly the shoulder in the Cu L_α spectrum for La_2CuO_4 quenched from a high temperature suggests excess Cu and the absence of $(CuO)^+$ states (the spectrum is a little narrower towards low x-ray energies than in the case of $YBa_2Cu_3O_{6+\delta}$). The fact that the emission from the annealed La_2CuO_4 sample is identical to that from $La_{1.85}Sr_{0.15}CuO_4$ indicates that it is superconducting at least over the thickness probed by the experiment (Raveau et al., 1988).

Finally we confront our data with band structure calculations. Amongst the many highly precise local density calculations already implemented we have chosen those using the full potential linearized augmented plane wave (FLAPW) method because they have been used to obtain theoretical x-ray spectra. The calculated Cu L_α and O K_α spectra for La_2CuO_4 (Redinger et al., 1987a), $YBa_2Cu_3O_7$ (Redinger et al., 1987b) and $Bi_2Sr_2CaCu_2O_8$ (Marksteiner et al., 1988) are presented Fig. 8 along with the experimental data set on a binding energy scale.

The overall agreement between the calculated and experimental O K_α emission is satisfactory as might be expected for a broad band regime. This is especially true of the binding energies and the shape of the spectra for $YBa_2Cu_3O_7$ and $Bi_2Sr_2CaCu_2O_8$. Some discrepancy in shape is apparent for La_2CuO_4 but it should be pointed out that, in-plane and out-of-plane oxygens are involved, so the relative positions of the O(1) and O(2) partial DOS have been shifted by ≈ 0.7 eV to take into account the estimated difference in the O 1s core level binding energies from each type of site. This may be somewhat overestimated. If so, agreement with experiment could be improved. We can conclude that the usefulness of the local density approximation as a means of interpreting the O 2p spectroscopic data seems to be ascertained.

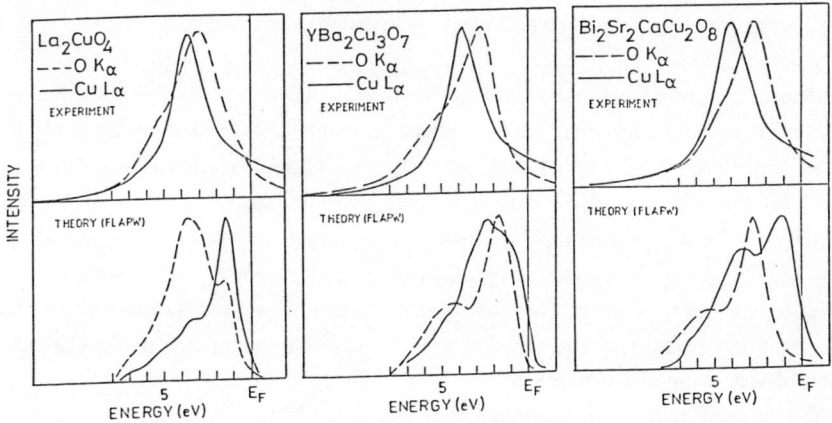

Fig. 8. Comparison between calculated (FLAPW) and experimental Cu L_α and O K_α spectra.

The situation concerning the more localized Cu 3d states is, as expected, less straightforward. We have explained above that using the $2p^53d^{10}\underline{L}$ XPS peak to set the binding energy for the XES experiment only implies that we can now estimate the binding energy ε_V of the $3d^9\underline{L}$ configuration. From XPS and x-ray absorption spectroscopy (XAS) it is possible to obtain ε_L, the difference in energy between $3d^n\underline{L}$ and $3d^n$ configurations because the energies of the well-screened core hole states $2p^53d^{10}$ and $2p^5 3d^{10}\underline{L}$ are known from the Cu L_3 XAS and Cu $2p_{3/2}$ XPS data respectively. According to Bianconi et al. (1987), $\varepsilon_L = 2.4$ eV for CuO and 1.9 eV for $YBa_2Cu_3O_7$. Thus the one-hole $3d^9$ state binding energy in $YBa_2Cu_3O_7$ can be estimated to be ε_V -1.9 ≈ 2.1 eV. This agrees rather well with the d-DOS peak position, whatever the compound, using FLAPW or other approximations (see e.g., Temmerman et al., 1987, Mattheiss and Hamann, 1988).

It should be pointed out that we have not used the L_3 photoabsorption peak energy to set the binding energy scale for Cu L_α because physically this would represent an excited initial state i.e., an initial state where a 2p core level electron had been promoted to the 3d band ($2p^5 3d^{10}$), the x-ray transition leaving a $3d^9$ final state. (This would have been the case had the spectra been produced by monochromatized radiation tuned to the L_3 excitation threshold).

We note that the calculated Cu d DOS shape, which is very dependent on hybridization with O p states over the whole width of the oxygen band, is completely different from that observed experimentally. This is a clear indication that the observable Cu 3d-related structure in high energy spectra cannot be compared to the rigidly-shifted calculated DOS.

In summary, therefore, x-ray emission spectra contain unique information on the electronic properties of perovskite-related cuprates because the component Cu 3d and O 2p electron states can be separated out from the total DOS. More detailed work is required to elucidate the origin of small differences in the O 2p DOS between different superconducting oxides but because they form extended bands confrontation with calculated ground-state DOS can be expected to be useful. The Cu 3d signals are very similar from one high T_c superconductor to the next but are very sensitive to the presence of Cu^I valence states present in their non-superconducting phases. There are good reasons to believe that the O - Cu charge transfer models which have successfully accounted for much of the XPS data can also be invoked to explain more delocalized d states in the high T_c compounds compared to CuO.

Preliminary results on non-superconducting three-dimensional oxides of the $La_4BaCu_5O_{13+\delta}$ type suggest, however, that here too the Cu states are less localized than in CuO so that enhanced charge transfer cannot be concluded to be a sufficient condition for the superconducting state.

REFERENCES

Barnole, V., J.-M. Mariot, C.F. Hague and H.-J. Güntherodt (1988). Cu 3d soft x-ray emission spectra from $MBa_2Cu_3O_7$ (M=Y, Eu and Er). *Physica C*, 153-155, 125-6.

Barnole, V., J.-M. Mariot, C.F. Hague, C. Michel and B. Raveau (1989). Cu L_α and O K_α emissions from high T_c superconductors. (to be published)

von Barth, U. and G. Grossman (1979). The effect of the core hole on x-ray emission spectra in simple metals. *Solid State Commun.*, 32, 645-9.

Beille, J., R. Cabanel, C. Chaillout, B. Chevalier, G. Demazeau, F. Deslandes, J. Etourneau, P. Lejay, C. Michel, J. Provost, B. Raveau, A. Sulpice, J.-L. Tholence et R. Tournier (1987). Supraconductivité en-dessous de 40K de La_2CuO_4 de structure orthorhombique.

C.R. Acad. Sc. Paris, 304-II, 1097-101.

Bianconi, A., (1980). Soft x-ray absorption spectroscopy: surface EXAFS and surface XANES. Appl. Surf. Sci., 6, 392-418.

Bianconi, A., A. Clozza, A. Congiu-Castellano, S. Della Longa, M. De Santis, A. Di Cicco, K. Garg, P. Delogu, A. Gargano, R. Giorgi, P. Lagarde, A. M. Flank and A. Marcelli (1987). Experimental evidence of itinerant Cu $3d^9$-oxygen hole many body configuration in the high-T_c superconductor $YBa_2Cu_3O_{\approx 7}$. Int. J. Mod. Phys. B., 1, 853-63.

Bonnelle, C. (1966) Contribution à l'étude des métaux de transition du premier groupe, du cuivre et de leurs oxydes par spectroscopie X dans le domaine 13 à 22 Å. Ann. Phys. (Paris), 1, 439-81.

Dauth, B., Kachel, T., P. Sen, K. Fischer and M. Campagna (1987). Valence fluctuations and oxygen dimerization in high-temperature superconductors from x-ray photoemission spectroscopy. Z. Phys. B, 68, 407-10.

Durham, P. J., D. Ghaleb, B.L. Györffy, C.F. Hague, J.-M. Mariot, G.M. Stocks and W.M. Temmerman (1979). Soft x-ray emission and local densities of states in Cu - Ni alloys. J. Phys. F, 9, 1719-30.

Fuggle, J.C., J. Fink and N. Nücker (1988). The status of high energy spectroscopic studies of high T_c superconductors. Int. J. Mod. Phys. B (to be published).

Fujimori, A., E. Takayama-Muromachi and Y. Uchida (1987). Electronic structure of superconducting copper oxides. Solid State Commun., 63, 857-60.

Guennou, H., A. Sureau, G. Dufour, C.F. Hague and J.-M. Mariot (1981). Lα satellites of copper: a theoretical description by an SCF method including spin-orbit interaction. In: Inner-Shell and X-Ray Physics of Atoms and Solids (D.J. Fabian, H. Kleinpoppen and L.M. Watson, Eds.). Plenum, New York, pp. 797-800.

Hague, C.F. and D. Laporte (1980). Spectrometer for soft x-ray emission studies of liquid metals and metallic vapours.Rev. Sci. Instrum., 51, 621-5.

van der Laan, G., C. Westra, C. Haas and G.A. Sawatzky (1981). Satellite structure in photoelectron and Auger spectra of copper dihalides. Phys. Rev. B, 23, 4369-80.

Ley, L., S.P. Kowalczyk, F.R. McFeely, R.A. Pollak and D.A. Shirley (1973). X-ray photoemission from zinc: evidence for extra-atomic relaxation via semilocalized excitons. Phys. Rev. B, 8, 2392-402.

Mariot, J.-M., V. Barnole, C.F. Hague, V.Geiser and H.-J. Güntherodt (1987a). Experimental Cu 3d electronic structure in high T_c $La_{1.85}Ba_{0.15}CuO_4$ and $YBa_2Cu_3O_{8-x}$ superconductors. Solid State Commun ., 64, 1203-7.

Mariot J.-M., V. Barnole, C.F. Hague, V. Geiser and H.-J. Güntherodt (1987b). Cu valence d states in high T_c superconductors from soft x-ray fluorescence spectroscopy. J. Physique, 48-C9, 1203-6.

Mariot, J.-M., V. Barnole, C.F. Hague, G. Vetter and F. Queyroux (1989). Local densities of states of Cu_2O, CuO and $YBa_2Cu_3O_7$. (submitted for publication)

Marksteiner, P., Blaha and K. Schwarz (1986). Electronic structure and binding mechanism of Cu_2O. *Z. Phys. B*, 64, 119-27.

Marksteiner, P., S. Massidda, Jaejun Yu, A.J. Freeman and J. Redinger (1988). Calculated photoemission and x-ray emision spectra of $Bi_2Sr_2CaCu_2O_8$. *Phys. Rev. B*, 38, 5098-101.

Mattheiss, L.F. and D.R. Hamann (1988). Electronic band properties of $CaSr_2Bi_2Cu_2O_8$. *Phys. Rev. B*, 38, 5012-5.

Michel, C., F. Deslandes, J. Provost, P. Lejay, R. Tournier, M. Hervieu et B. Raveau (1987). L'oxyde $YBa_2Cu_3O_{8-y}$: une nouvelle perovskite de cuivre à valence mixte déficitaire en oxygène, supraconductrice en-dessous de 91K. *C.R. Acad. Sc. Paris*, 304-II, 1059-61.

Moruzzi, V.L., J.F. Janak and A.R. Williams (1978). *Calculated electronic properties of metals*. Pergamon, New York.

Raveau, B., C. Michel, M. Hervieu and J. Provost (1988). Crystal chemistry of perovskite superconductors. *Physica C*, 153-5, 3-8.

Redinger, J., Jaejun Yu, A.J. Freeman and P. Weinberger (1987a). Calculated local density theory of x-ray and photoemission spectra for superconducting $La_{2-x}M_xCuO_4$: localization of Cu-3d. *Phys. Lett. A*, 124, 463-68.

Redinger, J., A.J. Freeman, Jaejun Yu and S. Massida (1987b). Local density theory of x-ray and photoemission from $YBa_2Cu_3O_{7-\delta}$: the high T_c superconductors. *Phys. Lett. A*, 124, 469-73.

Sarma, D.D. and C.N.R. Rao (1987). Evidence for peroxide and Cu^{1+} species in $La_{1.85}Sr_{0.2}CuO_4$ from photoemission studies. *J. Phys.C*, 20, L659-63.

Sellmyer, D.J. (1978). Electronic structure of metallic compounds and alloys: experimental aspects. In: *Solid State Physics* (H. Ehrenreich, F. Seitz and D. Turnbull, Eds.). Academic, New York. Vol. 33, pp. 83-248.

Steiner, P., V. Kinsinger, I. Sander, B. Siegwart, S. Hüfner and C. Politis (1987). Photoemission on the high T_c superconductors Y-Ba-Cu-O. *Z. Phys. B*, 67, 13-23.

Steiner, P. (1988). XPS spectroscopy of high T_c oxides. *Int. Symp. on the Electronic Structure of high Tc Superconductors (Rome)*.

Stoffel, N.G., J.M. Tarascon, Y. Chang, M. Onellion, D.W. Wiles and G. Margaritondo (1987). Effects of oxygen stoichiometry on the electronic structure of $YBa_2Cu_3O_x$. *Phys. Rev. B*, 36, 3986-89.

Temmerman, W.M., G.M. Stocks, P.J. Durham and P.A. Sterne (1987). Electronic structure of La-Cu and Y-Ba-Cu oxides: ground-state properties and photoemission spectra. *J. Phys. F*, 17, L135-40.

Valjakka J., J. Utriainen, T. Aberg and J. Tulkki (1985). Direction-dependent initial-state relaxation in oxygen K x-ray emission. *Phys. Rev. B*, <u>32</u>, 6892-98.

Wassdahl, N., J.-E. Rubensson, G. Bray, J. Rindstedt, N. Martensson, J. Nordgren, R. Nyholm, S. Cramm, K.-L. Tsang, T.A. Calcott, D. L. Ederer and C.W. Clark (1987). Cu L and O K emission spectra of the $YBa_2Cu_3O_{7-x}$ superconductor excited by monochromatized synchrotron radiation. UUIP-1176 (unpublished).

SEP 0 6 1990